DICTIONARY OF
AGRICULTURE

DICTIONARY OF
AGRICULTURE

2nd Edition

Editor Alan Stephens

FITZROY DEARBORN PUBLISHERS
CHICAGO • LONDON

First published in Great Britain by
Peter Collin Publishing Ltd
1 Cambridge Road, Teddington, Middlesex, TW11 8DT

This edition published in the United States of America by
Fitzroy Dearborn, Publishers
70 East Walton Street, Chicago, Illinois 60611

Text computer typeset by Create Publishing Services Ltd, Bath
Printed and bound by WSOY in Finland

A Cataloging-in-Publication record for this book is available from the Library of Congress

ISBN 1-57958-076-9

First Published in the USA and UK 1998

Cover: Peter Aristedes, Chicago Advertising and Design

Preface to the first edition

In this dictionary we aim to provide the user with a comprehensive vocabulary of terms used in agriculture, horticulture and other related areas. The vocabulary ranges from the equipment used on a farm to breeds of livestock, varieties of crops, types of pesticide and herbicide, and farm building construction. It covers not only agriculture in temperate regions, but also makes particular reference to the agriculture of the tropics. The terms are defined in simple English in a way which makes them easily accessible to the student; examples of usage are given, especially in the form of quotations from newspapers and magazines published in many countries.

More extensive coverage of many topics is provided by the encyclopaedic comment sections which expand on the information given in the definitions, and give further details which may be of help to the student. At the back of the book, several pages of supplementary material offer information in the form of tables.

We are particularly grateful to Dr Jonathan Blackman of the University of Sussex and to Mr Neville Beynon of the Berkshire College of Agriculture, for helpful comments and suggestions for improvement which they made in the course of the compilation of the dictionary.

Preface to the second edition

New terminology has been added to this dictionary, to cover latest developments in various fields.

We are grateful to Simon Medaney of the Language Centre, Silsoe College, Cranfield University, for having commented on the proofs.

In this edition, we have also added phonetic pronunciation for the main entry words.

Pronunciation

The following symbols have been used to show the pronunciation of the main words in the dictionary:

Stress has been indicated by a main stress mark ('), but this is only a guide, as the stress of the word changes according to its position in the sentence.

Vowels		*Consonants*	
æ	back	b	bud
ɑː	hard	d	ditch
ɒ	fog	ð	weather
aɪ	fly	dʒ	jet
aʊ	plough	f	farm
aɪə	fire	g	gold
aʊə	shower	h	head
ɔː	coarse	j	yeast
ɔɪ	noise	k	coke
e	head	l	leaf
eə	fair	m	mixed
eɪ	main	n	nest
ə	absorb	ŋ	spring
əʊ	node	p	pond
əʊə	lower	r	rust
ɜː	bird	s	scale
iː	seep	ʃ	shell
ɪ	fit	t	teak
ɪə	clear	tʃ	chain
uː	pool	θ	thaw
ʊ	wood	v	value
ʌ	nut	w	work
		z	zone
		ʒ	fusion

Aa

A horizon ['eɪ hə'raɪzən] *noun* topsoil; *see also* HORIZON

Vitamin A ['vɪtəmɪn 'eɪ] *noun* (= *retinol*) vitamin which is soluble in fat and can be formed in the body, but which is mainly found in food, such as liver, vegetables, egg yolks and cod liver oil

COMMENT: lack of Vitamin A affects the body's growth and resistance to disease and can cause night blindness. The primary source of this vitamin is the green plant. It is of great importance for dairy cows: lack of the vitamin leads to retardation of growth in young stock and in adult animals appears to lower their resistance to infectious diseases

AA = ARBORICULTURAL ASSOCIATION

AAPP = AVERAGE ALL PIGS PRICE

abaca [ə'bækə] *noun* manila hemp

abattoir ['æbətwɑ:] *noun* place where animals are slaughtered and prepared for sale to the public as meat

abdomen ['æbdəmən] *noun* the space in an animal's body containing the stomach, intestines, liver and other vital organs

Aberdeen Angus [æbə'di:n 'æŋgəs] *noun* an early maturing breed of beef cattle, naturally hornless and usually black all over. Angus cattle can be rather small-headed with a long deep body. They are highly valued for quality beef

abomasal ulcer [æbəʊ'meɪsəl 'ʌlsə] *noun* disease common in both calves and adult cattle. Calves show poor growth and lose appetite. In rare cases, cows may bleed to death

abomasum [æbəʊ'meɪsəm] *noun* the fourth stomach of a ruminant; *see also* OMASUM, RETICULUM, RUMEN

abort [ə'bɔ:t] *verb* **(a)** to miscarry; to produce an embryo *or* foetus and to end a pregnancy before the foetus is fully developed **(b)** *(in plants and animals)* to remain underdeveloped; to wither away

◊ **abortion** [ə'bɔ:ʃn] *noun* situation where an unborn offspring leaves the womb before the end of a pregnancy; **contagious abortion** = brucellosis, an infectious disease, which is usually associated with cattle where it results in reduced milk yields, infertility and abortion; the UK is now a brucellosis-free area

ABRASSUCOS = ASSOCIACAO BRASILIERA DAS INDUSTRIAS DE SUCOS CITRICOS the Brazilian organization representing producers of juice from citrus plants

abreast parlour [ə'brest 'pɑ:lə] *noun* one of the four basic designs of milking parlour: the cows stand side by side with their heads facing away from the milker

ABRO = ANIMAL BREEDING RESEARCH ORGANISATION.

abscess ['æbses] *noun* collection of pus formed in the body, which may be acute or chronic

absinthe ['æbsɪnθ] *noun* perennial aromatic herb *(Artemisia absinthium)* used as medicine and (illegally) in making alcoholic drinks; also called 'wormwood'

absolute ['æbsəlu:t] *adjective* (i) complete; (ii) terminal point (not compared with anything else); **absolute humidity** = vapour concentration *or* ratio of the mass of water vapour in a given quantity of air to the amount of air

absorb [əb'zɔ:b] *verb* to take in any substance; *(of a solid)* to take in a liquid

◊ **absorption** [əb'zɔ:pʃən] *noun* action of absorbing; (i) taking nutrients from food into the bloodstream; (ii) the process by which chemicals gain entry into plant tissues

abstraction [æb'strækʃn] *noun* removal; **abstraction licence** = a licence issued by a Water Board to allow abstraction of water from a river or lake for domestic or commercial use. The licence is needed for irrigation

QUOTE a statutory limit on water abstraction from the Thames was set in 1911, in essence to maintain the quality of water in the river and provide for navigational needs
London Environmental Bulletin

abundant [ə'bʌndnt] *adjective* with very large numbers; giving a large crop; *the bush produces abundant red berries*

Acacia [ə'keɪʃə] *noun* species of tree, including the wattle, often grown as ornamental trees

acarid ['ækərɪd] *noun* mite or tick, a small insect which feeds on plants or animals by piercing the outer skin and sucking juices

◊ **acariasis** [ækə'raɪəsɪs] *noun* skin disease caused by ticks or mites

◊ **acaricide** *or* **acaridicide** [ə'kærɪsaɪd *or* ækə'rɪdɪsaɪd] *noun* poison used to kill mites and ticks

◊ **Acarida** *or* **Acarina** [ək'ærɪdə *or* əkə'riːnə] *noun* scientific name for the order of animals including mites and ticks

ACAS = ADVISORY CONCILIATION AND ARBITRATION SERVICE, AGRICULTURAL CHEMICALS APPROVAL SCHEME

ACC = AGRICULTURAL CREDIT CORPORATION, ASSOCIATION OF COUNTY COUNCILS

access ['ækses] *noun* **right of access** = (i) right of someone to be able to get to land by passing over someone else's property; (ii) right of the public to walk in areas of the countryside, providing they do not harm crops or farm animals; **access order** = court order which gives the public the right to go on private land

acclimatize [ə'klaɪmətaɪz] *verb* (i) to make something become used to a different sort of environment, usually a change in climate; (ii) to become used to a different sort of environment; *plants take some time to become acclimatized to tropical conditions;* **acclimatized sheep** = sheep which have become used to their local environment and as a result are less likely to stray

◊ **acclimatization** *or* **acclimation** [əklaɪmətaɪ'zeɪʃn *or* əklaɪ'meɪʃn] *noun* action of becoming acclimatized

COMMENT: when an organism such as a plant or animal is acclimatizing, it is adapting physically to different environmental conditions, such as changes in food supply, temperature or altitude

accommodation land [əkɒmə'deɪʃn 'lænd] *noun* land available for short-term tenancy

accredit [ə'kredɪt] *verb* to recognize officially; **accredited herd** = herd of cattle registered under a scheme as being free from Brucellosis; **accredited milk** = milk from a herd accredited as being free from Brucellosis

Acer ['eɪsə] *Latin name for* the maple

acetonaemia [æsətəʊ'niːmɪə] *noun* disease affecting cows, caused by ketone bodies accumulating. The animal loses appetite and the smell of acetone affects the breath, the urine and milk

achene [ə'kiːn] *noun* small dry one-seeded fruit, which does not split open to release its seed

acid ['æsɪd] *noun* chemical compound containing hydrogen, which dissolves in water and forms hydrogen, or reacts with an alkali to form a salt and water, and turns litmus paper red; *hydrochloric acid is secreted in the stomach and forms part of the gastric juices;* **inorganic acids** = acids which are derived from minerals, such as hydrochloric acid and sulphuric acid; **organic acids** = weak acids which contain carbon, some of which are pesticides

◊ **acidic** [ə'sɪdɪk] *adjective* referring to acids; *soil and vegetation in high altitude forests are directly exposed to an extremely acidic cloud base;* **acidic properties** = properties associated with acids

◊ **acidification** [əsɪdɪfɪ'keɪʃn] *noun* process of becoming acid *or* of making a substance more acid; *acidification of the soil leads to the destruction of some living organisms*

◊ **acidify** [ə'sɪdɪfaɪ] *verb* to make a

substance more acid; *acid rain causes acidified lakes with no fish population*

◊ **acidity** [ə'sıdıti] *noun* level of acid in a solution; *the alkaline solution may help to reduce acidity*

COMMENT: acidity and alkalinity are shown according to the pH scale on which pH7 is neutral; numbers above indicate alkalinity, while pH6 and below indicate acidity

acidophilus milk [æsı'dɒfıləs mılk] *noun* a cultured milk made from fresh milk which is allowed to go sour in a controlled way. One of the most popular types of cultured milk in Europe is yoghurt

acidosis [æsı'dəusıs] *noun* disease of cattle caused by excess concentrate feed

COMMENT: as acidity increases the rumen wall becomes inflamed. The animal dehydrates progressively, the blood turns more acidic and in extreme cases the animal may die

acid rain *or* **acid deposition** *or* **acid precipitation** [æsıd 'reın *or* æsıd depə'zıʃn *or* æsıd presıpı'teıʃn] *noun* rain (or snow) which contains a higher level of acid than normal

COMMENT: acid rain is mainly caused by sulphur dioxide, nitrogen oxide and other pollutants which are released into the atmosphere when fossil fuels containing sulphur (such as oil or coal) are burnt. Acid rain rarely falls near the source of the pollution, because the smoke from chimneys can be carried by air currents for many kilometres before it finally falls as rain. So Scandinavia receives acid rain which is caused by pollution from British and German factories; Canada receives acid rain from factories in the US. Acid soot, on the other hand, can fall relatively close to the source of pollution. It is caused when carbon combines with sulphur trioxide from sulphur-rich fuel to form particles of an acid substance which can damage the surfaces it falls on (such as stone buildings). The effects of acid rain are primarily felt by wildlife: the water in lakes becomes very clear as fish and microscopic animal life is killed. It is believed that it is acid rain that kills trees, especially conifers, which gradually lose their leaves and die

acid soil ['æsıd 'sɔıl] *noun* soil which has a pH value of 6 or less

COMMENT: most farm crops will not grow well if the soil is very acid. This can be cured by applying one of the materials commonly used for adding lime, such as ground chalk or limestone

ACP = ADVISORY COMMITTEE ON PESTICIDES

ACP states *see* LOME CONVENTION

ACPAT = ASSOCIATION OF CHARTERED PHYSIOTHERAPISTS IN ANIMAL THERAPY

ACR = AUTOMATIC CLUSTER REMOVAL

ACRE = ACTION WITH COMMUNITIES IN RURAL ENGLAND

acre ['eıkə] *noun* unit of measurement of land area, equal to 4,840 square yards, or 0.4047 hectares

◊ **acreage** ['eıkərıdʒ] *noun* area of land measured in acres; *US* **acreage allotment** = quota system operated in the USA, which limits the area of land which can be planted with a certain type of crop; **acreage reduction programme (ARP)** = American federal programme under which farmers are only eligible for subsidies if they reduce the acreage of certain crops planted (NOTE: the British equivalent is **set-aside)**

actinobacillosis [æktiːnəubæsı'ləusıs] *noun* disease of cattle affecting the tongue and throat; it also occurs in sheep as swellings on the lips, cheeks and jaws. It is also known as 'cruels'

◊ **actinomycosis** [æktınəumæı'kəusıs] *noun* disease of cattle and pigs, where the animal is infected with bacteria which form abscesses in the mouth and lungs, known as 'lumpy jaw'

activate ['æktıveıt] *verb* to make a chemical reaction take place

◊ **activator** ['æktıveıtə] *noun* substance which activates; **compost activator** = chemical added to a compost heap to speed up the decomposition of decaying plant matter

actuals ['æktʃuəlz] *noun* stocks of commodities, such as cotton or rice, which are available for shipping (as opposed to 'futures', commodities for delivery in the future, which have not yet been produced)

acute [ə'kjuːt] *adjective* (disease) which comes on suddenly and can be very serious; **acute toxicity** = level of concentration of a toxic substance, which makes animals seriously ill, or can even cause death

ADAS = AGRICULTURAL DEVELOPMENT AND ADVISORY SERVICE

COMMENT: ADAS became an independent executive agency in April 1992. Reorganised into 15 consultancy centres, 6 development centres and 12 research centres

additive ['ædɪtɪv] *noun* substance which is added, especially one which is added to animal feedingstuffs, which may include antibiotics, mineral supplements, vitamins and hormones, or one added to human food to improve the taste, smell, appearance or keeping

addled egg ['ædəld 'eg] *noun* rotten egg, an egg which produces no chick

ADHAC = AGRICULTURAL DWELLING HOUSE ADVISORY COMMITTEE

adlay ['ædleɪ] *noun* tropical millet with very large seeds grown in the Philippines and some other areas of South-East Asia; also known as 'Job's tears'

ad libitum *or* **ad lib feeding** [æd 'lɪbɪtəm *or* æd 'lɪb 'fiːdɪŋ] *noun* unrestricted supply of feed, day and night, usually in the form of dry feed for pigs and poultry

admixture [əd'mɪkstʃə] *noun* the proportion of a seed crop which is made up of weed seeds or other crop species

ADP = AGRICULTURAL DEVELOPMENT PROGRAMME

ADRA = ANIMAL DISEASES RESEARCH ASSOCIATION

adulterate [ə'dʌltəreɪt] *verb* to add water (to milk)

advanced register [əd'vɑːnst 'redʒɪstə] *noun* book which records breeding performance of outstanding livestock

adventitious [ædven'tɪʃəs] *adjective* (root) which develops from a plant's stem and not from another root

Adzuki beans [æd'zuːki biːnz] *noun* tropical legume grown in Japan and China

AEA = AGRICULTURAL ENGINEERS ASSOCIATION

aerate [eə'reɪt] *verb* to put air into a substance, especially to replace stagnant air in soil with fresh air; *worms are useful because they aerate the soil*

◊ **aeration** [eə'reɪʃn] *noun* action of aerating, the replacement of stagnant soil air with fresh air

COMMENT: the process of aeration of soil is mainly brought about by the movement of water into and out of the soil; rainwater drives out the air, and then as the water drains away or is used by plants, fresh air is drawn into the soil to fill the pore spaces. The aeration process is also assisted by changes in temperature, good drainage, cultivations and open soil structure. Sandy soils are usually well aerated; clay soils are poorly aerated

aerial ['eərɪəl] *adjective* which exists in the air; **aerial roots** = roots of certain plants, which hang in the air or wrap around other plants, and take up moisture from the air; **aerial spraying** = spraying of crops by plane or helicopter. This spraying mainly involves pesticides and fertilizers. Spraying is subject to various safety precautions and care must be taken to avoid drifting of harmful chemicals

QUOTE in 1986, there were about 50 operators licensed to spray from the air, now that figure has fallen to about 12. Spraying, in its heyday, accounted for less than 2% of the total area; now it is about 0.25%
Farmers Weekly

aerobic [eə'rəubɪk] *adjective* needing oxygen for its existence (NOTE: opposite is **anaerobic)**

affect [ə'fekt] *verb* to have a result on; *the restrictions cover a twenty mile radius, affecting thirty farms and smallholdings; the disease affects mainly young animals*

afforestation [əfɒrɪs'teɪʃn] *noun* (i) growing trees as a crop; (ii) planting of trees on land previously in another use; *there is likely to be an increase in afforestation of upland areas if the scheme is introduced*

aflatoxin [æflə'tɒksɪn] *noun* very toxic

substance formed by a fungus *Aspergillus flavus*, which grows on seeds and nuts, and is found in stored grain or in groundnut meal; it is harmful to cattle and pigs

AFRC = AGRICULTURAL AND FOOD RESEARCH COUNCIL an organization funded by the MAFF and the Department of Education

African buffalo ['æfrɪkən 'bʌfələʊ] *noun* undomesticated buffalo indigenous to Africa, including the Black Cape Buffalo and the Congo Buffalo; these buffalo can carry in their blood parasites that cause sleeping sickness

Africander [æfrɪ'kændə] *noun* breed of cattle found in parts of Southern Africa; originally used as draught animals, they are now used for beef production. The animals are red, with curved horns; they are resistant to heat and to disease; also breed of sheep

African fan palm ['æfrɪkən 'fæn pɑːm] *noun Borassus aethiopicum*, found in the savannahs of tropical Africa. The tree has a very tall smooth trunk with fan shaped leaves. The orange fruits are large and round or oval. The fruit is eaten raw, and the seeds used for palm wine. Leaflets used for thatching and mats

African swine fever ['æfrɪkən 'swaɪn fiːvə] *noun* a virus disease which is highly contagious among pigs. Animals suffer fever and high temperature followed by death; in Europe, it occurs in parts of Spain (NOTE: also known as **wart-hog disease)**

afrormosia [æfrɔː'məʊzɪə] *noun* hardwood from West Africa, resembling teak, now becoming scarce

afterbirth ['ɑːftbɜːθ] *noun* the remains of the placenta pushed out of the uterus of the dam at the birth of a young animal (NOTE: also called **cleansing)**

aftermath ['ɑːftəmɑːθ] *noun* grass which grows quickly after cutting for hay, and which will provide a second cut

agalactia [æɡə'læktɪə] *noun* disease of pigs, a form of post-farrowing shock. The sow does not secrete milk

agar *or* **agar agar** ['eɪɡə] *noun* culture

medium, the gelatin in which bacteria or tissue can be grown

agave ['æɡeɪv] *noun* succulent tropical plant which gives sisal

age [eɪdʒ] *verb* to treat flour to make the dough more elastic and whiter

agglutination [əɡluːtɪ'neɪʃn] *noun* action of grouping cells together (as of bacteria cells in the presence of serum *or* blood cells when blood of different types is mixed); **agglutination tests** = (i) tests to identify bacteria; (ii) tests used to detect Brucellosis in cattle

agist [ə'dʒɪst] *verb* to take another person's livestock to feed on your land

◊ **agistment** [ə'dʒɪstmənt] *noun* money paid for grazing stock on land owned by another person. The owner of the land is responsible for the feeding and care of the livestock

agitator ['ædʒɪteɪtə] *noun* part of a machine for harvesting root crops, such as potatoes, which shakes the earth off the crop after it has been lifted

agrarian [ə'greərɪən] *adjective* referring to matters of land tenure and problems arising from land ownership

Agregado [æɡreɪ'ɡɑːdəʊ] *noun* Brazilian term for a tenant farmer

agri- *or* **agro-** ['æɡrɪ *or* 'æɡrəʊ] *prefix* referring to agriculture *or* the cultivation of land

agribusiness [æɡrɪ'bɪznəs] *noun* the business of farming and making products and equipment used by farmers

COMMENT: the term is used to refer to large-scale farming businesses run along the lines of a conventional company, often involving the growing, processing, packaging and sale of farm products

agriculture ['æɡrɪkʌltʃə] *noun* cultivation of the land, including horticulture, fruit growing, crop and seed growing, dairy farming and livestock breeding; **Agriculture Acts** = Acts of Parliament, introduced to update legislation affecting agricultural policy; **'Agriculture in the United Kingdom'** = review undertaken each year by the British government, reporting on the state of the agricultural industry; **Agriculture Service** = a service provided by a branch of ADAS, which gave specialist advice to farmers, and ran several experimental stations in parts of England and Wales; now replaced by the Farm and Countryside Service and the Research and Development Service

COMMENT: agriculture in the United Kingdom has declined as an employer and will continue to do so. There are now, in 1996, about 250,000 full time farmers, compared with 600,000 in 1945. Up to 150,000 jobs are forecast to disappear from the countryside in the next five years

agricultural [ægrɪ'kʌltʃərl] *adjective* referring to farming; **agricultural holding** = a basic unit for agricultural production; **agricultural depopulation** = movement of people from rural areas; **agricultural waste** = waste matter produced on a farm such as manure from animals or excess fertilizers and pesticides which run off the fields

◊ **June Agricultural Census** in Great Britain, the annual farm returns on 4th June each year provide the basis for a census of agriculture carried out by the MAFF; the census covers crop areas, numbers of livestock, production and yields, number and size of holdings, numbers of workers, farm machinery, prices and incomes

◊ **Agricultural Chemicals Approval Scheme (ACAS)** scheme which gave advice to farmers on the use and efficiency of chemicals and which tested chemicals before use by farmers (operated by the Agricultural Chemicals Approved Organization). It was a voluntary scheme which has now been replaced by the FEPA legislation

◊ **Agricultural Development and Advisory Service (ADAS)** a branch of the MAFF, providing scientific, professional and technical advice to farmers. It has three divisions: Farm and Countryside Service, Veterinary Service, and the Research and Development Service

◊ **Agricultural Holdings Act 1984** an Act of Parliament which gives protection to tenants in questions of the fixing of rent and security of tenure. It makes provision for tenancies for a life time and for short-term lettings

◊ **Agricultural Improvement**

Scheme a scheme introduced in 1985 to provide capital grants to farmers

◊ **Agricultural Land Tribunal** a court established in 1947 to hear appeals against decisions affecting owners or tenants of agricultural land

◊ **Agricultural Mortgage Corporation** a corporation which makes loans available to borrowers on the security of agricultural land and buildings in England and Wales

◊ **Agricultural and Food Research Council (AFRC)** a council established to organize and provide funds for agricultural and food research

◊ **Agricultural Stabilization and Conservation Service (ASCS)** *US* service of the federal Department of Agriculture which operates the department's various schemes throughout the USA

◊ **Agricultural Training Board** a body established under the Industrial Training Act of 1966 to provide training for agricultural workers and farmers

◊ **Agricultural Wages Board** a board which fixes minimum wages and holiday entitlements for agricultural workers, and deals with terms and conditions of their employment

Agricultural Revolution *noun* the changes in agriculture which transformed Britain's countryside in the 18th and 19th centuries. The changes were in techniques such as crop rotation, selective breeding, new implements, and also in land tenure, with the removal of the open field system with common rights, and the gradual enclosure of land, and farm consolidation

agrochemicals [ægrəʊ'kemikəlz] *noun* pesticides and fertilizers developed artificially for agricultural use; **the agrochemical industry** = the branch of industry which produces pesticides and fertilizers used on farms

QUOTE more efficient agrochemicals applied at lower rates and which target pests weeds and diseases are reflected in the British Agrochemicals Association figures. Since 1983, the amount of active pesticide ingredients applied to UK soils and crops fell from 33,157+ to 23,504+. The area increased from 15.1m hectares to 19.6m hectares

Farmers Weekly

agroclimatology [ˌægrəʊklaɪməˈtɒlədʒi] *noun* study of climate and its effect on agriculture

◊ **agroforestry** [ˌægrəʊˈfɒrɪstri] *noun* growing farm crops and trees together as a farming unit

agronomy [əˈɡrɒnəmi] *noun* scientific study of soil management and the cultivation of crops

◊ **agronomist** [əˈɡrɒnəmɪst] *noun* person who studies crop cultivation; **pasture agronomist** = person who specializes in the study of types of grass grown in pastures

COMMENT: the use of land to raise crops for eating first started about 10,000 years ago. All plants grown for food have been developed over many centuries from wild plants, which have been progressively bred to give the best yields in different types of environment. Genes from wild plants are likely to be more hardy and resistant to disease, and are still kept in gene banks to strengthen new cultivated varieties

AHGS = AGRICULTURAL AND HORTICULTURAL GRANTS SCHEME

AHS = ARAB HORSE SOCIETY

COMMENT: Society founded in England in 1918 to promote the breeding and importation of pure-bred Arab horses

AHU = APPLICATION HAZARDS UNIT

AI [eɪˈaɪ] = ARTIFICIAL INSEMINATION **AI centre** = centre which keeps breeding bulls, boars and rams, and quantities of their semen for use in artificial insemination; **to AI** = to inseminate an animal artificially; *25 ewes were AI'd*

QUOTE AI centres not only supply semen from high-ranking sires, but generally offer a high degree of choice
Queensland Agricultural Journal

air pollution [ˈeə pəˈluːʃn] *noun* odour nuisance from livestock units and other farming activities is governed by the Environment Protection Act. Burning of agricultural crop residues is now banned

AIS = AGRICULTURAL INVESTMENT SCHEME

akee [əˈkiː] *noun* evergreen tree found in the West Indies; the fruit can be eaten raw, but is more usually cooked

alar [ˈeɪlə] *noun* chemical used by apple growers to make more buds set and grow into mature fruit

QUOTE millions of apples are being sprayed with a chemical said to cause cancer: the use of alar is now under review in the UK
Guardian

Ala Tau [ˈælə ˈtaʊ] *noun* breed of cattle found in the Soviet state of Kazakhstan; it is a dual-purpose breed, grey-brown in colour

albino [ælˈbiːnəʊ] *noun* animal, with little or no pigmentation in skin, hair or eyes

◊ **albinism** [ˈælbɪnɪzm] *noun* complete absence of pigmentation

albumens *or* **albumins** [ˈælbjumɪnz] *noun* bodies which are like the white of an egg, a constituent part of animal solids and fluids; they are soluble in water and coagulate when heated

alder [ˈɔːldə] *noun* fine European hardwood *(Alnus glutinosa)* which is waterproof

aldosterone [ælˈdɒstərəʊn] *noun* hormone which regulates the sodium balance in the body

aldrin [ˈɔːldrɪn] *noun* one of the organochlorine insecticides, used to control cabbage root fly, wireworms and leatherjackets. There is an agreement to restrict the use of this insecticide

alecost [ˈeɪlkɒst] *noun* perennial herb *(Chrysanthemum balsamita)* used in cooking; also known as 'mace'

aleurone [ˈælʊrəʊn] *noun* protein found in the outer skin of seeds

alfa [ˈælfə] *noun* North African name for esparto grass

alfalfa [ælˈfælfə] *noun* Arabic name for lucerne, a plant of the Leguminosae family, with clover-like leaves; grown to use as fodder; *see* LUCERNE

algae [ˈældʒiː] *plural noun* tiny plants living in water or in moist conditions, which contain chlorophyll and have no stems or roots or leaves; **algae poisoning** =

poisoning caused by the decomposition of algae which releases toxic substances. This may poison livestock if eaten; it occurs at the side of lakes

◊ **algaecide** ['æld₃ɪsaɪd] *noun* substance used to kill algae

> COMMENT: algae grow rapidly in water which is rich in phosphates. When the phosphate level increases, as when fertilizer runoff enters the water, the algae multiply greatly to form enormous floating mats (or blooms), blocking out the light and inhibiting the growth of other organisms. When the algae die, they combine with all the oxygen in the water so that other organisms suffocate

alimentary canal *or* **alimentary tract** [ælɪ'mentəri kə'næl *or* trækt] *noun* the digestive tract, the passage from the mouth to the rectum along which food passes and is digested

alkali ['ælkəlaɪ] *noun* one of many substances which neutralize acids and form salts (NOTE: British English plural is **alkalis,** but US English is **alkalies)**

◊ **alkaline** ['ælkəlaɪn] *adjective* containing more alkali than acid

> QUOTE in some countries alkaline diets containing calcium chloride, magnesium sulphate or ammonium sulphate are offered, but this is not common in Britain
> *Farmers Weekly*

alkalinity [ælkə'lɪnɪti] *noun* amount of alkali in something such as soil *or* water

> COMMENT: alkalinity and acidity are measured according to the pH scale. pH7 is neutral, and pH8 and upwards is alkaline

alkaloid ['ælkəlɔɪd] **1** *adjective* similar to an alkali **2** *noun* one of many poisonous substances (such as atropine *or* morphine *or* quinine) found in plants which use them as defence against herbivores; they are also useful as medicines

allele [ə'liːl] *noun* one of two or more alternative forms of a gene

allelopathy [ælɪ'lɒpəθi] *noun* the release of a chemical by a plant that inhibits the growth of nearby plants

allergy ['ælədʒi] *noun* being sensitive to

certain substances, which cause a physical reaction

◊ **allergen** ['ælədʒen] *noun* substance which produces a sensitive physical reaction

alleviation [əliːvi'eɪʃn] *see* FLOOD

Allium ['æliəm] *Latin name* family of plants including the onion, leek, garlic and chives

allo- ['æləʊ] *prefix* different

◊ **allogamy** [ə'lɒgəmi] *noun* fertilization by pollen from different flowers *or* from flowers of genetically different plants of the same species

> COMMENT: some fruit trees are self-fertile, that is, they fertilize themselves with their own pollen. Others need pollinators, usually different cultivars of the same species

allograft ['æləʊgrɑːft] *noun* homograft, graft of tissue from one specimen to another of the same species

◊ **allopatric** [ælə'pætrɪk] *adjective* (plants of the same species) which grow in different parts of the world and so do not cross-breed

allotment [ə'lɒtmənt] *noun* small area of land which is let on an annual basis by the municipality which owns it to an allotment-holder for the cultivation and production of vegetables and fruit for the consumption of the holder and his family. Under the 1947 Agriculture Act an allotment should not be larger than 0.25 acre, or 0.1 hectare

allspice ['ɔːlspaɪs] *noun* tropical tree grown in Jamaica; its dried berries are used as a flavouring

alluvium [ə'luːviəm] *noun* silt deposited by rivers when they flood, or by lakes; usually very fertile

◊ **alluvial** [ə'luːviəl] *adjective* referring to alluvium; **alluvial deposits** *or* **alluvial soils** = deposits of silt on the bed of a river *or* lake

alm [ælm] *noun* *(in German-speaking countries)* alpine pasture normally only grazed in summer

almond ['ɑːmənd] *noun* small tree *(Prunus dulcis)* grown for its edible nuts; it grows in most Mediterranean countries, and also in

California, South Australia and South Africa

◊ **almond oil** ['ɑːmənd 'ɔɪl] *noun* oil from almond seed used for toilet preparations and for flavouring

alp [ælp] *noun* a high mountain pasture, above the treeline

◊ **alpine** ['ælpaɪn] *adjective* referring to the Alps or to other high mountains; **alpine plants** = plants which grow on high mountains; *alpine vegetation grows above the treeline*

alpaca [æl'pækə] *noun* animal which is similar to the llama; a native of the Andes, it is domesticated and reared for its very soft and elastic wool

alpha acids ['ælfə 'æsɪdz] *noun* a number of related compounds found in hops, and which give them their bitter taste; the ability of hops to give beer a bitter taste is assessed by their alpha acid content

alpha amylase ['ælfə 'æmɪleɪz] *noun* an enzyme present in wheat seed, which changes some starch to sugar. Excessive amounts can result in loaves of bread with sticky texture. Alpha amylase activity can be measured by the Hagberg test or the Farrand test

alternate husbandry [ɔːl'tɜːnət 'hʌzbəndri] *noun* husbandry in which arable and grassland cultivation are alternated every few years

altiplano [æltɪ'plɑːnəʊ] *noun* high plateau land of Bolivia

altitude ['æltɪtjuːd] *noun* height above sea-level. This can affect climate in many ways: the temperature drops about 0.5°C for every 90m rise above sea-level. Every 15m rise in height usually shortens the growing season by two days and may check the rate of growth during the year. High land is likely to receive more rain than lowland areas

AMC = AGRICULTURAL MORTGAGE CORPORATION

American bison [ə'merɪkən 'baɪsən] *noun* North American animal of the cattle family, of which large numbers were slaughtered during the nineteenth century; it is now protected, and small numbers can be seen in some National Parks. The bison will breed with ordinary cattle

amino acid [ə'miːnəʊ 'æsɪd] *noun* chemical compound which is broken down from proteins in the digestive system and then used by the body to form its own protein; *proteins are first broken down into amino acids;* **essential amino acids** = amino acids which are essential for growth, but which cannot be synthesized and so must be obtained from food or medicinal substances

COMMENT: amino acids all contain carbon, hydrogen, nitrogen and oxygen, as well as other elements. Some amino acids are produced in the body itself, but others have to be absorbed from food. The number of amino acids varies between species and between infants and adults; eight of the essential amino acids are: isoleucine, leucine, lysine, methionine, phenylalanine, threonine, tryptophan and valine

ammonia (NH$_3$) [ə'məʊnɪə] *noun* gas with a strong smell, a compound of nitrogen and hydrogen, which is a normal product of organic metabolism, and is used in compounds to make artificial fertilizers; **ammonia treatment** = method of treating straw, using ammonia to make it more palatable and nutritious (NOTE: chemical symbol is **NH$_3$**)

◊ **ammoniacal** [ə'məʊnɪəkəl] *adjective* referring to ammonia; **ammoniacal nitrogen** = nitrogen derived from ammonia

COMMENT: ammonia is released into the atmosphere from animal dung. It has the effect of neutralizing acid rain, but in combination with sulphur dioxide it forms ammonium sulphate which damages the green leaves of plants

ammonium [ə'məʊnɪəm] *noun* ion formed from ammonia; **ammonium fixation** = the absorption of ammonium ions by the soil; **ammonium nitrate** = a popular fertilizer used as top dressing. It is available in a special prilled or granular form, and can be used both as a straight fertilizer and in compounds; **ammonium phosphate** = fertilizer which can be used straight, but is more often used in compounds. A certain amount of care is needed in its use, as applications may increase the acidity of the soil; **ammonium sulphate** *see* SULPHATE OF AMMONIA

amniotic fluid [æmnɪ'ɒtɪk 'fluːɪd] *noun*

fluid which surrounds the foetus in the uterus

amoeba [ə'miːbə] *noun* form of animal life made up of a single cell (NOTE: plural is **amoebae.** Note also the American spelling: **ameba)**

◊ **amoebic** [ə'miːbɪk] *adjective* caused by amoebae; **amoebic dysentery** = mainly tropical form of dysentery which is caused by *Entamoeba histolytica*, a parasite which enters the body through contaminated water

AMS = AGGREGATE MEASURE OF SUPPORT

amylase ['æmɪleɪz] *noun* enzyme which converts starch into maltose; *see also* ALPHA AMYLASE

anabolism [ə'næbəlɪzm] *noun* building of tissue, one of the processes in metabolism

◊ **anabolic steroids** [ænə'bɒlɪk 'stɪrɔɪdz] *noun* hormones which encourage growth and muscle building

anaemia [ə'niːmɪə] *noun* deficiency of red blood corpuscles. In young pigs it is caused by a shortage of iron in the sow's milk. Anaemia results in lack of condition and may cause death

anaerobic [ænə'rəʊbɪk] *adjective* not needing oxygen for existence; **anaerobic decomposition** = breaking down of organic material by microorganisms without the presence of oxygen (NOTE: opposite is **aerobic)**

◊ **anaerobically** [ænə'rəʊbɪkli] *adverb* without using oxygen; *slurry is digested anaerobically by bacteria*

◊ **anaerobism** [ænə'rəʊbɪzm] *noun* lack of oxygen (as in gley soils)

COMMENT: anaerobic digesters for pig, cattle and poultry waste feed the waste into a tank where it breaks down biologically to give off large amounts of methane. This gas is then used to generate electricity. The remaining slurry can be applied directly to the land

analyse ['ænəlaɪz] *verb* (i) to examine something in detail; (ii) to separate something into its component parts; *the laboratory is analysing the soil samples; when the water sample was analysed it was found to contain traces of bacteria*

◊ **analysis** [ə'nælɪsɪs] *noun* examination of a substance to find out what it is made of (NOTE: plural is **analyses)**

◊ **analyst** ['ænəlɪst] *noun* person who examines samples of substances *or* tissue, to find out what they are made of

COMMENT: chemical and electrical methods are used in soil analysis to determine the pH and lime requirements of a soil. Portable testing equipment, using colour charts, is sometimes used to test for pH

Ananas ['ænənæs] *Latin name for* pineapple

anaplasmosis [ænəplæz'məʊsɪs] *noun* infectious disease of cattle, characterized by anaemia

anchorage ['æŋkərɪdʒ] *noun* ability of plant roots to hold firm in the soil

Ancona [æn'kəʊnə] *noun* a laying breed of chicken, with white tips on black feathers

Andalusian [ændə'luːsɪən] *noun* **(a)** a dark red breed of cattle, used both as draught animals and for beef **(b)** a laying breed of chicken with blue feathers

angelica [æn'dʒelɪkə] *noun* plant with dark green stems, which are crystallized with sugar and used in confectionery

Angeln ['æŋgeln] *noun* German dual-purpose breed of cattle, red or brown in colour, with black hooves

Anglo-Nubian ['æŋgləʊ 'njuːbɪən] *noun* a hardy breed of goat with high milk yields; it has a brown coat with white patches

angora [æŋ'gɔːrə] *noun* (i) breed of rabbit, bred mainly for its fur; (ii) breed of goat, important as a source of mohair; **Texan angora goat** = breed of goat with very fine hair, imported from America

COMMENT: the original colour was white, but there are now grey, pale brown and other shades. The wool is extremely fine

Angus *see* ABERDEEN

anise ['ænɪs] *noun* annual herb *(Pimpinella anisum)* which produces small aromatic fruit (called aniseed) used for flavouring

Ankole [æŋ'kəʊli] *noun* East African breed of cattle, with very long horns

annual ['ænjuəl] **1** *adjective* **(a)** which happens each year *or* once a year; **Annual Review of Agriculture** = review (also called the Price Review) which was undertaken each year by the British government, reporting on the state of the agricultural industry, now called 'Agriculture in the United Kingdom'; **annual rings** = the rings of new wood formed each year in the trunk of a tree, and which can easily be seen when the tree is cut down. They are also used to estimate the age of a tree; *see also* DENDROCHRONOLOGY **(b)** (plant) whose life cycle is completed in a single year; **annual meadowgrass** = widespread weed *(Poa annua)* found in all arable and grass crops; **annual nettle** = a common weed *(Urtica urens)* **2** *noun* plant whose life cycle (germination, flowering, fruiting) takes place within the period of a year

COMMENT: typical examples of annuals are wheat and barley, which complete their life history in one growing season, that is, starting from seed, they develop roots, stem and leaves, then produce flowers and seed before dying. Plants which develop over two years are called 'biennials' and those which do not die at the end of the fruiting period are called 'perennials'. Annual rings are formed as a tree grows, because the wood formed in the spring has more open cells than that formed in later summer. The difference in texture forms the visible rings. Note that in tropical countries, trees grow all the year round and so do not form rings

anoestrus [æn'iːstrəs] *noun* situation where a female animal does not come on heat at the usual time

anopheles [ə'nɒfɪliːz] *noun* mosquito which carries the malaria parasite

antenna [æn'tenə] *noun* sensing organ which projects from an insect's head, shaped like a long thick hair (NOTE: plural is **antennae)**

anthelmintic [ænθəl'mɪntɪk] *noun* substance (such as thiabendazole) which is used as a treatment against parasites, as in worming livestock

anther ['ænθə] *noun* part of a stamen which produces pollen

anthesis [æn'θiːsɪs] *noun* the action of flowering, when the anthers emerge

anthrax ['ænθræks] *noun* a notifiable disease of animals, commonest among cattle and sheep. Pigs and horses can also be affected

COMMENT; anthrax is caused by the bacterium *Bacillus anthracis*, which is difficult to destroy and can stay in the soil and infect animals. Anthrax can be transmitted to humans by touching infected skin, meat, or other parts of an animal (including bone meal used as fertilizer). It causes pustules on the skin, or in the lungs (woolsorter's disease)

antibiotic [æntɪbaɪ'ɒtɪk] **1** *adjective* which stops the spread of bacteria **2** *noun* drug (such as penicillin) which is developed from living substances and which kills or stops the spread of microorganisms; **broad spectrum antibiotics** = antibiotics which are used to control many types of bacteria

COMMENT: penicillin is one of the commonest antibiotics, together with streptomycin, tetracycline, erythromycin and many others. Although antibiotics are widely and successfully used, new forms of bacteria have developed which are resistant to them. Antibiotics can also be used as feed additives to promote growth

QUOTE some antibiotics, such as penicillin and cloxacillin, are effective against staphs and streps, but have no effect on coliform bacteria. Others, like streptomycin, are effective against streptococci. A third group, the broad-spectrum antibiotics, which includes tetracycline, are effective against the common causes of mastitis
Practical Farmer

antibody ['æntɪbɒdi] *noun* substance which is naturally present in the body and which is produced in response to invasion by antigens (such as bacteria)

◊ **anticaking additive** [ænti'keɪkɪŋ 'ædɪtɪv] *noun* substance added to processed food to prevent it becoming solid when damp (in the EU, they have E numbers E530 - E578)

◊ **antifungal** [ænti'fʌŋgəl] *adjective* (substance) which kills *or* controls fungi

◊ **antigen** ['æntɪdʒən] *noun* substance such as a virus which makes the body produce antibodies to attack it

◊ **antioxidant** [ænti'ɒksɪdənt] *noun* substance which prevents oxidation, added to processed food to prevent it going bad

(in the EU, antioxidant food additives have the E numbers E300 - E321)

◊ **antiseptic** [æntɪˈseptɪk] **1** *adjective* (substance) which prevents germs spreading **2** *noun* substance which prevents germs growing or spreading and is applied as a disinfectant to wounds; *the wound was painted with antiseptic*

◊ **antiserum** [æntɪˈsɪːərəm] *noun* serum taken from an animal which has developed antibodies to bacteria, used to give temporary immunity to a disease (NOTE: plural is **antisera)**

◊ **antitoxin** [æntɪˈtɒksɪn] *noun* antibody produced by the body to counteract a poison in the body

◊ **antivenene** *or* **antivenom (serum)** [æntɪˈveniːn *or* æntɪˈvenəm] *noun* serum which is used to counteract the poison from snake *or* insect bites

antlers [ˈæntləz] *noun* branched horns of male deer

anus [ˈeɪnəs] *noun* opening at the end of the rectum, through which excreta are passed out of the body

anvil [ˈænvɪl] *noun* base or bed for shaping a horse shoe. Made of wrought iron

AONB = AREA OF OUTSTANDING NATURAL BEAUTY

APDC = APPLE AND PEAR DEVELOPMENT COUNCIL

apex [ˈeɪpeks] *noun* the main growing shoot of a plant; pinching out the apex will allow side shoots to develop and will help the plant develop a bushy form

aphid [ˈeɪfɪd] *noun* small insect (blackfly or greenfly) which sucks sap from plants and can multiply very rapidly; *analysis of the effects on the aphid population showed a 19% increase in the rate of production*

COMMENT: cereal aphids are various species of greenfly. Winged females are found feeding on cereal crops in May and June. The grain aphid causes empty or small grain by puncturing the grain in the milk ripe stage, letting the grain contents seep out. Aphids can carry virus diseases from infected plants to clean ones

Aphis [ˈeɪfɪs] *Latin name for* various species of aphid

apiary [ˈeɪpiəri] *noun* a place where bees are kept

◊ **apiarist** [ˈeɪpiərɪst] *noun* beekeeper, person who keeps bees

◊ **apiculture** [ˈeɪpɪkʌltʃə] *noun* the husbandry of bees for honey production

apple [ˈæpəl] *noun* edible fruit of the apple tree *(Malus domestica);* **apple blossom weevil** = insect which attacks apple flower buds, causing no fruit to develop

COMMENT: the apple is the most important UK fruit crop, growing mainly in Kent and Worcestershire. In 1957, England had about 26,000 hectares growing dessert apples, but by 1987, there were less than 14,000 hectares. 90% of all cooking apples grown are Bramleys. In the USA and Canada apples are mainly grown in Washington, Oregon, British Columbia and Nova Scotia. The apple is also grown extensively in temperate regions of Australia, South Africa and South America. 6,000 apple varieties once grew in Britain, and all of them are recorded in the UK's National Apple Register. Around 2,300 of these are growing at the National Fruit Trials at Brogdale in Kent. Of the recognized apple varieties the most important are Cox's Orange Pippin and Golden Delicious (dessert varieties) and Bramley's Seedling (cooking apple). Cider apples are grown mainly in Herefordshire and Somerset in the UK, and in Normandy in France

apply [əˈplaɪ] *verb* to put (a substance) on; *the fungicide should be applied in early spring*

◊ **application** [æplɪˈkeɪʃn] *noun* putting a substance on; *two applications of the pesticide should be enough to keep most pests off*

QUOTE nitrogen application is only part of an overall system of management aimed at getting the best out of the crop
Farmers Weekly

apportionment [əˈpɔːʃənmənt] *noun (in Scotland)* enclosure of part of common land for private farm use

appropriate technology [əˈprəʊprɪət tekˈnɒlədʒi] *noun* technology that is suited to the local environment, usually involving skills *or* materials that are easily available locally; *biomethanation seems a very appropriate technology for use in rural areas*

COMMENT: in many parts of world, devices to help the local population cultivate the land can be made out of simple pipes or pieces of metal. Expensive tractors may not only be unsuitable for the terrain involved, but also use fuel which costs more than the crops produced

approved bull [ə'pru:vd 'bʊl] *noun* bull which has MAFF approval for providing semen used in AI; **approved products** = chemicals used in agriculture and approved for use by the government

APR = ANNUAL PERCENTAGE RATE

APRS = ASSOCIATION FOR THE PROTECTION OF RURAL SCOTLAND

apricot ['eɪprɪkɒt] *noun* deciduous tree *(Prunus armeniaca)* bearing soft yellow fruit, similar to a small peach, but not as juicy

aqu- ['ækw] *prefix* meaning water

◊ **aquaculture** *or* **aquiculture** *or* **aquafarming** ['ækwəkʌltʃə *or* 'ækwɪkʌltʃə *or* 'ækwəfɑːmɪŋ] *noun* fish farming, breeding fish for food in special ponds

◊ **aquatic** [ə'kwætɪk] *adjective* (animals *or* plants) living in water

◊ **aqueous** ['eɪkwɪəs] *adjective* containing water; (solution) in water; **aqueous ammonia** = ammonia in solution, obtained from gas works and as a by-product; sometimes used as a fertilizer; **aqueous solution** = a solution of a substance in water

◊ **aquifer** ['ækwɪfə] *noun* porous rock *or* soil through which water passes and in which water gathers to supply wells. Aquifers can be used as a source of irrigation water

Arab ['ærəb] *noun* breed of horse, especially used for racing. Arabian horses are those in whose pedigree there is no other blood than Arabian. Of great antiquity, they are the purest of the equine races. Nearly every breed has had an infusion of Arab blood; *see also* AHS

Arabica [ə'ræbɪkə] *noun* coffee obtained from the coffea arabica shrub, originally grown in the southern parts of the highlands of Ethiopia, from where it was introduced into south-western Arabia.

Arabica coffee beans are generally considered to produce a milder and higher quality drink than those obtained from the Robusta coffee plant

Arabis Mosaic ['ærəbɪs mə'zeɪɪk] *noun* viral disease which infects a wide range of horticultural crops; leaves become stunted and plants die back

arable ['ærəbl] *adjective* (land) on which crops are grown; **arable crops** = crops which are cultivated on ploughed land; they are annual crops and include cereals, root crops and potatoes; **arable farming** = growing crops (as opposed to dairy farming, cattle farming, etc.); **arable land** = land suitable for ploughing; land which has been ploughed and cropped

COMMENT: the UK arable sector grows 3.5 million hectares of cereals with an estimated value of £2.5 billion (1994). The other important arable crops are sugar beet, potatoes and oilseed rape

arachidonic acid ['ærəkɪdɒnɪk 'æsɪd] *noun* one of the essential fatty acids

Arachnida [ə'ræknɪdə] *noun* class of animals with eight legs (such as spiders, mites, etc.)

COMMENT: Arachnida have pincers on the first pair of legs. Their bodies are divided into two parts, and they have no antennae

arbor- ['ɑːbɔː] *prefix* referring to trees

◊ **arboretum** [ɑːbə'reɪtəm] *noun* collection of trees from different parts of the world, grown for scientific study

◊ **arboriculture** [ɑː'bɒrɪkʌltʃə] *noun* cultivation of trees and shrubs

arctic ['ɑːktɪk] *adjective* referring to the region around the North Pole

ard [ɑːd] *noun* the scratch plough, made of a sharpened piece of wood which is dragged across the surface of the soil; this was the first type of plough to be used for cultivating the land (first used in South-West Asia)

Ardennes [ɑː'den] *noun* breed of horse; a short, hardy French breed found in Britain mainly on hill farms

are [eə] *noun* unit of metric land measurement; it equals 100 square metres

area ['eəriə] *noun* region of land; *the whole*

area has been contaminated by waste from the power station; **Area of Outstanding Natural Beauty (AONB)** = region in England and Wales which is not a National Park, but which is considered attractive enough to be preserved from being overdeveloped

Areca [ə'ri:kə] *see* BETEL NUT

arenaceous [ærɪ'neɪʃəs] *adjective* sandy; **arenaceous soils** = sandy soils, soils which have a high amount of sand particles

argente ['ɑ:ʒɒnt] *noun* French breed of rabbit

argillaceous [ɑ:dʒɪ'leɪʃəs] *adjective* clayey; **argillaceous soils** = soils with a high amount of clay particles

ARIA = ARABLE RESEARCH INSTITUTE ASSOCIATION

arid ['ærɪd] *adjective* (soil) which is very dry; (area of land) which has very little rain; **arid zone** = area in the tropics (between about 15° and 30° north and south) which is very dry and covered with deserts

◊ **aridity** [ə'rɪdɪtɪ] *noun* state of being extremely dry

aril ['ærɪl] *noun* an extra covering to the seed, found in certain plants

ark [ɑ:k] *noun* (i) small hut made of wood or corrugated metal, used to shelter sows kept outdoors in fields; (ii) also a mobile poultry house with wire or slatted floors, formerly used to house chickens at night and in bad weather

aromatic [ærə'mætɪk] *adjective* (plant) which has a strong pleasant smell; **aromatic herbs** = herbs, such as rosemary or thyme, which are used to give a particular taste to food

ARP = ACREAGE REDUCTION PROGRAM

arpent ['ɑ:pənt] *noun* old French measure of land, varying between 1¼ and 1⅜ of an acre (now only used occasionally in Quebec)

arrowroot ['ærəʊru:t] *noun* tropical plant *(Maranta arundinacea)* whose tubers yield starch. Used particularly in the preparation of invalid foods, since this form of starch is easily digested

arsenic ['ɑ:sənɪk] *noun* chemical element which forms poisonous compounds, such as arsenic trioxide, and is used to kill rodents (NOTE; symbol is **As;** atomic number is **33)**

◊ **arsenical** [ɑ:'senɪkl] *noun* one of the group of poisonous oxides of arsenic

arterial drain [ɑ:'tɪərɪəl 'dreɪn] *noun* main watercourse which carries water from smaller drains or ditches

artesian well [ɑ:'ti:zɪən 'wel] *noun* well which has been bored into a confined aquifer; the hydrostatic pressure is usually strong enough to force the water to the surface

arthropod ['ɑ:θrəpɒd] *noun* one of a very large group of animals with an external skeleton, including insects and crustaceans

artichoke ['ɑ:tɪtʃəʊk] *noun* vegetable grown as a specialized crop

COMMENT: there are two types of artichoke: the globe artichoke, a tall thistle-like plant (the leaves of the head are cooked and eaten), and the Jerusalem artichoke, a tall plant which develops tubers like a potato

artificial [ɑ:tɪ'fɪʃl] *adjective* which is made by man *or* which does not exist naturally; **artificial community** = plant community kept by man (as in a garden); **artificial manure** = fertilizer made by chemical processes or mined; **artificial rain** = rain which is made by scattering crystals of salt and other substances into clouds

◊ **artificial insemination (AI)** [ɑ:tɪ'fɪʃəl ɪnsemɪ'neɪʃn] *noun* way of breeding livestock by fertilizing females with sperm from specially selected males; AI is important in cattle breeding. Milk Marque operates a number of AI centres

QUOTE AI is obligatory for breeding modern strains of turkey because the size difference between males and female is such that natural mating is very difficult
Poultry Science Symposium 17

arvensis [ɑ:'vensɪs] *Latin word meaning* 'of the fields', used in many genetic names of plants

asbestos sheets *or* **asbestos-**

cement [æz'bestəs 'ʃiːts *or* sı'ment] *noun* material used to make prefabricated buildings; it is light and strong and can be cast into sheets which can be used as roofs for light farm buildings

Ascaris ['æskərıs] *noun* nematode worm which infects the small intestine

Ascochyta [æskəʊ'ʃiːtə] *noun* disease affecting pulses, such as peas and beans

ascorbic acid [ə'skɔːbık 'æsıd] *noun* Vitamin C, found in fresh fruit

ASCS = AGRICULTURAL STABILIZATION AND CONSERVATION SERVICE

asexual [eı'sekʃjuəl] *adjective* not sexual *or* not involving sexual intercourse; **asexual reproduction** = reproduction by taking cuttings of plants *or* by cloning, etc.

◊ **asexually** [eı'sekʃjuəli] *adverb* not involving sexual intercourse; *by taking cuttings it is possible to reproduce plants asexually*

ash [æʃ] *noun* **(a)** European hardwood tree *(Fraxinus excelsior)* **(b)** grey *or* black powder formed of minerals left after an organic substance is burnt; wood ash is particularly rich in potash and may be used as a fertilizer

asparagus [ə'spærəgəs] *noun* native European plant *(Asparagus officinalis)* of which the young shoots (called 'spears') are cut when they are about 25cm long and are eaten as a vegetable

◊ **asparagus pea** [ə'spærəgəs 'piː] *noun* plant *(Lotus tetragonolobus)* grown in southern Europe; the pods are edible

aspect ['æspekt] *noun* the direction in which land faces. It can affect the amount of sunshine (and heat) absorbed by the soil. In the UK, the temperature of north-facing slopes may be 1°C lower than on similar slopes facing south

aspen ['æspən] *noun* North American hardwood, a poplar tree *(Populus tremula)*

aspergillosis [æspɜːgı'ləʊsıs] *noun* infection of the lungs with *Aspergillus,* a type of fungus which affects parts of the respiratory system

ass [æs] *noun* long-eared animal of the horse family. Used as a draught animal and still important in Mediterranean countries, where it thrives better than the horse on scanty herbage

association [əsəʊsi'eıʃn] *noun* group of plants living together in a large area, forming a stable community; **Association of Agriculture** = a voluntary body formed in 1947 to promote a better understanding of farming and the countryside. It provides educational material to schools and arranges farm visits and farm open days for the general public

assortive mating [ə'sɔːtıv 'meıtıŋ] *noun* mating animals that have a similar appearance (NOTE: also called **mating likes)**

asulam ['æsjuləm] *noun* powerful herbicide, used to remove difficult weeds such as bracken

atavism ['ætəvızm] *noun* inheriting a disease from remote ancestors, the immediate parents of the diseased animal having not been affected

ATB = AGRICULTURAL TRAINING BOARD

at foot suckling ['æt fʊt 'sʌklıŋ] *noun* ewe with suckling lamb

atrazine ['ætrəziːn] *noun* a residual herbicide which acts on the soil

atrium ['eıtriəm] *noun* one of the two upper chambers of the heart

atrophic rhinitis [ə'trɒfık ræı'naıtıs] *noun* bacterial disease of young pigs causing inflammation of the nasal passages; can cause deformity of the snout

atropine ['ætrəpiːn] *noun* the alkaloid contained in the leaves and root of the deadly nightshade; it can cause atropine poisoning in animals which feed on the plant

attachment [ə'tætʃmənt] *noun* device which can be attached to a machine, such as a straw chopper which can be attached to a combine harvester

attenuated strains [ə'tenjuːeıtıd 'streınz] *noun (in vaccines)* pathogenic microorganisms (mainly bacteria and viruses) which have lost their virulence

attest [ə'test] *verb* to certify; **attested area** = an area declared to be free from a specific animal disease; **attested herd** = a herd tested and found to be free of bovine tuberculosis

◊ **attestation** [ætes'teɪʃn] *noun* action of attesting; **certificate of attestation** = certificate given to an attested herd

COMMENT: in 1960, the whole of the UK was declared an attested area, since bovine tuberculosis had by then been almost completely eradicated from all herds of cattle

attractant [ə'træktənt] *noun* substance emitted by an animal which attracts individuals of the opposite sex

ATV = ALL TERRAIN VEHICLE 4x4 lightweight vehicle with great mobility

aubergine ['əʊbədʒiːn] *noun* purple fruit of the eggplant *(Solanum melongena)* , used as a vegetable. A native of tropical Asia, it is sometimes called by its Indian name 'brinjal'

Aubrac ['ɔːbræk] *noun* rare breed of cattle from southern France; the animals are light yellow or brown in colour and are used for draught, milk and meat

auction ['ɔːkʃən] **1** *noun* selling of goods where people offer bids, and the item is sold to the person who makes the highest offer; **livestock auction** = auction sales where livestock are shown in a ring and sold to the highest bidder **2** *verb* to sell at an auction

◊ **auctioneer** [ɔːkʃə'niːə] *noun* person who conducts an auction

auger ['ɔːgə] *noun* **(a)** a tool for boring holes, having a long shank with a cutting edge and a screw point; soil augers are used to obtain samples of soil for analysis **(b)** device on a combine harvester (shaped like a large screw) which carries the grain up into the grain tank

Aujesky's disease [aʊ'jeskɪs dɪ'ziːz] *noun* a virus disease of animals, characterized by intense itching. The disease is most serious in young pigs, and often leads to death. A notifiable disease

auricle ['ɔːrɪkl] *noun* **(a)** small projection at the base of a leaf blade **(b)** tip of each atrium in the heart

Australorp ['ɒstrəlɔːp] *noun* Australian breed of chicken; the birds are black with red combs

auto- ['ɔːtəʊ] *prefix* meaning self

◊ **autogamy** [ɔː'tɒgəmi] *noun* pollination with pollen from the same flower, i.e. self-fertilization

automatic cluster removal (ACR) [ɔːtə'mætɪk 'klʌstə rɪ'muːvəl] *noun* automatic removal of the cluster of the milking machine from the udder at the end of milking

◊ **automatic pick-up hitch** *noun* mechanism on a tractor, operated by the hydraulic system, which allows the driver to hitch up to a trailer without leaving his seat

◊ **automation** [ɔːtə'meɪʃn] *noun* use of machinery to save manual labour

COMMENT: automation has considerably reduced the labour involved in dairying. It has now been adopted almost everywhere for the milking process (removing the cluster of the milking machine at the end of milking is done automatically) and for the making of conserved fodder

autotroph *or* **autotrophic organism** ['ɔːtətrɒf *or* ɔːtə'trɒfɪk] *noun* organism (such as a green plant) which takes its energy from the sun; *compare* HETEROTROPH

autumn fly ['ɔːtəm 'flaɪ] *noun* an irritating non-biting fly *(Musca autumnalis)*

auxin ['ɔːkzɪn] *noun* plant hormone which encourages tissue growth and so makes plants grow and fruit swell

COMMENT: some herbicides act as synthetic auxins by upsetting the balance of the plant's growth

available water [ə'veɪləbl 'wɔːtə] *noun* water which can be taken up by the plant roots; **available water capacity (AWC)** = the amount of water held by a soil between the amounts at field capacity and wilting point

Avena [æ'viːnə] *Latin name for* the oat family

Average All Pigs Price (AAPP) ['ævrɪdʒ ɔːl 'pɪgz praɪs] *noun* the average price for pigs, calculated each week and used in contracts for payment

Aves ['eɪviːz] *Latin name for* birds, the class of egg-laying feathered animals which are adapted to fly

aviary system ['eɪvjəri 'sɪstəm] *noun* a poultry housing system using a deep litter house with extra raised areas inside it for the birds to feed, roost and exercise in

avocado [ævə'kɑːdəʊ] *noun* pear-shaped green fruit of a tree *(Persea americana)* which is native of South and Central America; it is cultivated in Israel, Spain, the USA and elsewhere. The fruit has a high protein and fat content, making it very nutritious and an important food crop

awassi [ə'wæsi] *noun* breed of sheep found in Syria; it is suitable for semi-arid conditions

away-going crop [ə'weɪ 'gəʊɪŋ krɒp] *noun* crop sown by a tenant farmer before leaving the farm at the end of his tenancy; he is permitted to return and harvest the crop and remove it (NOTE: also known as **following crop** or **off-going crop**)

AWC = AVAILABLE WATER CAPACITY

awn [ɔːn] *noun* tip of a leaf which ends in a spine; in cereals awns are attached to the grains, and in barley each grain has a particularly long awn

AWT = ANIMAL WELFARE TRUST

axil ['æksɪl] *noun* upper angle between a leaf and the stem from which it springs

◊ **axillary** [æk'sɪləri] *adjective* referring to an axil; **axillary bud** = bud in the angle between a leaf and the main stem, producing a side shoot, as in the case of tomatoes

Aylesbury ['eɪlzbəri] *noun* a heavy table breed of duck, with white feathers

Ayrshire ['eəʃə] *noun* breed of dairy cow, originating in South-West Scotland. It is deep and broad at the hips and narrow at the shoulders. It was the main rival to the Friesian as a milk producer. It is hardy, and white and brown in colour

Azarole [æzə'rəʊl] *noun* small tree *(Crataegus azarolus)* bearing fruit which are used for making jellies; it is a native of the Mediterranean area

azobacter ['eɪzəʊbæktə] *noun* nitrogen-fixing bacteria present in the soil

Bb

B *chemical symbol for* boron

B horizon ['biː həˈraɪzn] *noun* the subsoil; *see* HORIZON

Vitamin B ['vɪtəmɪn 'biː] *noun* **Vitamin B complex** = group of vitamins which are soluble in water, including folic acid, pyridoxine, riboflavin and many others; **Vitamin B₁** = thiamine, vitamin found in yeast, liver, cereals and pork; **Vitamin B₂** = riboflavin, vitamin found in eggs, liver, green vegetables, milk and yeast; **Vitamin B₆** = pyridoxine, vitamin found in meat, cereals and molasses; **Vitamin B₁₂** = cyanocobalamin, vitamin found in liver and kidney, but not present in vegetables; *see note at* VITAMIN

BAA = BRITISH AGROCHEMICALS ASSOCIATION

Baars irrigator ['baːz 'ɪrɪɡeɪtə] *noun* a flexible hose with sixteen folding sprinklers, used for irrigating

baby beef ['beɪbi biːf] *noun* meat from cattle slaughtered around 12 months of age

BAC = BRITISH AGRICULTURAL COUNCIL

bacillus [bəˈsɪləs] *noun* bacterium shaped like a rod (NOTE: plural is **bacilli**)

◊ **bacillary** [bəˈsɪləri] *adjective* referring to bacillus; **bacillary dysentery** = dysentery caused by the bacillus *Shigella* in contaminated food; **bacillary white diarrhoea** = acute, infectious and highly fatal disease of chicks, caused by *Salmonella pullorum*

back [bæk] *verb (of wind)* to change direction, moving in an anticlockwise direction (NOTE: the opposite is **veer**)

back band ['bæk bænd] *noun* band of rope or chain over a cart saddle, to keep the shafts up

◊ **backboard** ['bækbɔːd] *noun* board at the back of a cart

◊ **backcrossing** ['bækkrɒsɪŋ] *noun* breeding a crossbred offspring back to one of its parents, usually a purebred

◊ **back end** ['bæk 'end] *noun* late autumn

background ['bækɡraʊnd] *noun* conditions which are always present in the environment, but are less obvious (or less important) than others; **background carboxyhaemoglobin level** = level of carboxyhaemoglobin in the blood of a person living a normal existence without being exposed to particularly high levels of carbon monoxide; **background radiation** = radiation which comes from natural sources like rocks *or* the earth *or* the atmosphere and not from a single man-made source

COMMENT: background radiation can depend on the geological structure of the area. Places above granite are particularly subject to high levels of radiation. Other sources of background radiation are: cosmic rays from outer space, radiation from waste products of nuclear power plants which has escaped into the environment, radiation from TV and computer screens

bacon ['beɪkn] *noun* cured back and sides of a pig; bacon may be green or smoked

◊ **baconer** ['beɪknə] *noun* pig bred and reared for bacon

COMMENT: bacon is cured in brine for several days; some bacon is also smoked by hanging in smoke, which improves its taste; unsmoked bacon is also known as 'green' bacon

bacteria [bækˈtɪəriə] *plural noun* submicroscopic organisms, which help in the decomposition of organic matter, some of which are permanently present in the intestines of animals and can break down food tissue; many of them can cause disease; **bacteria beds** = filter beds of rough stone, forming the last stage in the treatment of sewage (NOTE: the singular is **bacterium**)

◊ **bacterial** [bæk'tɪərɪəl] *adjective* referring to bacteria *or* caused by bacteria

> COMMENT: bacteria can be shaped like rods (bacilli), like balls (cocci) or have a spiral form (such as spirochaetes). Bacteria, especially bacilli and spirochaetes, can move and reproduce very rapidly

bacterial canker [bæk'tɪərɪəl 'kæŋkə] *noun* disease affecting the plant genus *Prunus*. Stems of the plant swell and exude a light brown gum and can cause browning of foliage (*Pseudomonas morsprunorum*)

bacteriophage [bæktɪərɪəʊ'feɪdʒ] *noun* virus which affects bacteria

Bactrian camel ['bæktrɪən 'kæməl] *noun* Asian camel, found in the dry deserts of Mongolia and Turkestan; it has two humps

badger ['bædʒə] *noun* white-faced grey-coated carnivore, which is harmless as far as crops are concerned and useful as it kills insects, slugs and mice. It is possible that bovine tuberculosis is transmitted by badgers

badlands ['bædlænz] *noun* (*in North and South America*) intensely dissected area of land which is unsuitable for agriculture

BAEC = BRITISH AGRICULTURAL EXPORT COMMITTEE

bag [bæg] **1** *noun* (**a**) unit of measurement used in the coffee trade (one bag contains 60 kilos) (**b**) udder of a cow **2** *verb* to cut wheat with a sickle

bagasse [bæ'gæs] *noun* fibres left after sugar cane has been crushed

BAGMA = BRITISH AGRICULTURAL AND GARDENING MACHINERY ASSOCIATION

Bagot ['bægət] *noun* rare breed of goat; the preferred colouring is black head, neck and shoulders, with the rest of the body white. The hair is long. Both sexes have horns

bail [beɪl] *noun* (**a**) a bar separating horses in a shed (**b**) milking shed which can be moved from place to place

bailiff ['beɪlɪf] *noun* person employed to manage a farm on behalf of the owner

bait [beɪt] *noun* feed of hay and oats for horses

bajra ['bɑːdʒrə] *noun* Indian name for pearl millet

bake [beɪk] *verb* to cook in an oven; **baking powder** = powder formed of dry acid which reacts with water to produce carbon dioxide, and so aerate dough; **baking quality** = ability of flour to retain carbon dioxide bubbles until it has become hard; *see also* STRENGTH

◊ **bakers** ['beɪkəz] *noun* large potatoes, graded by size, and used to cook as baked potatoes

Bakewell ['beɪkwel] Sir Robert Bakewell (1735-1795), an important Leicestershire farmer whose experiments on longhorn cattle showed that the way to achieve rapid improvement in stock was by carefully controlled inbreeding; he also originated the improved Leicester sheep

Baladi [bə'lɑːdi] *noun* type of Egyptian cattle; *see also* EGYPTIAN

balance ['bæləns] *noun* state where two sides are equal; **the balance of nature** *or* **ecological balance** = situation where different organisms live in a stable state in the same ecosystem; **to disturb the balance of nature** = to make a change to the environment which has the effect of putting some organisms at a disadvantage against others; **water-salt balance** = situation where the level of water in the soil balances the amount of salt in it

◊ **balanced diet** ['bælənsd 'daɪət] *noun* diet which provides the animal with all the nutrients it needs in the correct proportions. Deficiency in the diet of animals leads to acute nervous disorders

bald faced [bɔːld 'feɪst] *adjective* (animal) with white on the face

bale [beɪl] **1** *noun* (**a**) package of hay or straw, square, round or rectangular in shape, usually tied with twine; **big bale** = a large bale, weighing from 600 to 900 kilos; **big bale silage (BBS)** = silage stored in big bales; **bale buster** = machine for unrolling big bales of straw, hay and silage, for bedding and feeding; **bale loader** = attachment to the front loader arms of a tractor, which allows it to lift bales onto a trailer or into a barn; **bale sledge** = sledge pulled behind a baler to collect the finished bales; **bale wrapper** = machine mounted on a tractor used to wrap film around a bale

(b) standard pack for cotton **2** *verb* to take loose straw, grass or hay and press it into bales

◊ **baler** ['beɪlə] *noun* machine that picks up loose hay or straw and compresses it into bales of even size and weight, and then ties them with twine. The completed bale is dropped onto the ground at the back of the baler; **baler twine** = cord used to tie bales, now often made of rot-proof polypropylene

COMMENT: there are two types of big bale, the round and the rectangular. The baling is slower with the round bale, but the completed bale is more weatherproof and can be safely left out in the field for longer periods of time. One of the main advantages of the big baler is that the number of bales per hectare is very low. This makes handling from the field to the barn very much simpler, provided the right equipment is used

QUOTE big bale silage is much easier to handle or to feed than either hay or conventional silage
Practical Farmer

balk [bɔːlk] *see* BAULK

balling ['bɔːlɪŋ] *noun* dosing of larger animals, by rolling the drug into a ball which is then swallowed; **balling gun** = instrument used to force balls of drugs down the throats of animals

balm [bɑːm] *noun* an aromatic herb *(Melissa officinalis)* grown for its lemon scented leaves

Bambarra groundnut [bæm'bærə 'graʊndnʌt] *noun* an exotic legume *(Vaandzeia subterranea)*. The seeds have a low oil content and are grown for their protein and starch. Found mainly in West Africa, and also in Zambia

bamboo [bæm'buː] *noun* plant *(Bambusa vulgaris)* growing mainly in temperate and subtropical regions of China, Japan and Korea; bamboo produces very long straight stems which are cut and dried and used in construction work; in horticulture, thin bamboo stems are widely used as supports for plants; **bamboo shoots** = young shoots of the bamboo plant which are cut and cooked as vegetables

Bampton Nott ['bæmptən nɒt] *noun* ancient breed of sheep, the ancestor of the Devon Longwool

banana [bə'nɑːnə] *noun* fruit of a large tropical plant *(Musa sapientum)* and its varieties

COMMENT: The banana is grown on plantations, especially in South and Central America, the West Indies, the Canary Islands and the Philippines. All varieties of banana need high temperatures, a great deal of moisture and deep soil. The bananas grow in 'hands' of 12 to 16 fingers along a main stem, together forming a bunch. Another species of *Musa* yields the plantain, which is grown in the tropics for food

BANC = BRITISH ASSOCIATION OF NATURE CONSERVATIONISTS

banding ['bændɪŋ] *noun* placing greasebands round the trunks of fruit trees to trap insects

band sprayer ['bænd 'spreɪə] *noun* crop sprayer that applies chemicals in narrow strips, mostly used with precision seeders

bane [beɪn] = LIVER FLUKE

bantam ['bæntəm] *noun* small breed of domestic fowl, most kinds being about half the size and weight of the common fowl

Banteng ['bænteŋ] *noun* breed of cattle found in Burma, Thailand, Malaysia and Indonesia. The cows are bright reddish-brown in colour with white patches; mature bulls are black. The animals are used for draught and for meat

banyan ['bænjæn] *noun* Indian fig tree

Bapedi [bə'piːdi] *noun* breed of black and white dairy cattle found in Eastern Transvaal

barb [bɑːb] *noun* breed of horse originating in Morocco, influential in the development of the thoroughbred

barban ['bɑːbən] *noun* herbicide for use on many varieties of winter wheat and barley to control wild oats

bar code ['bɑː kəʊd] *noun* data represented as a series of printed stripes of different widths, which can be read by a bar-code reader

COMMENT: bar codes are found on most goods and their packages; the width and position of the stripes can be recognized by the reader and give information about the goods, such as price, stock quantities, etc. Many packaged foods, even fresh foods, are bar-coded to allow quicker data capture in the supermarket

bare [beə] *adjective* (land) with no plants growing on it; **bare fallow** = land left fallow and ploughed several times during the course of a year, the aim being to get rid of weeds

◊ **bareback** ['beəbæk] *adjective* riding an unsaddled horse

bark [bɑːk] *noun* protective hard outer layer covering the trunk and branches of a tree, formed of dead tissue

◊ **barking** ['bɑːkɪŋ] *noun* removing the bark of a tree (as in harvesting cork from cork oaks); **ring-barking** = the cutting of a strip of bark from a tree as a means of making the tree more productive; it restricts growth and encourages fruiting

barley ['bɑːli] *noun* common cereal crop *(Hordeum sativum)*, grown in temperate areas; **barley yellow dwarf virus (BYDV)** = virus spread by aphids, which causes poor root development in plants. Red and yellow colour changes occur in leaves and yields are much reduced. Also affects wheat and grass

COMMENT: barley has six spikelets at each node. If each of these develops it forms a six-rowed barley; in other varieties the lateral spikelets are much smaller than the central one, and these form four-rowed or two-rowed barley. Barley is widely grown in northern temperate countries, with the largest production in Germany and France; it is a very important arable crop in the UK. The grain is mainly used for livestock feeding (six-rowed barley) and for malting (two-rowed barley) for use in producing alcoholic drinks. It is rarely used for making flour. Barley straw is used mainly for litter, but some is fed to cattle

barn [bɑːn] *noun* farm building used for storing hay or grain; in the USA, barns are also used for keeping animals; **barn-dried hay** = hay dried indoors by blowing air through it; usually more nutritious than field-dried hay; *see also* DUTCH BARN

◊ **barnyard** ['bɑːnjɑːd] *noun* farmyard, the area round a barn

Barnevelder ['bɑːnveldə] *noun* breed of fowl

bar pig ['bɑː pɪg] *noun* castrated male pig (NOTE: also called **barrow** or **hog**)

bar point ['bɑː pɔɪnt] *noun* spring-loaded type of a plough body, useful when boulders occur below the surface of the land being ploughed

barren ['bærən] *adjective (animals)* which cannot bear young; *(plants)* which do not produce fruit; *(land)* which is not fertile; **barren brome** = widespread weed *(Bromus sterilis)* which affects winter cereals

◊ **barrener** ['bærənə] *noun* female animal unable to produce young

Barrosa [bə'rəusə] *noun* breed of cattle found in northern Portugal; used both for draught and for meat. The animals are reddish-brown with large curved horns

barrow ['bærəu] *noun* male pig after castration, while a suckler or weaner

basal metabolism ['beɪsəl me'tæbəlɪzm] *noun* energy used by a body at rest (i.e. energy needed to keep the body functioning and the temperature normal); this can be calculated while an animal is in a state of complete rest, by observing the amount of heat given out or the amount of oxygen taken in and retained; **basal metabolic rate (BMR)** = rate at which a body uses energy when at rest

BASC = BRITISH ASSOCIATION FOR SHOOTING AND CONSERVATION

Bashi ['bæʃi] *noun* East African breed of cattle, small in size and short horned; the colour varies from red to fawn or black

basic ['beɪsɪk] *adjective* **(a) basic price** = a support price fixed each year by the EU for certain fruit and vegetables; **basic slag** = waste from furnaces, which is used as a fertilizer because of its phosphate content **(b)** (chemical substance) which reacts with an acid to form a salt

basidiomycetes [bəsɪdɪəumaɪ'siːtiːz] *noun* large group of fungi, including mushrooms and toadstools

basil ['bæzəl] *noun* aromatic herb *(Ocimum basilicum)*, used in cooking, particularly with fish. Although probably native to India, basil is widely grown in Europe, especially in Italy. It is frost-tender, and usually grown as an annual in northern areas

basin irrigation ['beisin iri'geiʃn] *noun* form of irrigation where the water is trapped in basins surrounded by low mud walls (as in rice paddies)

BASIS 1 = BRITISH AGROCHEMICAL SUPPLY INDUSTRY SCHEME LTD scheme for registering distributors of chemicals used to protect crops, so that dangerous chemicals are stored and used correctly **2** = BRITISH AGROCHEMICALS STANDARDS INSPECTION SCHEME

basmati [bæs'mɑːti] *noun* common variety of long-grain rice

bastard fallow ['bɑːstəd 'fæləu] *noun* land left fallow for the time between harvesting and sowing, usually ploughed to control weeds

batch drying ['bætʃ 'draɪŋ] *noun* a process for drying bales of hay in batches; bales are placed on a platform and heated air is blown up between the bales

batt [bæt] *noun* long wooden handle of a scythe

battery ['bætri] *noun* system of raising chickens for egg production, where thousands of birds are kept in small cages; **battery hens** = laying birds reared and kept in multi-bird cages arranged in rows one above the other

COMMENT: a method of egg production which is very energy-efficient. It is criticized, however, because of the quality of the eggs, the possibility of disease and the polluting substances produced, and also on grounds of cruelty because of the stress caused to the birds. See also FREERANGE EGGS

baulk [bɔːlk] *noun* narrow strip of land left unploughed to mark the boundary between fields (NOTE: also written **balk**)

bawon [bə'wɒn] *noun* system used in Java for harvesting rice: the villagers who harvest the rice each receive a share of the crop

bay [bei] *noun* **(a)** stall in a stable **(b)** horse of a reddish-brown colour **(c) bay (laurel)** =

evergreen shrub *(Laurus nobilis)* with aromatic leaves used for flavouring

COMMENT: not to be confused with the common laurel *(Kalmia)* whose leaves are highly poisonous

BBS = BIG BALE SILAGE

BCPC = BRITISH CROP PROTECTION COUNCIL

BDM = BLEU DE MAINE

BDPS = BRITISH DEER PRODUCERS SOCIETY

BEA = BRITISH EGG ASSOCIATION

beak [biːk] *noun* hard parts forming the mouth of a bird; also called the 'bill'

beam [biːm] *noun* the main frame of a plough, to which the soil-engaging parts are attached

beans [biːnz] *noun* various varieties of legumes with edible seeds; **French bean** = *Phaseolus vulgaris,* a common green vegetable grown for sale fresh or for canning and freezing (its dried ripe seeds are known as haricot beans); **runner bean** = *Phaseolus coccineus,* a common vegetable crop grown almost only for the fresh trade; **bean aphid** = blackfly *(Aphis fabae).* Very small, oval-bodied, black or dark green fly; on summer host plants, colonies of aphids make the plants wilt; **bean stem rot** = fungus disease of beans; *see also BLACKEYED, BROAD, BUTTER, GRAM, HARICOT, KIDNEY, NAVY, PEA, PIGEON, RUNNER, SNAP, STRING*

COMMENT: bacteria on the roots of bean crops can fix nitrogen and crops following will benefit from the nitrogen left in the soil by the beans. Beans normally have a single square hollow stem bearing a large number of compound bluish-green leaves. Black and white flowers appear in spring, developing into green pods. The main types of beans are field beans *(Vicia faba),* used for stock feeding, or for producing broad beans, which are the immature seeds used for human consumption. Field beans are usually grown as a break crop. Winter beans are sown in October and spring beans in February/March. Spring beans include tick beans (with small seeds), minor (very small seeds, used to feed racing pigeons) and horse beans (large seeds)

beastings ['biːstɪŋz] *see* COLOSTRUM

beaters ['biːtəz] *noun* **(a)** steel bars on the drum of a combine harvester **(b)** people who rouse game in hunting or shooting

Beaumont period ['bəʊmɒnt 'pɪəriəd] *noun* period of 48 hours during which temperatures do not fall below 10°C (50°F) and relative humidity remains above 75%; potato blight is likely to occur within 21 days of this, though spraying may prevent the blight from appearing

bed [bed] **1** *noun* specially planted area of land, such as an asparagus bed, flower bed or a strawberry bed **2** *verb* **to bed out** = to plant a flower bed with plants, especially in such a way as to give a decorative effect; **bedded set** = young hop plant rooted from a cutting

◊ **bedding** ['bedɪŋ] *noun* **(a)** materials such as straw, shavings or sand, used as litter for animals to lie on **(b)** planting out small flower plants into a flower bed; **bedding plants** = small annual flower plants which are used for bedding out

bee [biː] *noun* four-winged stinging insect; honey bees are kept in hives for commercial production of honey; *see also* APIARY

◊ **beehive** ['biːhaɪv] *noun* housing for a colony of bees, containing the frames in which bees store honey

◊ **beekeeper** ['biːkiːpə] *noun* person who keeps bees for honey

beech [biːtʃ] *noun* common temperate hardwood tree *(Fagus sylvatica)*. The timber is used in furniture making

beef [biːf] *noun* meat of bull or cow; **beef bull** = bull reared to produce beef; **beef cow** = cow kept for rearing calves for beef production; **beef shorthorn** = compact short-legged breed of beef cattle, colour may be red, red and white, white or roan

◊ **beefalo** ['biːfələʊ] *noun* American breed of cattle, brown in colour; it is derived from crossing bison, Charolais and Hereford breeds

◊ **beef carcass classification scheme** *or* **beef carcass service** a scheme, set up in the UK in 1989, to provide authentication of carcass weight and dressing specification; classification, labelling, sexing and identification of breed are options. The service is offered in slaughterhouses by the Meat and Livestock Commission

◊ **Beef Premium Scheme** an EU scheme that applies only in the UK. In 1988, the UK variable premium scheme was replaced by a single premium of £27 per head, paid once for each male adult and restricted to 90 animals per holding per year. There will be a restriction on intervention buying by limiting the total amount of beef put into intervention each year, with some 4 million cattle marketed each year. The value of home production in 1994 was £2 billion, around 15% of the total value of agricultural output

QUOTE the variable Premium Scheme (Beef) is replaced by the Special Beef Premium
Beef Year Book 1989

beer [biːə] *noun* alcoholic drink made by fermenting malted barley with large quantities of water

beet [biːt] *noun* **(a)** plant with a succulent root, a source of sugar; **weed beet** = wild beet which can be self-sown in beet crops **(b)** *US* = BEETROOT

◊ **beetroot** ['biːtruːt] *noun* salad vegetable with a bright red root

beet flea beetle [biːt 'fliː 'biːtl] *noun* sugar beet pest *(Chaetocnema concinna)*

beetle ['biːtl] *noun* any insect whose upper wings form hard wing-cases

beggary ['begəri] *noun* popular name for the common fumitory, a widespread weed *(Fumaria officinalis)*

BEIC = BRITISH EGG INDUSTRY COUNCIL

Belgian Blue ['beldʒən 'bluː] *noun* dual-purpose breed of cattle, the result of crossing Friesians and Shorthorns with native Belgian stock. The animals are coloured blue and white; there is a British breed, derived from this, called British Belgian Blue. Cannot give birth normally and have to be caesared

◊ **Belgian hare** ['beldʒən 'heə] *noun* breed of rabbit

Belgian Heavy Draught ['beldʒən hevi 'drɔːft] *noun* breed of horse of great weight and traction power

belladonna [belə'dɒnə] *noun* another name for the deadly nightshade

bellows ['beləʊz] *noun* instrument for blowing air into a fire, used by blacksmiths

bellwether ['belweðə] *noun* sheep with a bell hung round its neck, which leads a flock

belly ['beli] *noun* underside of an animal; **pork bellies** = part of a pig which is processed to produce bacon

belt [belt] *verb* to clean out a sheep's fleece with shears

◊ **belt drive** ['belt 'draɪv] *noun* transmission of power from an engine or electric motor to another machine, by means of a belt

Belted Galloway ['beltɪd 'gæləweɪ] *noun* breed of beef cattle, coloured black with a white belt round the body. The animals are long-haired with a dense undercoat. It is a medium-sized polled breed used for cross-breeding with White Shorthorn bulls to produce blue-grey suckler cows

Beltsville ['beltsvɪl] *noun* breed of large turkey, with white feathers

benazolin [bə'næzəliːn] *noun* herbicide for controlling broadleaved weeds

Beneder ['benɪdə] *noun* African breed of large goat, found in Somalia; the animals are red or black-spotted

bennet ['benɪt] *see* BENT

benniseed ['benɪsiːd] *noun* West African name for sesame, an erect herb which grows to two metres tall. The seeds yield a valuable oil. Also called 'sim-sim' in East Africa; *see also* SESAME

COMMENT: the crop is of African origin, but is now grown in tropical, subtropical and Mediterranean areas. It is important in India, China, Sudan and Nigeria. The seeds are crushed to make an oil used in cooking, and the sesame cake which remains after the oil has been extracted is an important protein-rich stock feed

benomyl ['benəmiːl] *noun* fungicide used against eyespot in cereals

bent [bent] *noun* (i) stiff-stemmed grasses of the genus *Agrostis*, found in hill pastures and tolerant of poor conditions; (ii) old dry stalks of dead grass (NOTE: also called **bennet**)

benzene hexachloride (BHC) ['benziːn heksə'klɔːraɪd] *noun* insecticide used as a dust or spray against pea and bean weevil, and as a seed dressing against wireworm (NOTE: chemical formula is $C_6H_6Cl_6$)

Berkankamp scale ['bɜːkəkæmp 'skeɪl] *noun* scale used to describe the growth stages in oilseed rape crop

Berkshire ['bɑːkʃə] *noun* breed of small pig, dark coloured with a compact body, short head and prick ears; the snout, feet and tail are white; **Berkshire Knot** = local breed of sheep crossed with Southdown to develop the Hampshire Down breed

Berrichon du Cher ['berɪʃɒn dju ʃeə] *noun* French breed of sheep, now imported into the UK

berry ['beri] *noun* small fleshy seed-bearing fruit of a bush; there are usually many seeds in the same fruit, and the seeds are enclosed in a pulp (as in a tomato or gooseberry)

berseem [bɜː'siːm] *noun Trifolium alexandrinum*, a form of kohlrabi, a common fodder crop grown in Egypt and popular also in India

best-before date ['bestbɪ'fɔː deɪt] *noun* date stamped on foodstuffs sold in supermarkets, which is the last date when the food is guaranteed to be in good quality; *similar to* SELL-BY DATE

betel ['biːtl] *noun* shrub of which the leaves are used along with areca nut and other ingredients to make the popular stimulant chewing mixture used in India

◊ **betel nut** ['biːtl nʌt] *noun* seed of the betel palm *(Areca catechu)* small pieces of which are chewed alone or mixed with other ingredients

Beulah speckle face ['bjuːlə 'spekəl feɪs] *noun* breed of sheep, with a black and white speckled face; it is native to the hills of Wales

BFSS = BRITISH FIELD SPORTS SOCIETY

BGH = BOVINE GROWTH HORMONE

BGS = BRITISH GRASSLAND SOCIETY

BHC = BENZENE HEXACHLORIDE insecticide used as a dust or spray against pea and bean weevil, and as a seed dressing against wireworm; also called HCH; *see also* LINDANE

bhp = BRAKE HORSEPOWER the power of an engine shown as the resistance of the brake to the vehicle's movement

bible ['baɪbl] *noun* name given to the omasum, the third stomach of ruminants

bid [bɪd] *noun (at an auction)* offer to buy something at a certain price; **opening bid** = first bid; **closing bid** = last bid, which is successful

◊ **bidder** ['bɪdə] *noun* person who makes a bid at an auction

biddy ['bɪdi] *noun* popular name for a hen

bident ['baɪdənt] *noun* two-year-old sheep

biennial [baɪ'enɪəl] **1** *adjective* happening every two years; (plant) which takes two years to complete its life cycle **2** *noun* plant which completes its life cycle (germination, flowering, fruiting) over a period of two years

COMMENT: biennial plants spend the first year producing roots, stem and leaves. In the following year the flowering stem and seeds are produced, after which the plant dies. Sugar beet and swedes are typical biennials, although they are grown as annuals, and are harvested at the end of their first year when the roots are fully developed

biffin ['bɪfɪn] *noun* variety of cooking apple

bifoliate [baɪ'fəʊlɪət] *adjective* (plant) which has two leaves only

big bale (BB) ['bɪg beɪl] *see* BALE

◊ **big bud** ['bɪg bʌd] *noun* swelling of black currant buds caused by gall mites

bigg [bɪg] *noun* four-rowed barley

bighorn ['bɪghɔːn] *noun* American breed of sheep found in the Rocky Mountain region

Bilharzia [bɪl'hɑːtsɪə] *noun* Schistosoma, fluke which enters the blood stream and causes bilharziasis

◊ **bilharziasis** [bɪlhɑːts'aɪəsɪs] *noun* schistosomiasis, a disease of man and domestic animals, caused by parasites in the blood stream

bill [bɪl] *noun* hard part round the mouth of a bird, also called the 'beak'

billhook ['bɪlhʊk] *noun* cutting implement with a curved blade, used for trimming hedges

billy goat ['bɪli gəʊt] *noun* male goat or buck

bin [bɪn] *noun* large container for storage, such as a maize bin, or a bin for holding wool in a shearing shed

bind [baɪnd] *verb* **(a)** to cut corn and attach it in sheaves **(b)** to stick together; **binding agent** = additive which makes prepared food remain in its proper form and not disintegrate

◊ **binder** ['baɪndə] *noun* machine formerly used to cut and bind corn, drawn by a tractor; now replaced by the combine harvester

◊ **bindweed** ['baɪndwiːd] *noun* perennial weed with creeping roots, which climbs by twisting round other plants; *see* BLACK BINDWEED, FIELD BINDWEED

bine [baɪn] *noun* **(a)** new shoot of a hop plant which is twisted round the strings up which it will start to grow **(b)** the stem of a climbing plant, such as the runner bean

bing [bɪŋ] *noun* feed passage

binomial classification [baɪ'nəʊmɪəl klæsɪfɪ'keɪʃn] *noun* way of naming living organisms, devised by Linnaeus, using two Latin names

COMMENT: the first name is the name of the genus or generic name, and the second or specific name is the name of the species; so the species man is *Homo sapiens*, the house sparrow is *Passer domesticus*, and the redwood is *Sequoia sempervirens*. A third name can be added to give a subspecies

bio- ['baɪəʊ] *prefix* referring to living organisms

biochemical oxygen demand (BOD) [baɪəʊ'kemɪkl 'ɒksɪdʒən dɪ'mɑːnd] *noun* amount of pollution in water, shown as the amount of oxygen which will be needed to oxidize the polluting substances; *the main aim of sewage treatment is to reduce the BOD of the liquid*

COMMENT: diluted sewage passed into rivers contains dissolved oxygen which is absorbed by bacteria as they oxidize the pollutants in the sewage. The oxygen is replaced by oxygen from the air. Diluted sewage should not absorb more than 20ppm of dissolved oxygen

biocide ['baɪəʊsaɪd] *noun* substance which kills living organisms; *biocides used in agriculture run off into lakes and rivers*

◊ **biocontrol** [baɪəʊkən'trəʊl] *noun* = BIOLOGICAL CONTROL

biodegradable [baɪəʊdɪ'greɪdəbl] *adjective* (substance) which can easily be decomposed by organisms

biogas ['baɪəʊgæs] *noun* gas (partly methane and partly carbon dioxide) which is produced from fermenting waste, such as animal refuse; *farm biogas systems may be uneconomic unless there is a constant demand for heat; the use of biogas systems in rural areas of developing countries is increasing*

◊ **bioinsecticide** [baɪəʊɪn'sektɪsaɪd] *noun* insecticide developed from natural toxins

biological control *or* **biocontrol** [baɪə'lɒdʒɪkl] kən'trəʊl] *noun* control of pests by using predators to eat them

COMMENT: biological control of insects involves using bacteria, viruses, parasites and predators to destroy the insects. Plants can be controlled by herbivorous animals such as cattle

biomass ['baɪəʊmæs] *noun* all living organisms in a given area

COMMENT: about 2,500 million people, half the world's population, rely on biomass for virtually all their cooking, heating and lighting. Most of these people live in rural areas of developing countries. Concern over the global environment could provide the impetus for determined research and development of biomass

biomethanation [baɪəʊmeθə'neɪʃn] *noun* system of producing biogas for use as fuel or light

QUOTE biomethanation is attractive for use in rural areas for several reasons: it is an anaerobic digestion process, which is the simplest, safest way that has been found for treating human excreta and animal manure
Appropriate Technology

biopesticide [baɪəʊ'pestɪsaɪd] *noun* pesticide made from biological sources, that is from toxins which occur naturally

COMMENT: biopesticides have the advantage that they do not harm the environment as they are easily inactivated and broken down by sunlight. This is, however, a practical disadvantage for the farmer who uses them, since they may not be as efficient as artificial chemical pesticides in controlling pests

biotechnology [baɪəʊtek'nɒlədʒi] *noun* the use of technology to manipulate and combine different genetic materials to produce living organisms with particular characteristics; *artificial insemination of cattle was one of the first examples of biotechnology*

COMMENT: biotechnology offers great potential to increase farm production and food processing efficiency, to lower food costs, to enhance food quality and safety and to increase international competitiveness

birch [bɜːtʃ] *noun* common hardwood tree (*Betula pendula*) found in northern temperate zones; the tree has thin bark; it bears catkins

bird [bɜːd] *noun* animal which lays eggs, and has wings; **bird of prey** = bird which kills and eats animals; **bird song** = singing calls made by birds, to communicate with each other

◊ **birdsfoot trefoil** ['bɜːdsfʊt 'triːfɔɪl] *noun* plant with small yellow flowers, grown for fodder

COMMENT: all birds are members of the class Aves. They have feathers, and the front limbs have developed into wings, though not all birds are now able to fly. Birds are closely related to reptiles, and have scales on their legs. Birds (rooks, pigeons, pheasants) can cause very serious damage to crops: various controls can be used such as shooting, bird-scarers and destruction of nests. Birds also destroy many pests, such as wireworms, leatherjackets and caterpillars. In general, birds are more helpful than harmful, though this will depend on the type of farming undertaken

biscuit-making quality ['bɪskɪt meɪkɪŋ 'kwɒləti] *noun* a hard wheat grain which produces a weak flour (the dough does not rise when cooked)

bison ['baɪsən] *noun* large wild cattle

> COMMENT: the European bison (*Bos bonasus*) is now only found in zoos and some reserves. The American bison (*Bos bison*) is still found in reserves in the USA and Canada. Both are very large animals with massive heads and shoulders, and coats of short curly brown hair

bit [bɪt] *noun* the metal part of a bridle, placed in a horse's mouth to give the rider control over the animal

bite [baɪt] *noun* grazing; **early bite** = grazing in the spring, provided by new growths of grass which sprouts when the weather gets warmer

bitter ['bɪtə] *adjective* which has a sharp taste, and is not sweet; **bitter pit** = a disease of apples

black [blæk] *adjective* **black bean** = type of very hard tropical wood, resistant to termites; **black bent** = grassweed plant (*Agrostis gigantea*) which affects cereals; **black disease** = liver disease of sheep and cattle, rarely found in pigs and horses; **black earth** = chernozem, dark fertile soil, rich in organic matter, found extensively on the plains of Russia and North America; **black eye bean** *or* **blackeyed bean** = common name for the cow pea (*Vigna unguiculata*) in the USA and West Indies; **black quarter** *see* BLACKLEG; **black spot** = fungus which attacks plants, making black spots appear on the leaves

◊ **blackberry** ['blækbəri] *noun* soft black fruit of the bramble

◊ **black bindweed** ['blæk 'baɪndwiːd] *noun* common weed (*Bilderdykia convulvulus*) which is widespread in spring arable crops

◊ **blackcurrant** [blæk 'kʌrənt] *noun* soft fruit (*Ribes*) grown for its small black berries, used also in making soft drinks

blackface *or* **blackfaced sheep**

[blækfeɪs *or* 'blækfeɪst ʃiːp] *noun* one of several breeds of sheep with black faces (such as the Scottish Blackface or the Suffolk); **Blackface mountain** = common breed of sheep found in Ireland

blackfly ['blækflaɪ] *noun* small black aphid (*Aphis fabae*) a very small, oval-bodied, black or dark green fly

◊ **black gram** [blæk 'græm] *see* URD

◊ **blackgrass** ['blækgrɑːs] *noun* weed (*Alopecurus myosuroides*) which is widespread among winter cereals, especially on heavy soils; also known as slender foxtail

◊ **blackhead** ['blækhed] *noun* common and fatal disease of young turkeys

◊ **blackleg** ['blækleg] *noun* **(a)** bacterial disease of potatoes, affecting the base of the stem which turns black and rots **(b)** bacterial disease of sheep and cattle; it causes swellings containing gas on the shoulders, neck and thigh, and can cause death within 24 hours of the appearance of the symptoms

◊ **Black Leghorn** [blæk 'leghɔːn] *noun* one of the several varieties of leghorn chicken; the breed is black all over, with a large red single comb; it has yellow legs and flesh and lays white eggs

◊ **black mould** ['blæk 'məʊld] *noun* fungus growth on cereals (*Cladosporium herbarum*)

◊ **black mustard** ['blæk 'mʌstəd] *noun* variety of rape (*Brassica nigra*) sown in spring and harvested by combine in August. The oil and powder from the seeds are important ingredients of table mustard

◊ **black pod** [blæk 'pɒd] *noun* serious fungal disease affecting cacao trees

◊ **black rust** ['blæk 'rʌst] *noun* disease of cereals (*Puccinia graminis*)

◊ **Blacksided Trondheim** ['blæksaɪdɪd 'trɒndhaɪm] *noun* Norwegian breed of polled cattle; the animals are small and white, with black patches

◊ **blacksmith** ['blæksmɪθ] *noun* person who makes things, especially horseshoes, from wrought iron

◊ **blackthorn** ['blækθɔːn] *see* SLOE

◊ **Black Welsh mountain** ['blæk welʃ 'maʊntən] *noun* breed of sheep with dark brown wool; a completely black strain has been isolated in separate flocks to supply the specialist demand for black wool

◊ **blackwood** ['blækwʊd] *noun* Australian hardwood tree

blade [bleɪd] *noun* **(a)** thin flat leaf (such as a leaf of grass) **(b)** sharp flat cutting part of a knife, scythe, mowing machine, etc.

blanch [blɑːnʃ] *verb* **(a)** to make plants white, by covering them up; celery is blanched either by putting dark tubes round the stems or by earthing the plants up **(b)** to partly cook vegetables by putting them in boiling water for a short time; this is done before preserving

blast [blɑːst] *see* BLOAT

◊ **blast freezing** ['blɑːst 'friːzɪŋ] *noun* method of quick-freezing oddly-shaped food, by subjecting it to a blast of freezing air

bleach [bliːtʃ] *verb* to remove all colour from a substance (as in the preparation of white flour)

bleat [bliːt] **1** *noun* sound made by a sheep or goat **2** *verb (of sheep or goat)* to make a sound; *the bleating of lambs in the field showed that a fox was near see also* GRUNT, LOW, NEIGH

Blenheim orange ['blenɪm 'ɒrɪnʃ] *noun* apple which ripens late in season; the skin is golden in colour

Bleu du Maine (BDM) [blɜː də 'meɪn] *noun* breed of sheep, originating in France and introduced into the UK to produce crossbred ewes with good conformation

blight [blaɪt] **1** *noun* fungus disease of plants; *they are taking steps to eradicate the epidemic of potato blight;* **bacterial pea blight** = fungal disease attacking peas (it is a notifiable disease) **2** *verb* to ruin *or* to spoil the environment; *the landscape was blighted by open-cast mining*

blind gut ['blaɪnd 'gʌt] = CAECUM

blind quarter ['blaɪnd 'kwɔːtə] *noun* nonfunctional udder, usually of sheep

blinkers ['blɪŋkəz] *noun* a head covering with leather eyeshields which allows a horse to see only to the front

bloat [bləʊt] *noun* swelling of a cow's rumen due to gas from lucerne and clovers which is unable to escape; the gas presses on the diaphragm and the animal may die of asphyxiation (NOTE: also known as **blast**)

Blonde (d'Aquitaine) [blɒnd dækɪ'teɪn] *noun* breed of cattle, originating in South-West France, and now established in Britain; it produces large calves which develop into good beef animals. The colour varies from off-white to light brown

blood [blʌd] *noun* red liquid in an animal's body; **blood cell** *or* **blood corpuscle** = cell (red blood cell *or* white blood cell) which is one of the components of blood; **blood chemistry** = substances which make up blood, which can be analysed in blood tests, the results of which are useful in diagnosing disease; **blood count** = test to count the number of different blood cells in a certain quantity of blood; **blood plasma** = watery liquid which forms the greatest part of blood; **blood serum** = watery liquid which separates from coagulated blood; **blood sports** = sport in the countryside involving the killing of animals, such as foxes and hares; **blood sugar level** = amount of glucose in the blood; **blood typing** = classifying the blood group of an animal and establishing parentage from this

◊ **bloodlines** ['blʌdlaɪnz] *noun* general term used to describe relationships between animals, the pedigree lines in a flock or herd

◊ **bloodmeal** ['blʌdmiːl] *noun* protein-rich feedstuff

◊ **bloodstock** ['blʌdstɒk] *noun* collective term for thoroughbred horses

◊ **bloodstream** ['blʌdstriːm] *noun* blood as it passes round the body

◊ **blood sucker** ['blʌdsʌkə] *noun* insect or parasite which sucks blood from an animal

COMMENT: blood is formed of red and white corpuscles, platelets and plasma. It circulates round the body, going from the heart and lungs along arteries and returns to the heart through the veins. As it moves round the body it takes oxygen to the tissues and removes waste material from them. Waste material is removed from the blood by the kidneys or exhaled through the lungs. It also carries hormones produced by glands to the various organs which need them

bloom [bluːm] **1** *noun* **(a)** flower of a plant; *the blooms on the orchids have been ruined* *by frost;* **tree** *or* **field in bloom** = a tree *or* field covered with flowers **(b)** powdery substance on the surface of a fruit such as grapes (in fact a form of yeast) **(c)** fine hairy covering on some fruit, such as peaches **2** *verb* to flower; *the plant blooms at night; some cacti only bloom once every seven years*

blossom ['blɒsəm] **1** *noun* flower or flowers which come before an edible fruit (such as apple blossom or cherry blossom); **trees in blossom** = trees covered with flowers **2** *verb* to open into flower

blotch [blɒtʃ] *see* LEAF BLOTCH, NET BLOTCH

blower ['bləuwə] *noun* machine which blows; **snow blower** = machine which picks up snow from the ground and blows it away to clear a path; **blower unit** = part of an agricultural machine which blows out waste

blowfly ['bləuflaɪ] *noun* name for a number of species of fly, such as *Lucila cuprina,* which deposit their eggs in flesh; also called the meat fly (NOTE: a sheep which has been attacked by blowfly is said to be **blown)**

blue [blu:] *see* BLUE CROSS

◊ **blueberry** ['blu:bəri] *noun* wild plant *(Vaccinium)* now cultivated for its blue berries

bluebottle ['blu:bɒtl] *noun* two-winged fly, whose maggots live in decomposing flesh, but are sometimes also found on living sheep

blue cross ['blu: 'krɒs] *noun* term used for blue and white pigs, usually produced by crossing saddleback and landrace breeds

◊ **blue-ear** ['blu:ɪə] *noun* disease of pigs; animals may farrow prematurely. Made notifiable in the UK in 1991

◊ **Blue-faced Leicester** ['blu:feɪst 'lestə] *noun* middle-sized longwool breed of sheep, with dense curly fleece. The ram is excellent for crossing with smaller types of ewe; it is the sire of the 'mule' and the so-called 'Welsh mule'

◊ **Blue grey** ['blu: greɪ] *noun* crossbred type of cattle, resulting from mating a white shorthorn bull with a Galloway cow

◊ **blue nose** ['blu:nəuz] *noun* disease affecting horses

◊ **blue tongue** ['blu:tʌŋ] *noun* viral disease of cattle

BMR = BASIC METABOLIC RATE

boar [bɔ:] *noun* male uncastrated pig; **wild boar** = *Sus scrofa,* species of feral pig, common in parts of Europe, but extinct in the UK. Wild boars are preserved for hunting, but are now bred on farms; their meat is dark, with very little fat, and is of high value

board [bɔ:d] *noun* the floor of a shearing shed

◊ **boarding** ['bɔ:dɪŋ] *noun* tilting a plough towards the ploughed land so as to increase the pressure on the mouldboards

bobby calf ['bɒbɪ 'kɑ:f] *noun* unwanted male calf in extreme dairy breeds, such as the Channel Island breeds

bocage [bɒ'kɑ:ʃ] *noun* type of rural landscape with small fields, hedges and trees; used to describe the landscape of north-west France

BOD = BIOCHEMICAL OXYGEN DEMAND

Boer [bɔ:] *noun* South African breed of goat, white with red head markings. The animals are raised both for meat and for their skins

bog [bɒg] *noun* soft wet land, usually with moss growing on it, which does not decompose, but forms a thick layer of acid peat; *bog mosses live on nutrients which fall in rain;* **blanket bog** = wide area of bog

◊ **boggy** ['bɒgi] *adjective* soft and wet (soil)

◊ **bogland** ['bɒglænd] *noun* area of bog

boil [bɔɪl] *verb* to cook in water at 100°C

◊ **boiler** *or* **boiling fowl** ['bɔɪlə *or* 'bɔɪlɪŋ faul] *noun* hen sold for the table, when no longer used for laying

bole [bəul] *noun* base of a tree trunk

boll [bɒl] *noun* seed pod of the cotton or flax plant; **boll number** = number of bolls per cotton plant; **boll weevil** = American beetle *(Anthonomus grandis)* which attacks cotton plants

---

QUOTE the plant height is about 150cm; boll number varies from 20- 25 per plant with average boll weight of 3.0g and above

Indian Farming

bolt ['bəʊlt] *verb* **(a)** *(of a vegetable)* to produce flowers and seeds too early (as in the case of beetroot or lettuce) **(b)** *(of biennial plants)* to behave as an annual

◊ **bolter** ['bəʊltə] *noun* plant which grows too fast, producing a low seed yield

COMMENT: bolting occurs as a response to low temperatures and results in the plant failing to build up food reserves in the root, so producing seed in its first year; bolting is highly undesirable in many plants, especially in the sugar beet, where the roots become woody and have a low sugar content

bolt-on payment ['bəʊltɒn 'peɪmənt] *noun* cash incentive paid for a special purpose; *increased payments for set-aside schemes will include bolt-on payments which allow farmers to encourage wildlife and flowers*

bolus ['bəʊləs] *noun* ball of partly digested food regurgitated by ruminants

boma ['bəʊmə] *noun (East Africa)* small enclosure for keeping cattle

Bon Chretien ['bɒŋ 'kretjæŋ] *see* PEAR

bonding ['bɒndɪŋ] *noun* important time when a newly-born young animal becomes 'tied' to its mother in the period just after birth; the young then begins to imitate its mother and learn from her

bone [bəʊn] *noun* one of the calcified pieces of connective tissue which make a skeleton

COMMENT: bones are formed of a hard outer layer (compact bone) which is made up of a series of layers of tissue and a softer inner part which contains bone marrow

bonemeal ['bəʊnmiːl] *noun* fertilizer made of ground bones or horns, reduced to a fine powder. It is also used as a fine meal for growing animals, used for the calcium, phosphorus and magnesium it contains

boost [buːst] *noun* instrument for marking sheep, usually with hot tar, often bearing the owner's initials

boot [buːt] *noun* characteristic swelling in the stem of a cereal plant, produced as the developing ear moves up the stem

borage ['bɒrɪdʒ] *noun* herb used as an oilseed break crop

Boran ['bɔːræn] *noun* East African breed of beef cattle, found especially in Kenya; the animals are usually white

Bordeaux mixture [bɔː'dəʊ 'mɪkstʃə] *noun* mixture of copper sulphate, lime and water, used as a fungicide on vines and fruit trees

border disease ['bɔːdə dɪ'ziːz] *noun* disease affecting sheep, caused by a virus. Ewes may abort, or lambs may be born weak and may show muscle tremors. Infection enters the flock when diseased sheep are brought in and an infected ewe excretes the virus

Border Leicester ['bɔːdə 'lestə] *noun* breed of longwool sheep, derived from the English Leicester; its head is bare of wool and has a pronounced 'Roman' nose; the breed is used to produce ram lambs for breeding

bore cole ['bɔːkəʊl] *see* KALE

Boreray ['bɔːreɪ] *noun* rare breed of horned sheep, with grey or cream coloured fleece

-borne [bɔːn] *suffix meaning* carried by; **foodborne diseases** = diseases which are transmitted from feedstuff

Bornu White ['bɔːnuː 'waɪt] *noun* West African goat, a Nigerian breed

borogluconate [bɒrəʊ'gluːkəneɪt] *see* CALCIUM BOROGLUCONATE

boron ['bɔːrɒn] *noun* chemical element present in borax, and essential for healthy plant growth as a trace element in soils; boron deficiency is a cause of heart rot in sugar beet (NOTE: chemical symbol is **B**; atomic number is **5)**

Bos [bəʊs] *noun* the genus (part of the family Bovidae) to which cattle, buffalo, bison, yaks and gaur belong

bosk [bɒsk] *noun* small wood

botany ['bɒtəni] *noun* scientific study of plants

◊ **botanical** [bɒ'tænɪkl] *adjective* referring to botany; **botanical insecticide** = insecticide made from a substance extracted from plants (the best-known botanical insecticides are pyrethrum, derived from the chrysanthemum, and nicotine, derived from the tobacco plant)

◊ **botanist** ['bɒtənɪst] *noun* scientist who studies plants

bot fly ['bɒt flaɪ] *noun* fly of the *Oestrus* family; a two-winged fly whose maggots are parasitic on sheep and horses

◊ **bots** [bɒts] *noun* the larval stage of the bot fly which lays its eggs on horses' forelegs in summer. Cause of restlessness

bothy ['bɒθi] *noun* sparsely furnished farm worker's cottage in Scotland

Botrytis [bɒ'traɪtɪs] *noun* fungal disease affecting plants, especially young seedlings; *see also* CHOCOLATE SPOT

bottle [bɒtl] **1** *noun* glass container for liquids or preserves; **bottle feeding** = feeding young animals with liquids from a bottle with a teat **2** *verb* to keep in a bottle; to preserve (food) by heating it inside a glass jar with a suction cap

◊ **bottle gourd** ['bɒtl gʊəd] *noun* annual climbing plant found in tropical Africa. The dried seedcases are used as food containers and to make percussion instruments *(Lagenaria siceraria)*

botulism ['bɒtjuːlɪzm] *noun* type of food poisoning, caused by a toxin of *Clostridium botulinum* in badly canned or preserved food

boulder ['bəʊldə] *noun* large often rounded rock; **boulder clay** = stony clay, found in glacial deposits covering large parts of the Midlands and Northern England; now usually called 'till'

boundary ['baʊndəri] *noun* line marking the edge of a property; **boundary stone** = stone used to mark the edge of a property

Bovidae ['bɒvɪdiː] *noun* largest class of even-toed ungulates, including cattle, sheep and goats

◊ **bovine** ['bəʊvaɪn] *adjective* referring to cattle; **bovine immunodeficiency virus** = disease of cattle a version of cattle aids; **bovine leucosis** = cancerous disease in cattle; **bovine tuberculosis** = bacterial disease of cattle which has been almost completely eliminated from the UK by the attested herds scheme

◊ **bovine somatotropin (BST)** *noun* natural hormone found in cows, which has been produced artificially by genetic engineering; it is used to increase milk yields and is said to increase them by up to 40%. It is not licensed for use in the UK, but is being used in trials

◊ **bovine spongiform encephalopathy (BSE)** *noun* a fatal brain disease of cattle, also called 'mad cow disease'

COMMENT: caused by the use of ruminant-based additives in cattle feed, by which 'scrapie' (the disease affecting sheep) infects cattle. BSE-infected meat is believed to be connected to a new strain of Creuzfeldt-Jacob disease in humans

QUOTE BSE first appeared on English dairy farms in 1987. By December, 1988, 1,677 cattle had been slaughtered after contracting the infection. BSE is a new addition to a group of animal viruses known for about 200 years. The similarity between BSE and scrapie suggests that scrapie has been transmitted from sheep to cattle.
Guardian

QUOTE an extra £6.3 million is to be given to the agriculture and food research council to investigate the cause of BSE
Guardian

bowel ['baʊwəl] *noun* an animal's large intestine

◊ **bowel oedema** ['baʊwəl ɪ'diːmə] *noun* bacterial infection of pigs associated with sudden changes of diet and management. Most common in pigs 8-14 weeks old. Pigs may throw fits, stagger and become paralysed. In acute cases piglets are found dead

box [bɒks] *verb (in Australia)* to mix different flocks of sheep

BPA = BRITISH PIG ASSOCIATION

brace [breɪs] *noun* pair of dead game birds

bracken ['brækən] *noun* fern *(Pteridium aquilinum)* which grows widely on acid soils; a perennial, found mainly on moors and upland areas; **bracken poisoning** = poisoning of livestock from eating bracken; it can cause serious illness or death

> COMMENT: bracken is hard to eradicate as it sprouts easily after being burnt. The spores of bracken contain a carcinogenic substance

bract [brækt] *noun* small green leaf at the base of a flower

Bradford worsted count ['brædfəd 'wɜːstɪd kaʊnt] *noun* a method of measuring wool quality

brahman [brɑːmən] *noun* cattle of Indian origin; they have a large hump over the neck and shoulders; they tolerate high temperatures and are resistant to insects and tropical diseases, and are important for beef production in hot and humid regions. They are often called 'zebu'

braird [breɪrd] *noun* fresh shoots of corn or other crops

brake [breɪk] *noun* mechanism used to slow down a moving vehicle or engine; **brake harrow** = harrow used for breaking up large clods of earth; **brake horsepower (bhp)** = power developed by a tractor's engine, calculated as the force applied to the brakes

bramble ['bræmbl] *noun* wild blackberry bush *(Rubus fruticosus)*, with edible black fruits

Bramley's seedling ['bræmlɪz 'siːdlɪŋ] *noun* a common variety of cooking apple

bran [bræn] *noun* (a) outside covering of the wheat seed, removed when making white flour, but an important source of roughage and some vitamin B (NOTE: also called **hull**) (b) feedingstuff, used especially for cattle and poultry

branch [brɑːnʃ] *noun* long growth on a tree, growing from the main trunk, and carrying smaller shoots and twigs

◊ **branched** ['brɑːnʃt] *adjective* with branches

brand [brænd] **1** *noun* mark burnt with a hot iron on an animal's hide, to show ownership **2** *verb* to mark an animal's hide with a hot iron to show the owner

Brangus ['bræŋgəs] *noun* crossbreed of cattle, obtained by crossing a Brahman with an Aberdeen Angus. It is a black polled breed, and is found in Central and South America, the USA, Canada, Australia and Zimbabwe

brank [bræŋk] *see* BUCKWHEAT

brash [bræʃ] *noun* soil containing many stone particles or fragments of rock

brasics ['bræsɪks] *noun* common name for the charlock *(Sinapsis arvensis)*

brassica ['bræsɪkə] *noun* generic term for members of the cabbage family, such as broccoli, Brussels sprouts, cauliflowers, kales, savoys, swedes and turnips

bratting ['brætɪŋ] *noun* putting a jacket on sheep, usually hoggs, to protect from cold in winter

brawn [brɔːn] *noun* (a) meat dish prepared from cut, boiled and pickled pig's head (b) obsolete term for a stag boar (NOTE: also called **a brawner)**

braxy ['bræksɪ] *noun* bacterial disease of sheep; the affected animals show loss of appetite, dullness, difficult breathing, although these signs of the disease are rarely seen since death occurs rapidly, after five or six hours

breadcorn ['bredkɔːn] *noun* corn from which the flour for bread is obtained

breadfruit ['bredfruːt] *noun* starchy fruit of a tree *(Artocarpus communis)* grown in the Pacific Islands. The fruit is used as a vegetable

break [breɪk] *verb* (a) to train a horse to wear a saddle (b) to mill flour, by passing it through break rolls

◊ **break crop** ['breɪk krɒp] *noun* crop grown between periods of continuous cultivation of a main crop; on a cereal-growing farm, crops such as sugar beet or potatoes may be introduced to give a 'break' from continuous cereal growing

◊ **break feeding** ['breɪk 'fiːdɪŋ] *noun (in Australia)* method of rationing animal feed by moving animals from one grazing area to another

◊ **break in** ['breɪk 'ɪn] *verb* (a) to train a wild horse (b) *(in Australia)* to bring uncultivated land into cultivation

◊ **break rolls** ['breɪk 'rəʊlz] *noun* heavy

rollers with grooves in them, through which the grain passes when being milled; as the grain goes through it is broken into pieces and finally reduced to flour

breast [brest] *noun* front part of the body of a bird; lower part of the body of a sheep (similar to the belly on a pig)

Brecknock hill cheviot ['breknɒk hɪl 'tʃeviət] *noun* breed of sheep found in the Welsh border counties for producing ideal halfbreds, and also in lowlands for prime lamb production

breeching ['briːtʃɪŋ] *noun* a leather strap round the hindquarters of a shaft horse, which allows the horse to push backwards and so make the cart go backwards

breed [briːd] **1** *noun* group of organisms of a certain species, which have been developed by people over a period of time to stress certain characteristics; *a hardy breed of sheep; two new breeds of rice have been developed* **2** *verb* **(a)** *(of organisms, both animals and plants)* to reproduce; *rabbits breed very rapidly* **(b)** to encourage to develop; *insanitary conditions help to breed disease* **(c)** to raise a certain type of animal *or* plant, by crossing one variety with another to produce a new variety where the characteristics which the breeder wants to keep are strongest; *farmers have bred new hardy forms of sheep; his father breeds pedigree Friesians*

◊ **breed in** ['briːd 'ɪn] *verb* to introduce a characteristic into an animal breed or plant variety, by breeding until it is a permanent characteristic of the breed

◊ **breeder** ['briːdə] *noun* person who breeds new forms of animals *or* plants; *a cat breeder; a rose breeder*

◊ **breeding** ['briːdɪŋ] *noun* the action of raising a certain type of animal or plant; **breeding crate** = box construction designed to take the weight of a bull, when mating with a heifer or cow; **breeding stock** = prime animals kept for breeding purposes to maintain and improve quality of stock; *see also* CROSSBREEDING, INBREEDING, LINE BREEDING, OUTBREEDING

> QUOTE Milk Marketing Board AI figures show a swing away from traditional British beef breeds. Between April and December 1988, Limousin accounted for 13.8%, Charolais 10.8%, Hereford 6.8%, Belgian Blue 5.4%, Simmental 4.4% of all inseminations
>
> *Farmers Weekly*

breeze fly ['briːz 'flaɪ] *noun* large two-winged blood-sucking fly, similar to a gadfly

brew [bruː] *verb* to make beer by fermenting water and malted barley

◊ **brewers' grain** ['bruːəz 'greɪn] *noun* by-product of brewing; the residue of barley after malting. It is a valuable cattle food, fed wet or dried, and is a source of fibre and protein

brickearth ['brɪkɜːθ] *noun* common name for loess; soils developed in brickearth are fertile and well drained; they are used for intensive crop production

bridle ['braɪdl] *noun* part of a horse's harness which goes over the head, including the bit and rein

◊ **bridlepath** *or* **bridleway** ['braɪdlpɑːθ *or* 'braɪdlweɪ] *noun* track along which horses can be ridden

Brie [briː] *noun* soft French cheese, made in large flat round shapes

brindled ['brɪndəld] *adjective* brownish coloured, with spots or marks of another colour

brine [braɪn] *noun* solution of salt in water, used for preserving food

> COMMENT: some meat, such as bacon, is cured by soaking in brine; some types of pickles are preserved by cooking in brine; some foodstuffs are preserved in brine in jars

brinjal ['brɪndʒəl] *noun* Indian name for aubergine

brisket ['brɪskɪt] *noun* (i) lower front part of the body of a cow, between the front legs; (ii) the meat from this part; *when a cow is down for a length of time it should be supported on the brisket;* **brisket board** = board across the front of a cubicle to prevent a cow from lying down too near the front

British alpine ['brɪtɪʃ 'ælpaɪn] *noun* large goat of the Swiss type; it is black with white markings on the face and legs. Males have a large beard. It is a good milking breed

◊ **British Belgian Blue** *noun* breed of cattle in the UK, originating from the Belgian Blue and registered by the breed society

◊ **British Dane** *noun* dairy breed of cattle

◊ **British Holstein** *noun* breed of dairy cattle developed in the UK from the Holstein

◊ **British Lop** *noun* rare breed of pig; very large white, with lop ears. The breed is found mainly in South-West England

◊ **British Milksheep** *noun* a medium polled sheep, with white face and legs. Mainly used as a dairying ewe

◊ **British Oldenburg** *noun* breed of sheep giving a high yield of wool of good quality

◊ **British Romagnola** *noun* breed of beef cattle developed in the UK from stock originating in north Italy

◊ **British Saddleback** *noun* breed of pig, derived from the Wessex and Essex breeds; it is coloured black with a white band round the front of the body. Used for breeding hardy crosses

◊ **British Saanen** *noun* large white breed of goat, with a high milk yield; the breed originated in Switzerland

◊ **British Sugar Corporation** a corporation which buys sugar beet each year from farmers; the corporation is responsible for processing and marketing the sugar

◊ **British Toggenburg** *noun* dark brown coloured goat, developed from the Swiss Toggenburg breed

◊ **British White** *noun* rare breed of cattle, white with coloured points. The breed is hornless, short-legged and medium-sized. It is still used as a suckling breed for beef production

◊ **British Wool Marketing Board (BWMB)** government body responsible for the buying, grading and marketing of home-produced wool

> COMMENT: the British Wool Marketing Board is responsible for the specifications by which wool is graded. These are based on the length of staple, lustre, softness, springiness, strength, colour and fineness. The diameter of the wool fibre (measured in microns) is used to measure fineness, and in Britain ranges from 28 (the thickest) to 58 (the thinnest)

◊ **British Veterinary Association** professional body which represents veterinary surgeons

broad bean ['brɔːd 'biːn] *noun* a common

bean *(Vicia faba)* with large flattish pale green seeds; it is grown throughout the world. Also called 'field bean'

broadcast ['brɔːdkɑːst] *verb* to scatter seed freely over an area of ground, as opposed to sowing in drills

◊ **broadcaster** ['brɔːdkɑːstə] *noun* machine for sowing seeds broadcast

◊ **broadcast fertilizer distributor** *noun* machine which scatters fertilizer; the most common makes have a hopper for the fertilizer and a horizontal spinning disc below it. Another type has a horizontal spout which moves from side to side. Most broadcasters are mounted with hopper capacities of up to 600kg; trailed hoppers may have capacities of up to five tonnes

broadleaved ['brɔːdliːvd] *adjective* (plants) with large leaves, such as broadleaved weeds; **broadleaved evergreens** = evergreen trees with large leaves, such as rhododendrons, tulip trees, etc.

◊ **broadleaf** *or* **broadleaved trees** ['brɔːdliːf *or* 'brɔːdliːvd 'triːz] *noun* deciduous trees (such as beech, oak, etc.) which have wide leaves, as opposed to the needles on conifers; *the plan is to plant 12,000 hectares of agricultural land a year, one third of which will be broadleaves*

◊ **broadshare** ['brɔːdʃeə] *noun* cultivator used on heavy ground; used for stubble clearing and land reclamation work

◊ **broad spectrum antibiotics** [brɔːd'spektrəm æntɪbaɪ'ɒtɪks] *noun* antibiotics used to control many types of bacteria

broccoli ['brɒkəli] *noun* one of several types of brassica; one is the winter-grown type of cauliflower; other varieties include sprouting broccoli, of which there are purple and white kinds, the plants making their curds on numbers of side shoots. Perennial broccoli grows into large plants and forms a number of quite large cauliflower-like heads each season for two or three years; *see also* CALABRESE

broiler ['brɔɪlə] *noun* chicken raised for the table, usually under intensive conditions; it is usually less than 3 months old

broken mouthed ['brəukən 'mauðd] *adjective* (sheep) which has lost some of its teeth

◊ **broken-winded** ['brəʊkən 'wɪndɪd] *adjective* (horse) disabled by ruptured cells in the lungs

brome grass ['brəʊm 'grɑːs] *noun* weed grass found in arable fields and short leys; there are many species *(Bromus)* and none are of any value; *see also* BARREN BROME

bronchi ['brɒŋkiː] *plural noun* air passages leading from the throat into the lungs

◊ **bronchitis** [brɒŋ'kaɪtɪs] *noun* inflammation of the membranes in the bronchi, often caused by viruses; *see also* HUSK

Bronze [brɒnz] *noun* breed of large table turkey

brood [bruːd] **1** *noun* group of offspring produced at the same time, especially group of young birds; *the territory provides enough food for two adults and a brood of six or eight young;* **brood mare** = female horse kept for breeding **2** *verb* to raise chicks after they have hatched

◊ **brooder** ['bruːdə] *noun* container with a source of heat, used for housing newly hatched chicks

◊ **brooding time** ['bruːdɪŋ 'taɪm] *noun* length of time a bird sits on its eggs to hatch them out

◊ **broody hen** ['bruːdi 'hen] *noun* (i) hen which persists in sitting on its eggs to hatch them; (ii) the stage in a fowl's productive life when it has finished laying and sits on the eggs

brown [braʊn] *adjective* **brown earth** *or* **brown forest soil** = good fertile soil, slightly acid, containing well-incorporated humus; **brown flour** = wheat flour (such as wholemeal flour) which contains some bran and has not been bleached; **brown foot rot** = disease affecting wheat seedlings *(Fusarium avanaceum)* **brown podzolic soil** = brown earth from which humus and/or iron and aluminium have been leached from the topsoil, but in which there is no bleached horizon; *compare* PODZOL **brown rice** = rice grain that has had the husk removed, but has not been milled and polished to remove the bran; **brown rot** = fungal disease, of which the spores damage flower stalks and young shoots of apples, pears, and plums. The infected fruits shrivel up and can remain on the tree, creating a new source of infection; **brown**

rust = fungal disease *(Puccinia)* of barley and wheat, causing the grain to shrivel; **brown scale** = disease *(Parthenolecanium corni)* causing stunted growth and leaf defoliation. Attacks vines, currants and, in greenhouses, peaches

◊ **Brown Swiss** *noun* medium-sized Swiss dual-purpose breed of grey-brown cattle, now found in many parts of the world including Canada, Mexico and Angola

browse [braʊz] *verb (of animals)* to eat the above-ground parts of plants; *compare* GRAZE

◊ **browser** ['braʊzə] *noun* animal which browses; *goats are useful browsers*

Brucella [bruː'selə] *noun* type of rod-shaped bacterium

◊ **brucellosis** [bruːsɪ'ləʊsɪs] *noun* infectious disease, also known as contagious abortion, which is usually associated with cattle where it results in reduced milk yields, infertility and abortion; the United Kingdom is now a brucellosis-free area. In humans, brucellosis often takes the form of undulant fever. Agglutination tests are used to detect brucellosis

bruise [bruːz] *verb* to harm the flesh under the skin, usually by hitting; *ripe pears have to be handled carefully to avoid bruising*

brush [brʌʃ] *noun* trimmings from small trees and shrubs

◊ **brush drill** ['brʌʃ 'drɪl] *noun* old type of seed drill, in which a rotating brush moves the seed into the drill tube

◊ **brushwood** ['brʌʃwʊd] *noun* low undergrowth or thicket

Brussels sprouts [brʌselz 'spraʊts] *noun* variety of cabbage with a tall main stem, which develops buds which are picked and eaten as fresh vegetables or may be kept frozen

BSE = BOVINE SPONGIFORM ENCEPHALOPATHY a fatal brain disease of cattle, the great majority of cases can be shown to have had access to potentially contaminated feed. Cattle become agitated and aggressive

QUOTE Britain's BSE epidemic at last appears to be under control. Over the past two years there has been a big fall in the number of cases
Farmers Weekly

BSI = BRITISH STANDARDS INSTITUTION

BST = BOVINE SOMATOTROPIN natural hormone found in cows, which has been produced artificially by genetic engineering; it is used to increase milk yields and is said to increase them by between 12% and 20%. It is not licensed for use in the UK, but is being used in trials

QUOTE the USA plans to approve its use. The EU is uncertain and in making its decision will need to consider 1) milk surpluses; 2) that BST use will favour large-scale dairy operations over the small farmers; 3) possible consumer opposition; 4) US trade retaliation; 5) BST's effects on animal health
New Scientist

buck [bʌk] *noun* male goat or rabbit

bucket feeding [ˈbʌkɪt ˈfiːdɪŋ] *noun* feeding a young animal with milk from a bucket

buckrake [ˈbʌkreɪk] *noun* machine for collecting green crops cut for silage; the machine has a number of steel tines, usually rear-mounted in a tractor linkage system. A buckrake is used by going backwards along a swath with the tines pointing downwards

buckwheat [ˈbʌkwiːt] *noun* grain crop *(Fagopyrum esculentum)* which is not a member of the grass family; it can be grown on the poorest of soils. In the USA, when buckwheat is ground into flour, it is used to make grits (NOTE: also called **brank)**

bud [bʌd] **1** *noun* young shoot on a plant, which will later become a leaf or flower; **tree in bud** = tree which has buds which are swelling and about to produce leaves or flowers **2** *verb* to propagate plants by grafting a bud from one in place of a bud on the stock

◊ **budding** [ˈbʌdɪŋ] *noun* way of propagating plants, where a bud from one plant is grafted in place of a bud on the stock (used, for example, to propagate roses) (NOTE: also called **bud grafting)**

QUOTE bud grafting is the accepted technique of vegetative propagation in *Hevea brasiliensis* in commercial practice
The Planter

buffalo [ˈbʌfələu] *noun* **(a)** animal of the cow family, a common domestic animal in tropical countries, where it is used both for milk and also as a draught animal; dairy buffaloes contribute substantially to income and employment in India (NOTE: also called the **water buffalo) (b)** *US* type of large wild cattle, also called 'bison'

QUOTE Egypt's main milk producing animal is still the water buffalo. For milk production, female buffalo calves growing from 200kg to the proper mating weight of 350kg are reared totally on forages, with minimum concentrates, berseem and rice straw in winter, followed by berseem hay, rice straw and green maize in summer
Middle East Agribusiness

buffer stock [ˈbʌfə ˈstɒk] *noun* stock of a commodity (such as coffee) held by an international organization and used to control movements of the price on international commodity markets

Buff Orpington [ˈbʌf ˈɔːpɪŋtən] *noun* dual-purpose breed of chicken, brown in colour

bug [bʌg] *noun* any small insect

buisted [ˈbjuːstɪd] *noun* sheep marked after shearing

bulb [bʌlb] *noun* fleshy stem like an onion, formed of layers of tissue, which can be planted and which will produce leaves, flowers and seed

bulk milk collection [bʌlk ˈmɪlk kəˈlekʃn] *noun* collection of large quantities of milk by tankers from farms

◊ **bulk storage** [ˈbʌlk ˈstɔːrɪdʒ] *noun* storing of fertilizer or grain in dry covered barns, rather than in bags

bull [bʊl] *noun* uncastrated adult male ox

◊ **bull beef** [ˈbʊl ˈbiːf] *noun* beef from a bull, which is leaner than meat from a steer

◊ **bull calf** [ˈbʊl ˈkɑːf] *noun* male calf

bullets [ˈbʊlɪts] *noun* doses of mineral

given to cattle and sheep by means of a dosing gun; bullets provide the animal with a long-lasting supply of minerals to overcome deficiency diseases

bulling ['buliŋ] *noun* heifer cow of right inclination and body condition to be served

◊ **bulling hormone** ['buliŋ 'hɔːməʊn] *see* OESTROGEN

bullock ['buləl] *noun* **(a)** young male ox **(b)** castrated bull, also called a 'steer'; bullocks are more docile than bulls and 'finish' quicker than entire animals. They are used as draught animals in many parts of the world

bulrush millet ['bulrʌʃ 'mɪlɪt] *noun Penniseteum typhoides,* one of a group of cereals used as a staple food in the Sahel and Sudan zones of Africa

bumble foot ['bʌmbl 'fʊt] *noun* condition of the feet of poultry, characterized by abscesses, causing lameness

bund [bʌnd] *noun* soil wall built across a slope to hold back water *or* to hold waste in a sloping landfill site

QUOTE farmers have traditionally relied on a system of bunding to grow sorghum in the area's semi-arid soil
New Scientist

bundle ['bʌndl] *noun* twigs or stems of plants, tied together

bunt [bʌnt] *noun* disease of wheat caused by the smut fungus *(Tilletia caries)*

◊ **bunt order** ['bʌnt 'ɔːdə] *noun* the order of social dominance established by cattle and pigs which is the order in which the animals feed and drink (NOTE: the equivalent in birds is called the **pecking order**)

bur *or* **burr** [bɜː] *noun* seed case of the cleavers plants, with tiny hooked hairs, which catch on the coats of animals, and help to disperse the seed

burdizzo [bɜːˈdiːtsəʊ] *noun* implement used in 'bloodless' castration by crushing the spermatic cord

burn [bɜːn] **1** *noun (in Scotland)* small stream **2** *verb* to destroy by fire; *several hundred hectares of forest were burnt in the fire see also* SLASH AND BURN

◊ **burnt lime** ['bɜːnt 'laɪm] *noun* form of lime produced by burning lumps of limestone or chalk. It is usually ground up ready for spreading; also called 'quicklime', 'lump lime', and 'shell lime'

burrow ['bʌrəʊ] *noun* hole made in the earth by an animal, such as a rabbit or fox

bush [bʊʃ] *noun* **(a)** low shrub *or* small tree; *a coffee bush;* **bush fruit** = fruit from bushes (such as gooseberries and red currants), as opposed to fruit from trees **(b)** *(in semi-arid regions)* **the bush** = wild land covered with bushes and small trees; **bush-fallow** = cultivation system where trees and bushes are cleared from virgin land which is then left fallow for a period before cultivation starts

bushel ['bʊʃl] *noun* measure of capacity for corn and fruit, equivalent to eight gallons; **bushel weights** = the weight of average bushels of a product

COMMENT: the amount contained in a bushel varies: in the UK it is equivalent to 36.4 litres, but in the USA it is equivalent to 35.3 litres. The weight of a bushel varies with the crop: a bushel of wheat weighs 60lb; of barley 48lb; and of rice 45lb

buster ['bʌstə] *see* BALE BUSTER

butter ['bʌtə] *noun* fatty substance made from cream by churning; butter is high in saturated fat; **butter beans** = leguminous plant *(Phaseolus lunatus)* with large white seeds; also called 'Lima bean'

◊ **buttercup** ['bʌtəkʌp] *noun Ranunculus,* a common weed with yellow flowers

◊ **butterfat** ['bʌtəfæt] *noun* fatty substance contained in milk; **butterfat content** = the amount of butterfat in milk; this will vary with the breed of cow. Jersey cow milk contains 5.14% butterfat, and is much used for butter-making; the Gerber test is used to measure butterfat content

◊ **butterfly** ['bʌtəflaɪ] *noun* breed of English rabbit

◊ **buttermilk** ['bʌtəmɪlk] *noun* the liquid which remains after the churning of cream, when making butter

◊ **butter mountain** ['bʌtə 'maʊntən] *noun* popular term for vast quantities of dairy produce in the form of butter, which has been paid for by EU governments and put into cold store

BVA = BRITISH VETERINARY ASSOCIATION

BWGS = BROADLEAVED WOODLAND GRANT SCHEME

BWMB = BRITISH WOOL MARKETING BOARD

BYDV = BARLEY YELLOW DWARF VIRUS

byre [baɪə] *noun* cow house

Cc

C horizon ['si: hə'raɪzən] *see* HORIZON

C 1 *abbreviation for* Celsius **2** *chemical symbol for* carbon

◊ **C₃** metabolic pathway in plants, which uses three-carbon compounds to fix CO_2 from the atmosphere

◊ **C₄** metabolic pathway for CO_2 fixation, which uses four-carbon compounds. Plants with this mechanism, called C_4 plants, such as maize, are adapted to high sunlight and arid conditions; they have low photorespiration

◊ **Vitamin C** ['vɪtəmɪn 'siː] *noun* ascorbic acid, vitamin which is soluble in water and is found in fresh fruit (especially oranges and lemons) and in raw vegetables, liver and milk

COMMENT: lack of Vitamin C can cause anaemia and scurvy

Ca *chemical symbol for* calcium

caatinga [kɑːˈtɪŋgə] *noun* term used in South America for semi-arid scrub vegetation

cab [kæb] *noun* housing for the driver on a tractor, usually protected by anti-roll bars

cabbage ['kæbɪdʒ] *noun* cultivated vegetable (*Brassica oleracea*) with a round heart or head, a useful food for stock; other suitable varieties are grown for human consumption; **cabbage root fly** = *Erioischia brassicae*, a fly whose larvae attack the roots of Brassica seedlings, the plants turn bluish in colour and wilt and die; **cabbage white butterfly** = common white butterfly (*Pieris brassicae*) which lays eggs on the leaves of plants of the cabbage family; the caterpillars cause much damage to the plants

caboclo [kæˈbɒkləu] *noun* subsistence farmer in South America

cacao [kəˈkaʊ] *noun* term used instead of cocoa, when referring to the growing cocoa tree (*Theobroma cacao*) or its products

cactus ['kæktəs] *noun* succulent plant with a fleshy stem, often protected by spines; found wild in the deserts of North and Central America (NOTE: plural is **cacti** or **cactuses**)

cade lamb ['keɪd 'læm] *noun* lamb reared from a bottle because of the death of its mother

cadmium ['kædmɪəm] *noun* metallic element which is naturally present in soil and rock in association with zinc (NOTE: chemical symbol is **Cd**; atomic number is **48**)

COMMENT: cadmium is used for making rods for nuclear reactors. It is also present in tobacco smoke, and is found in fish and shellfish such as oysters. It is often found in sewage sludge from industrial sources

CAE = CAPRINE ARTHRITIS-ENCEPHALITIS

caecum ['siːkəm] *noun* wide part of the intestine leading to the colon

Caerphilly [kəˈfɪli] *noun* hard white cheese, originally made in South Wales

caesar ['siːzə] *verb* to perform a surgical intervention to enable an animal to give birth (as in Belgian Blue cattle)

caesium ['siːzɪəm] *noun* metallic element which is one of the main radioactive pollutants taken up by fish (NOTE: chemical symbol is **Cs**; atomic number is **55**)

cafezal ['kæfeɪzæl] *noun* a coffee plantation in Brazil

caffeine ['kæfiːn] *noun* alkaloid present in kola nuts which gives them their stimulating properties. An essential ingredient of Coca-Cola™

cage [keɪdʒ] *noun* housing made of wood or metal, used for battery hens; **laying cage** = specially built cage for laying hens; the cages are arranged in tiers and each cage should allow the birds to stand comfortably, allow the eggs roll forward and permit access to food and water, easy cleaning and easy handling of the birds; **cage rearing** = method of rearing poultry in which birds are taken right through from day-old chicks to placement in laying cages. Chicks are started in the top tier and then spread through all the tiers of cages within a few weeks. Cages are designed so that chicks can be brooded in them and yet they may also be used as laying cages

◊ **cage wheel** ['keɪdʒ 'wiːl] *noun* metal wheel fitted to the outside of a normal tractor wheel in order to reduce ground pressure

cake [keɪk] *noun* form of cattle food, made of compressed seeds of various plants, such as groundnut and soya bean; used generally to refer to most livestock feeding stuffs

calabash ['kæləbæʃ] *noun* fruit of the *Cresentia cujete* tree; large fruit with woody outer shells used as food containers and percussion instruments

calabrese ['kæləbriːz] *noun* variety of sprouting broccoli, grown as a vegetable for human consumption; it produces a large central head, and after this is cut, sprouts are produced for several months. Large quantities are grown for quick-freezing or canning

calcareous [kæl'keəriəs] *adjective* (soil) containing calcium; chalky (soil); (rock such as chalk or limestone) containing calcium

calcicole *or* **calciphile** *or* **calcicolous plant** ['kælsɪkəʊl *or* 'kælsɪfaɪl *or* kæl'sɪkələs plɑːnt] *noun* plant which grows well on chalky *or* alkaline soils

calcification [kælsɪfɪ'keɪʃn] *noun* hardening by forming deposits of calcium salts

◊ **calcified** ['kælsɪfaɪd] *adjective* made hard

◊ **calcifuge** *or* **calciphobe** ['kælsɪfjuːdʒ *or* 'kælsɪfəʊb] *noun* plant which prefers acid soils and cannot exist on chalky *or* alkali soils

◊ **calcimorphic soil** [kælsɪ'mɔːfɪk 'sɔɪl] *noun* soil which is rich in lime

◊ **calciphile** ['kælsɪfaɪl] *see* CALCICOLE

◊ **calciphobe** ['kælsɪfəʊb] *see* CALCIFUGE

calcium ['kælsiəm] *noun* metallic chemical element which is naturally present in limestone, chalk and is essential to biological life (it is a major component of bones and teeth); **calcium borogluconate** = chemical which is given in the form of injections to cows suffering from milk fevers, as a result of calcium deficiency; **calcium carbonate** = chalk, white mineral found widely in many parts of the world, formed from animal organisms; **calcium cycle** = cycle of events by which calcium in the soil is taken up into plants, passed to animals which eat the plants, and then passed back to the soil again when the animals die and decompose; **calcium deficiency** = lack of calcium in an animal's bloodstream; **calcium hydroxide** = slaked lime, mixture of calcium oxide and water, used on soils to improve their quality; **calcium oxide** = quicklime, chemical used in many industrial processes and also spread on soil to reduce acidity; **calcium sulphate** = gypsum, sometimes used to improve salt-affected soils; **calcium supplement** = addition of calcium to the diet, or as injections, to improve the level of calcium in the bloodstream; **calcium uptake** = the taking of calcium into an animal's bloodstream as it eats (NOTE: chemical symbol is **Ca**; atomic number is **20**)

COMMENT: calcium is essential for various body processes such as blood clotting, and is an important element in a correct diet. Milk, cheese, eggs and certain vegetables are the main sources of calcium for human consumption. In birds, calcium is responsible for the formation of strong egg shells. Lack of calcium in the bloodstream of cattle may lead to a disorder known as milk fever

calf [kɑːf] *noun* young of a cow, less than one year old; **calf diphtheria** = disease affecting the mouth and throat of a calf; **calf enteric disease** = disease of calves causing severe diarrhoea; **calf pneumonia** = disease caused by a virus, and affecting dairy-bred and suckled calves; *see also* CALVE

COMMENT: a male calf is known as a bull calf; a female calf is a heifer calf. The meat of calves fed on a milk diet is known as veal

calomel ['kæləmel] *noun* mercury chloride, used to kill moss on lawns and to treat pinworms in the intestine

caloric [kə'lɒrɪk] *adjective* referring to calories; **caloric requirement** = amount of energy (shown in calories) which an animal such as a human needs each day

◊ **calorie** *or* **gram calorie** ['kæləri *or* 'græm 'kæləri] *noun* unit of measurement of heat *or* energy (NOTE: the **joule** is now more usual)

◊ **Calorie** *or* **large calorie** ['kæləri] *noun* kilocalorie *or* 1,000 calories (NOTE: spelt with a capital. Also written **cal** after figures: **250 cal**)

◊ **calorific value** [kælə'rɪfɪk 'væljuː] *noun* heat value of a substance, the number of Calories which a certain amount of a substance (such as a certain food) contains

COMMENT: one calorie is the amount of heat needed to raise the temperature of one gram of water by one degree Celsius. A Calorie or kilocalorie is the amount of heat needed to raise the temperature of a kilogram of water by one degree Celsius. The Calorie is also used as a measurement of the energy content of food and to show the caloric requirement, or amount of energy needed by an average person

calve [kɑːv] **1** *verb* to give birth to a calf; **calving box** = special pen in which a cow is put to calve; **calving interval** = period of time between one calving and the next; **calving time** = time when a cow is ready to calve **2** *noun* **calves** *see* CALF

◊ **calver** ['kɑːvə] *noun* cow which has had calves; **first calver** = cow which has produced her first calf

QUOTE freshly calved cows should not enter the milking herd until 3-5 days after calving. If the cow is eating well and has normally expelled the placenta, she should then be moved from the calving pen to the milking herd
Indian Farming

calyx ['keɪlɪks] *noun* part of a flower shaped like a cup, made up of the green sepals which cover the flower when it is in bud (NOTE: the plural is **calyces)**

Cambridge ['keɪmbrɪdʒ] *noun* breed of sheep

◊ **Cambridge roller** ['keɪmbrɪdʒ 'rəʊlə] *noun* heavy roller with a ribbed surface, consisting of a number of heavy iron wheels or rings, each of which has a ridge about 4cm high; the ribbed soil surface left by the roller provides an excellent seedbed for grass and clover seeds

camel ['kæməl] *noun* animal found in North Africa, and the Near and Middle East, used mainly as a means of transport, but also for ploughing, lifting water and threshing

COMMENT: there are two types of camel: the Arabian dromedary, found in the Middle East, India and North Africa, and the Bactrian camel (which has two humps) and is found in Asia in the deserts and colder dry steppes of Mongolia and Turkestan

Camellia [kə'miːlɪə] *noun* family of semi-tropical evergreen plants, including the tea plant

Camembert ['kæməmbɜː] *noun* soft French cheese, produced in Normandy

campos ['kæmpɒs] *noun* important tropical grasslands in Brazil, situated south of the Equatorial forests of the Amazon basin

campylobacter [kæm'pɪləbæktə] *noun* bacteria found in the gut of chickens. Dairy cattle also carry the organism. It is a cause of food poisoning in humans

can [kæn] **1** *noun* metal container for preserving food **2** *verb* to preserve food by sealing it in special metal containers; **canning factory** = factory where food is canned

Canadian Holstein [ke'neɪdɪən 'hɒlstiːn] *see* HOLSTEIN

canal [kə'næl] *noun* waterway made by people to take water to irrigate land

Canary grass [kə'neəri 'grɑːs] *noun* weed which produces a large cylindrical-shaped flower not unlike Timothy heads; troublesome in cereals

candling ['kændlɪŋ] *noun* checking process where eggs are passed over a source of light which detects blood spots in the egg or cracks in the shell

cane [keɪn] *noun* stem of large grasses, such as the sugar cane, and of other plants such as blackberries and raspberries; **cane fruits** = fruits from plants belonging to the genus *Rubus* including raspberry, blackberry and loganberry. The canes need a post and wire system for support. Fruit is sold on the fresh market, as well as being used for processing, and in recent years there has been an increased interest in the PYO outlets; **sugar cane** = tall plant of the family of grasses; the seed or grain is of no commercial value and the plant is cultivated for the sake of the juice extracted from its stem. Sugar cane is mainly grown in the tropics, especially in Cuba and the West Indies; **cane sugar** = sugar which is processed from the juice which is extracted from the stems of sugar cane in crushing mills

canker ['kæŋkə] *noun* disease causing infected areas on the outer surface of a plant

COMMENT: the chief victims of canker are fruit trees, especially apples. Cankers appear as sunken areas on the bark, or near a wound. Fungus spores infect the wound edges, laying bare the wood. The disease spreads from the infected area, causing fruit spurs to wilt

cannabis ['kænəbɪs] *noun* Indian hemp plant *(Cannabis sativa)* which produces an addictive drug (marijuana)

cannibalism ['kænɪbəlɪzm] *noun* the practice of an animal which feeds on its own species. In poultry, this may follow on from featherpecking, and may be caused by the crowded conditions in which birds are housed. Sows may eat the young of other sows in intensive breeding conditions

cannula ['kænjʊlə] *noun* tube through which a trocar is inserted, used to puncture an animal's rumen to allow an escape of gas

canola [kə'nəʊlə] *noun (in Canada)* oilseed rape

canopy ['kænəpi] *noun* cover provided by a crop over the soil

Canterbury hoe ['kæntəbəri 'həʊ] *noun* hoe which does not have a blade, but is like a three-pronged fork, with the prongs set at right angles to the handle

◊ **Canterbury lamb** ['kæntəbəri 'læm] *noun* lamb reared in New Zealand, mainly for export

cantle ['kæntl] *noun* rear bow of a saddle

CAP *see* COMMON AGRICULTURAL POLICY

capability class [keɪpə'bɪlɪti 'klɑːs] *noun* classification of the usefulness of land for agricultural purposes

capacity [kə'pæsɪti] *noun* amount which something can contain; **field capacity** = the amount of water which is retained by soil after free drainage has taken place; **moisture-holding capacity (MHC)** = the amount of water held between field capacity and permanent wilting point

caper ['keɪpə] *noun* Mediterranean shrub *(Capparis spinosa)* the flower buds of which are used as a flavouring

capercaillie [kæpə'keɪli] *noun* large game bird *(Tetrao urogallus)* found in northern coniferous forests

capillary [kə'pɪləri] *noun* (a) tiny tube carrying a liquid in an organism, such as a very small blood vessel (b) **capillary flow** = movement of a liquid upwards inside a narrow tube *or* upwards through the soil

◊ **capillarity** [kæpɪ'lærəti] *noun* = CAPILLARY FLOW

COMMENT: capillary flow is important in water in soil, as it does not drain away. It moves through the soil by capillary action, i.e. by the surface tension between the water and the walls of the fine tubes or capillaries. It is a very slow movement, and may not be fast enough to supply plant roots in a soil which is drying out

capital items ['kæpɪtəl 'aɪtəmz] *noun* items such as machinery, buildings, fences and drains used in farm production. UK agriculture currently accounts for just under 2% of the country's total spending on capital goods

capon ['keɪpɒn] *noun* castrated edible cockerel; a cockerel which has been treated with a sex-inhibiting hormone grows and increases in weight more rapidly than a bird which has not been treated

cappie ['kæpi] *noun* disease of sheep, mainly of older lambs and young sheep, associated with thinning of the skull bones. In severe cases the animal cannot eat or close its mouth

capping ['kæpɪŋ] *noun* hard crust which sometimes forms on the surface of soil, often only about 2-3cm thick. It can be caused by heavy rain on dry soil; also by tractors and other heavy farm machinery

caprine ['kæpraɪn] *adjective* referring to goats; **caprine arthritis-encephalitis** *or* **caprine arthritic encephalitis (CAE)** = disease of goats, where the animal suffers loss of condition, swollen joints and pneumonia, leading eventually to death. The disease is spread by contact with saliva and milk. There is no cure. The first case in the UK was reported in April 1990

capsicum ['kæpsɪkəm] *noun* group of plants (also called 'peppers') grown for their pod-like fruit, some of which are extremely pungent, such as the chilli and Cayenne peppers. Others, including the red, green or sweet peppers are less pungent and are used as vegetables

capsid bug ['kæpsɪd bʌg] *noun* tiny insect which sucks the sap of plants, one of a large family of sucking insects that chiefly attack fruit trees

capsular ['kæpsjulə] *adjective* referring to a capsule

◊ **capsule** ['kæpsjuːl] *noun* **(a)** membrane round an organ *or* egg, etc. **(b)** dry seed case which bursts open to allow the seeds to shoot out (the seed cases of poppies are capsules)

captan ['kæptæn] *noun* fungicide, used to fight apple and pear scab; it is also used in seed dressings for peas and other vegetables

caracole ['kærəkəʊl] *noun* half turn performed by a horse

carambola [kærəm'bəʊlə] *noun* yellow fruit of a tropical tree *(Averrhoa carambola)* found in Indonesia; the fruit are used in preserves and drinks

caraway ['kærəweɪ] *noun* seeds of a herb *(Carum carvi)* used as a flavouring in bread and cakes

carbamate ['kɑːbəmeɪt] *noun* type of pesticide *or* insecticide (such as Zectran and Zineb) developed to replace organochlorines such as DDT

COMMENT: the advantage of carbamates over DDT is that they are not persistent and do not enter the human food chain

carbohydrates [kɑːbəʊ'haɪdreɪts] *noun* organic compounds which derive from sugar and which are the main ingredients of many types of food

COMMENT: carbohydrates are compounds of carbon, hydrogen and oxygen. They are found in particular in sugar and starch from plants, and provide the body with energy. Plants build up valuable organic substances from simple materials. The most important part of this process, which is called photosynthesis, is the production of carbohydrates such as sugars, starches and cellulose. They form the largest part of food of animals

carbon ['kɑːbən] *noun* one of the common non-metallic elements, an essential component of living matter and organic chemical compounds; **carbon cycle** *or* **circulation of carbon** = carbon atoms from carbon dioxide are incorporated into organic compounds in plants during photosynthesis. They are then oxidized into carbon dioxide again during respiration by the plants or by herbivores which eat them and by carnivores which eat the herbivores, so releasing carbon to go round the cycle again; **carbon sink** = part of the biosphere (such as a tropical forest) which absorbs carbon, as opposed to animals which release carbon into the atmosphere in the form of carbon dioxide (NOTE: chemical symbol is **C**; atomic number is **6**)

◊ **carbonate** ['kɑːbəneɪt] *noun* compound formed from a base and carbonic acid; **calcium carbonate** = chalk, a white mineral found widely in many parts of the world, formed from animal organisms

◊ **carbonic acid (H_2CO_3)** [kɑː'bɒnɪk 'æsɪd] *noun* weak acid formed by dissolving CO_2 in water, as in rainwater

carbon dioxide (CO_2) ['kɑːbən daɪ'ɒksaɪd] *noun* colourless gas produced when carbon is burnt with oxygen

COMMENT: Carbon dioxide exists naturally in air and is produced by burning or rotting organic matter. In animals, the body's metabolism makes the tissues burn carbon, which is then breathed out by the lungs as waste carbon dioxide. Carbon dioxide is removed from the atmosphere by plants when it is split by chlorophyll in photosynthesis to form carbon and oxygen. It is also dissolved from the atmosphere in sea water. The increasing release of carbon dioxide into the atmosphere, especially from burning fossil fuels, adds to the greenhouse effect. Carbon dioxide is used in solid form as a means of keeping food cold

carbon monoxide (CO) ['kɑːbən mən'ɒksaɪd] *noun* poisonous gas found in fumes from car engines, from burning gas and in cigarette smoke

carboxyhaemoglobin [kɑːbɒksɪhiːmə'gləubɪn] *noun* compound of carbon monoxide and haemoglobin formed when a person breathes in carbon monoxide; **background carboxyhaemoglobin level** = level of carboxyhaemoglobin in the blood of a person living a normal existence without exposure to particularly high levels of carbon monoxide

carcass ['kɑːkəs] *noun* dead body of an animal; especially in the meat trade, the body of an animal after removing head, limbs and offal; **carcass classification scheme** = system of judging the thickness of flesh on a carcass, and the fat cover over a carcass; *see also* CONFORMATION (NOTE: also written **carcase**)

COMMENT: the fat classes in the Beef Carcass Classification scheme range from Class 1 with the least fat cover through Class 3, which is average, to 5H which has the highest fat cover

cardamom ['kɑːdəməm] *noun Elettaria cardamomum,* a crop grown in India and Sri Lanka for its seeds, which are used in the preparation of curry powder

cardoon [kɑː'duːn] *noun* vegetable *(Cynara cardunculus)* grown in Mediterranean areas, similar to the globe artichoke

cargo ['kɑːgəʊ] = BROWN RICE

carnivore *or* **carnivorous animal** ['kɑːnɪvɔː *or* kɑː'nɪvərəs] *noun* animal (such as the cat or dog) which lives on flesh of other animals

◊ **carnivorous** [kɑː'nɪvərəs] *adjective* (animal) which eats the flesh of other animals

carob ['kærəb] *noun* long flat dried pod of the carob tree, used in food preparations

and in animal feed, where it is called 'locust bean'

carotene ['kærətiːn] *noun* orange or red pigment in carrots, egg yolk and some natural oils, which is converted by an animal's liver into vitamin A

carpel ['kɑːpəl] *noun* female part of a plant, formed of an ovary and stigma

carr [kɑː] *noun* type of fen which supports some trees, such as the alder, in East Anglia

carrier ['kæriə] *noun* animal which has recovered from a disease or in which the disease is latent, and which is capable of passing on the infection to another animal

carrot ['kærət] *noun* vegetable root crop *(Daucus carota)* grown for human consumption; most are grown for fresh sale, some for canning and freezing. Damaged roots which cannot be sold can be fed to cattle; **carrot fly** = a small fly *(Psila rosae)* with a reddish brown head and yellowish wings; it lays its eggs in the soil surface near plants and the larvae burrow into the root. The foliage wilts and dies, and in cases of severe infestation, the whole crop is lost

carry ['kæri] *verb* to keep (cattle) on farmland; **carrying capacity** = maximum number of productive livestock that can be supported by a given area of land, so that the vegetation eaten by the grazing animals does not exceed the rate at which it grows again

cart [kɑːt] *noun* farm vehicle, used for carrying produce

◊ **carter** ['kɑːtə] *noun* person in charge of a cart being used to move produce

◊ **carthorse** ['kɑːthɔːs] *noun* farm work horse

cartilage ['kɑːtɪlɪdʒ] *noun* thick connective tissue which lines the joints and acts as a cushion, and which forms part of the structure of an organ. The chief diseases which affect cartilage in animals are necrosis, resulting in the death of cells of the cartilage, and caused by injury, and ossification of the cartilage (also caused by injury), where the cartilage turns into bone

caryopsis [kærɪ'ɒpsɪs] *noun* seed of cereals and grasses

CAS = CENTRE FOR AGRICULTURAL STRATEGY

case [keɪs] *noun* **(a)** hard outer cover; **seed case** = hard outside cover which protects the seeds in certain plants; **case hardening** = formation of a hard surface on a piece of food, by deposition of sugar or salt **(b)** quantity of produce (one dozen bottles of wine, thirty dozen eggs)

casein ['keɪsiːn] *noun* protein found in milk

COMMENT: casein is precipitated when milk comes into contact with an acid, and so makes milk form cheese

cash crops ['kæʃ krɒps] *noun* crops which are grown to be sold rather than eaten by the person who grows them, or for feeding to livestock on the farm

QUOTE in the Philippines as a whole, more than 30% of the total cultivated land area was given over to cash crop production for export in 1980, mainly bananas, pineapples and sugar cane
The Ecologist

cashew nut [kə'ʃuː 'nʌt] *noun* nut from a tropical tree (*Anarcardium occidentale*), popular when roasted as dessert nuts; grown extensively in India and East Africa; native to Central and South America

cashmere ['kæʃmiːə] *noun* very fine down undercoat on a goat, less than 18 microns

cassava [kə'sɑːvə] *noun* starchy root crop; there are two varieties, the sweet cassava (*Manihot dulcis*) and the bitter cassava or manioc plant (*Manihot utilissima*)

COMMENT: cassava is an important world crop. The root of the bitter cassava is poisonous, and has to be washed before being made into flour for human consumption. The flour is called tapioca in Asia, in Brazil it is mandioca, and in Spanish Latin America it is yuca; cassava is also used in animal feedingstuffs. Brazil is the largest producer, followed by Zaire and Nigeria

cast [kɑːst] **1** *noun* **worm cast** = waste earth rejected by an earthworm **2** *verb* **(a)** to bear an offspring prematurely; *a cast calf;* **cast ewes** = old breeding ewes which are sold off by hill farmers to farmers on lower ground

who take one more crop of lambs from them **(b)** to place an animal on its side on the ground; **cast sheep** = sheep lying on its back and unable to get up again

◊ **casting** ['kɑːstɪŋ] *noun* method of ploughing, in which the area is ploughed in a circle, going in an anticlockwise direction

castor oil ['kɑːstə 'ɔɪl] *noun* oil derived from the seeds of the castor oil plant (*Ricinus communis*), used as a common purgative for fowls and calves

castrate [kæ'streɪt] *verb* to remove the testicles of a male animal

◊ **castration** [kæ'streɪʃn] *noun* removal of the essential sex organs, testes and ovaries, from male or female animals; this allows bullocks and heifers to be housed together, as castrated animals are more docile

COMMENT: castration may be by cutting the scrotum, as in pigs, or by atrophication which follows ringing, as in sheep and cattle. In fowls, a sex-inhibiting hormone is used

casual ['kæʒuəl] *adjective* **casual labour** *or* **casual workers** = workers who are hired for short periods from time to time; **casual labourer** *or* **casual worker** = worker who can be hired for a short period, usually paid on a piecework basis (for example students hired to pick soft fruit)

catabolism [kə'tæbəlɪzm] *noun* breaking down of complex chemicals into simple chemicals, as a part of metabolism; it includes the breaking down of nutrients from food to provide energy

CAT = CENTRE FOR ALTERNATIVE TECHNOLOGY

catch [kætʃ] **1** *noun* amount of fish caught **2** *verb* to hunt and take animals (usually fish); *(in Australia)* **catching pen** = pen into which sheep are put while waiting to be sheared

◊ **catch crop** ['kætʃ 'krɒp] *noun* **(a)** fast-growing crop grown in the time interval between two main crops **(b)** fast-growing crop planted between the rows of a main crop

◊ **catch drain** *or* **catch-water drain** [kætʃ *or* 'kætʃwɔːtə 'dreɪn] *noun* type of drain designed to take rainwater from sloping ground

catchment (area) ['kætʃmənt] *noun* area of land which collects and drains the water which falls on it (such as the area round a lake or the basin of a river)

caterpillar ['kætəpɪlə] *noun* soft-bodied larva of many species of butterflies and moths; caterpillars feed mainly on foliage, but can also attack roots, seeds and bark of crops

cation ['kætaɪən] *noun* ion with a positive electric charge; **cation exchange** = exchange which takes place when the ions of calcium, magnesium and other metals found in soil replace the hydrogen ions in acid

cattle ['kætl] *noun* domestic farm animals of the class Bovidae, raised for their milk and meat; **cattle grid** = type of grill made of parallel bars, covering a hole dug in the road; it prevents stock from crossing the grid and leaving their pasture, but allows vehicles and humans to pass; **cattle plague** = disease of cattle, eradicated from the UK in 1877, but still found in parts of Asia and Africa; **cattle rustler** = person who steals cattle; **cattle rustling** = stealing of cattle

◊ **cattleman** ['kætlmæn] *noun* person who looks after cattle

COMMENT: domesticated cattle belong to the genus Bos, of the family Bovidae, and within the genus there are a number of different species. The European domesticated cattle are *Bos taurus;* the Indian and African domesticated cattle are *Bos indicus;* the buffalo is *Bos bubalus.* The main breeds of European dairy cattle are: Ayrshire, Dairy Shorthorn, Friesian, Guernsey, Holstein, and Jersey. The dairy cow is one of the most valuable domesticated animals, providing high-quality human foods. Milk is the main product, although cull cows are an important source of meat. The main breeds of beef cattle are: Aberdeen Angus, Beef Shorthorn, Charolais, Devon, Galloway, Hereford. The main dual-purpose breeds (i.e. breeds which provide both beef and dairy products) are Dexter, South Devon and Welsh Black

cauliflower ['kɒlɪflaʊə] *noun* plant of the cabbage family, with a large white head made up of a mass of curds; it may be eaten raw, but more often is cooked

cavings ['keɪvɪŋz] *noun* broken pieces of straw from a threshing machine

cayenne pepper [keɪ'en 'pepə] *noun* plant *(Capsicum frutescens)* producing a pungent red pepper from ground dried pods; it is grown in the tropics and in warm temperate areas

Cayuga [kæ'juːgə] *noun* breed of duck, producing dark green eggs; the plumage is shiny green-black

CC = COUNTRYSIDE COMMISSION

Cd *chemical symbol for* cadmium

CEC = CATION EXCHANGE CAPACITY

cedar ['siːdə] *noun* American red wood *(Thuya plicata),* soft, but resistant to water and now becoming scarce. It is used mainly for outdoor construction work

celeriac [sə'lerɪæk] *noun* *Apium graveolens,* a variety of celery with a thick edible root; the root is used as a vegetable in soups and salads

celery ['selərɪ] *noun* vegetable plant *(Apium graveolens),* with thick edible leaf stalks; the plant is grown in trenches to help growth and to blanch the stems; some varieties are self-blanching

cell [sel] *noun* tiny unit of matter surrounded by membrane, which is the base of all plant and animal tissue; **cell division** = way in which a cell reproduces itself by mitosis; **cell membrane** = outside wall of all cells, both plants and animals; **cell wall** = outside wall of a plant cell, formed of cellulose (does not exist in animal cells)

◊ **cellular** ['seljʊlə] *adjective* **(a)** referring to cells *or* formed of cells **(b)** made of many similar parts connected together; **cellular tissue** = form of connective tissue with large spaces

COMMENT: the cell is a unit which can reproduce itself. It is made up of a jelly-like substance (cytoplasm) which surrounds a nucleus, and contains many other small organisms or organelles which vary according to the type of cell. Cells reproduce by division (mitosis). Some microorganisms are formed of a single cell, but most organisms are formed by the division and reproduction of many millions of cells

cellulose ['seljʊləuz] *noun* **(a)**

carbohydrate which makes up a large percentage of plant matter, especially cell walls **(b)** chemical substance processed from wood, used for making paper, film and artificial fibres

COMMENT: cellulose is not digestible, and is passed through the digestive system as roughage

Celsius ['selsiəs] *noun* scale of temperature where the freezing and boiling points of water are 0° and 100° (NOTE: used in many countries, but not in the USA, where the Fahrenheit system is still preferred. Normally written as a **C** after the number: **52°C** (say: 'fifty-two degrees Celsius'). Used to be called **centigrade)**

COMMENT: to convert Celsius temperatures to Fahrenheit, multiply by 1.8 and add 32. So 20°C is equal to 68°F

Celtic field system [keltɪk 'fiːld 'sɪstəm] *see* FIELD

CEMP = CONFEDERATION OF EUROPEAN MAIZE PRODUCERS

census ['sensəs] *see* AGRICULTURAL CENSUS

centigrade ['sentɪgreɪd] *noun* scale of temperature where the freezing and boiling points of water are 0° and 100°; *see note at* CELSIUS

centrifugal [sentrɪ'fjuːgəl] *adjective* which goes away from the centre; **centrifugal sugar** = raw sugar containing 96% to 98% sucrose, which has been isolated from sugar beet or cane by standard extraction processes

◊ **centrifugation** *or* **centrifuging** [sentrɪfjuˈgeɪʃn *or* 'sentrɪfjuːdʒɪŋ] *noun* separation of solids from a liquid by spinning

◊ **centrifuge** ['sentrɪfjuːdʒ] *noun* device which spins round very fast, and separates the various components of a liquid mixture or removes solids

cep [sep] *noun* edible mushroom-like fungus *(Boletus edulis)*

cereals ['sɪəriəlz] *noun* specialized types of grasses that are cultivated for their large seeds or grains, especially to make flour for breadmaking or for animal feed; *the EU grows large quantities of cereals;* **cereal stands** = fields of standing cereal crops

COMMENT: cereal plants are all members of the Graminales family or grasses. The commonest are oats, wheat, barley, maize and rye in colder temperate areas, and rice and sorghum in warmer regions. Other cereals of local importance are millet, teff in Ethiopia and adlay in India. Cereal production has considerably expanded and improved with the introduction of better methods of sowing, combine harvesters, driers, bulk handling and chemical aids such as herbicides, fungicides, insecticides and growth regulators

cerebrocortical necrosis [serəbrəʊˈkɔːtɪkl nɪˈkrəʊsɪs] *noun* disease of sheep caused by thiamine deficiency; the animal appears blind and fits may follow

cerrado [seˈrɑːdəʊ] *noun* tropical savanna grassland of Brazil

certificate [səˈtɪfɪkət] *noun* official paper which states something, such as the National Certificate in Agriculture; **certificate of bad husbandry** = certificate issued to a tenant farmer by an Agricultural Land Tribunal if the tenant is inefficient and unable to farm to a satisfactory standard

◊ **certification** [sɜːtɪfɪˈkeɪʃn] *noun* official granting of a certificate

◊ **certify** ['sɜːtɪfaɪ] *verb* to state officially that something is correct; **certified seed** = seed which has been successfully tested for purity, disease and weed contamination and is granted certification for sale; **certified stock** = stock of grain which has been approved for delivery

cesspool *or* **cesspit** ['sespuːl *or* 'sespɪt] *noun* septic tank *or* tank in the ground which collects sewage from a house, and from which the sewage gradually seeps away, not connected to a mains drainage system

chaff [tʃɑːf] *noun* husks of corn which separate from the grain during threshing; short lengths of cut straw, used as feed for ruminants and as a component of manures

chain [tʃeɪn] *noun* **(a)** (i) number of metal rings attached together to make a line; (ii) measure of length equal to 22 yards (originally measured with a chain, called Gunter's chain); **chain harrow** = type of harrow built in a similar way to a piece of chain-link fencing. The links may be plain or spiked, the spiked type being used to aerate grassland; **chain-link fencing** = material for fencing, made of an open web of thick wire links, twisted together; supplied on a roll, it is one of the easiest forms of fencing to put in place **(b)** number of parts linked together; **food chain** = series of organisms which pass energy from one to another as each provides food for the next (the first organism in the food chain is the producer, and the rest are consumers)

chalaza [kəˈleɪzə] *noun* coil of fibrous protein which holds the yolk in the centre of the egg

chalk [tʃɔːk] *noun* fine white limestone rock formed of calcium carbonate

◊ **chalky** [ˈtʃɔːki] *adjective* (soil) full of chalk

COMMENT: chalk is found widely in many parts of Northern Europe. Formed from animal organisms it is also used as an additive (E170) in white flour. Ground chalk is used for liming soils. The sharp-edged flints of various sizes found in soils overlying some of the chalk formations in Europe, are very wearing on farm implements

challenge feeding [ˈtʃæləndʒ ˈfiːdɪŋ] *noun* feeding of dairy cows with concentrates to provide extra nourishment

chamomile [ˈkæməmaɪl] *see* MAYWEED

chandler [ˈtʃɑːndlə] *noun* person who deals in goods, such as a seed chandler or corn chandler

channel [ˈtʃænəl] **1** *noun* bed of a river *or* the ground across which a river or stream flows; **drainage channel** = small ditch made to remove rainwater from the soil surface **2** *verb* to send (water) in a particular direction

◊ **channelize** [ˈtʃænəlaɪz] *verb* to make straight a stream which has many bends, so as to make the water flow faster

◊ **Channel Island breeds** the Guernsey and Jersey breeds of dairy cattle

chapati [tʃəˈpɑːti] *noun* flat unleavened bread made from cereal flour and water, eaten as a staple food throughout the Indian subcontinent

chaptalization [tʃæptəlaɪˈzeɪʃn] *noun* addition of sugar to wine which has not reached the standard level of alcohol, in order to increase the level of alcohol (the additive can be ordinary white sugar or some form of alcohol)

charlock [ˈtʃɑːlɒk] *noun* a widespread weed *(Sinapis arvensis)* mainly affecting spring cereals and other spring crops; also commonly called brasics or wild mustard

Charmoise [ˈʃɑːmwɑːz] *noun* breed of sheep found in central France

Charolais [ˈʃærəleɪ] *noun* breed of beef cattle originating in central France; creamy white in colour, valued for their fast growth and lean meat. They tend to have large hindquarters which, when passed on to an unborn calf, can lead to difficulties at birth. They are large heavy cattle, and have been exported all over the world in recent years

Charollais [ˈʃærɒleɪ] *noun* breed of sheep originating in France and having a characteristic 'red' face; used as an alternative to Suffolk and Texel rams on lowland crossbred or mule flocks

chats [tʃæts] *noun* small potatoes, separated from larger (or ware) potatoes during the grading process

chayote [tʃaɪˈəʊti] *noun* pear-shaped vegetable *(Sechium edule)* grown in the West Indies

Cheddar [ˈtʃedə] *noun* hard yellow cheese, originally made in the West Country

cheese [tʃiːz] *noun* food made from cow's milk curds. Cheese is also made from goat's milk, and more rarely from ewe's milk. The curd is pressed and left to mature for a period of time (longer in the case of hard cheese). There are many varieties of both hard and soft cheese: the British Caerphilly, Gloucester, Cheddar and Cheshire are all hard cheeses; the French Brie and Camembert are soft

chelates [ˈkiːleɪts] *noun* compounds of trace elements and organic substances which are water-soluble and may be safely applied as foliar sprays or to the soil

chem- [kem] *prefix* referring to chemistry *or* chemicals

◊ **chemical** [ˈkemɪkl] **1** *adjective* **chemical compound** = substance formed from two or more chemical elements, in which the proportions of the elements are always the same; **chemical elements** = the different substances which exist independently and cannot be broken down to simpler substances; **chemical food poisoning** = poisoning by chemical substances in food (such as toxic substances naturally present in some plants, or insecticides in processed food); **chemical score** = comparison of the relative protein values of particular foodstuffs, tested in laboratory experiments; **chemical symbol** = letter (or letters) used to indicate an element (used especially in formulae, C for carbon, Co for cobalt, etc.) **2** *noun* substance produced by a chemical process *or* formed of chemical elements; *the widespread use of chemicals in agriculture*

◊ **chemistry** ['kemɪstri] *noun* study of substances, elements and compounds and their reactions with each other

Cher [ʃeə] *see* BERRICHON

chernozem ['tʃɜːnəʊzem] *noun* dark fertile soil, full of organic matter containing abundant calcium carbonate in its lower horizons

COMMENT: chernozem is found in the temperate grass-covered plains of the Soviet Union and North and South America

cherry ['tʃeri] *noun* (a) *Prunus,* a tree with many small-stoned fruit (b) used to refer to any small fruit; **cherry plum** = cooking plum *(Prunus cerasifera)* which is small and usually bright red; **cherry tomato** = variety of tomato *(Lycopersicon esculentum)* with very small fruit; **coffee cherry** = the red fruit of the coffee plant, which contains the coffee beans

chervil ['tʃɜːvɪl] *noun* herb *(Anthriscus cerefolium)* used as a garnish and also in salads and soups

Cheshire ['tʃeʃə] *noun* crumbly hard white British cheese

chestnut ['tʃesnʌt] **1** *noun* **sweet chestnut** = European hardwood tree *(Castanea sativa)* important for its nuts and timber **2** *adjective* reddish-brown colour, used to describe horses

Cheviot ['tʃiːviət] *noun* large, hardy breed of sheep; white-faced and usually hornless. The short, thick wool is of good quality, of middle length and fairly dense

Chewings fescue ['tʃuːwɪŋz 'feskjuː] *noun* common variety of grass used for pastures in New Zealand

Chianina [kiæ'niːnə] *noun* breed of beef cattle, originating in Tuscany in Italy; one of the largest breeds of cattle, it is white in colour with black hooves, muzzle and horn tips. It is a dual-purpose breed, used both for beef production and as draught animals

chick [tʃɪk] *noun* young, newly hatched bird, up to the time when it is weaned from the hen or brooder

◊ **chicken** ['tʃɪkɪn] *noun* (a) a young bird of a domestic fowl (b) the meat of domestic poultry

COMMENT: chicken manure is now being used as fuel in programmes to set up renewable energy power stations

◊ **chickpea** ['tʃɪkpiː] *noun* *Cicer arietinum,* one of the pulses; important in India and Pakistan as a source of protein; also called gram, and in South America garbanzos

◊ **chickweed** ['tʃɪkwiːd] *noun* a widespread choking weed *(Stellaria media)* found in cereals and grass; it germinates and flowers all the year round. Also called 'white bird's-eye'

Chickasaw ['tʃɪkəsɔː] *noun* American variety of plum

chicle ['tʃɪkl] *noun* gum used in making chewing gum, made from the sap of the sapodilla *(Achras zapota)*

chicory ['tʃɪkəri] *noun* blue-flowered plant *(Cichorium intybus)* cultivated for its salad leaves and roots; dried chicory root is used as a substitute for coffee

chill [tʃɪl] **1** *noun* coldness; **wind chill factor** = way of calculating the risk of exposure in cold weather by adding the speed of the wind to the number of degrees of temperature below zero **2** *verb* to preserve by cooling to a temperature just above freezing; *see also* COOK CHILL

◊ **chillshelter** ['tʃɪlʃeltə] *noun* form of feedlot, where a small area of land for the intensive fattening of cattle is surrounded by a high embankment to protect the animals from cold

◊ **chill starvation** ['tʃɪl stɑː'veɪʃn] *noun* disease affecting very young lambs, caused by loss of body heat during severe weather conditions

chilli ['tʃɪli] *noun* very pungent spice *(Capsicum frutescens)*, with bright red seed pods

Chinchilla [tʃɪn'tʃɪlə] *noun* small rabbit important for its soft grey fur

Chinese goose [tʃaɪ'niːz 'guːs] *noun* breed of goose with a lighter carcass, raised for meat production

◊ **Chinese pig breeds** *see* MEISHAN, TAIHU

chinook [tʃɪ'nuːk] *noun* warm wind which blows from the Rocky Mountains down onto the Canadian plains in winter; *similar to* FOEHN

chip basket ['tʃɪp 'baːskɪt] *noun* basket woven from thin strips of wood

chisel plough ['tʃɪzl 'plaʊ] *noun* plough with a heavy-duty frame, with tines bolted to it. Rigid tines are normally used, and each tine has a point which can be replaced. The working depth is deeper than with normal ploughing and is hydraulically controlled

chit [tʃɪt] **1** *noun* shoot *or* sprout **2** *verb* to promote germination in seed before sowing; setting up seed potatoes to sprout before planting them; *about half of the potato crops in the UK are grown from chitting tubers*

◊ **chitting house** ['tʃɪtɪŋ 'haʊs] *noun* storage building for trays of potatoes, where they are kept to sprout before planting

chitterlings ['tʃɪtəlɪŋz] *noun* small intestines of pigs, used for food

chives [tʃaɪvz] *noun* onion-like herb *(Allium schoenoprasum)* of which the leaves are used as a garnish or in soups and salads

chlamydiosis [kləmɪdɪ'əʊsɪs] *noun* viral bacterial infection which is transmitted by infected birds, such as ducks and pigeons

COMMENT: symptoms can be similar to those of flu, or in bad cases, pneumonia and hepatitis. The disease comes from birds and their feathers, and affects particularly poultrymen. It can also affect sheep and lambs

chlor(o)- ['klɔːrəʊ] *prefix* referring to (i) chlorine; (ii) green

◊ **chloride** ['klɔːraɪd] *noun* a salt which is a compound of chlorine; **sodium chloride** = common salt

◊ **chlorinated** ['klɒrɪneɪtɪd] *adjective* treated with chlorine; **chlorinated hydrocarbon insecticide** = organochlorine, a type of insecticide made synthetically as a compound of chlorine

COMMENT: chlorinated hydrocarbon insecticides include DDT, aldrin and lindane. These types of insecticide are very persistent, with a long half-life of up to 15 years, while organophosphorous insecticides have a much shorter life. Chlorinated hydrocarbon insecticides not only kill insects, but also enter the food chain and kill small animals and birds which feed on the insects

chlorination [klɒrɪ'neɪʃn] *noun* sterilizing by adding chlorine

◊ **chlorinator** ['klɒrɪneɪtə] *noun* apparatus for adding chlorine to water

COMMENT: chlorination is used to kill bacteria in drinking water, in swimming pools, and sewage farms, and has many industrial applications such as sterilization in food processing

chlorine ['klɔːriːn] *noun* an element, a powerful greenish gas occurring in nature as various salts or chlorides. It is a nutrient of major importance both in the animal diet and in the growth of plants (NOTE: symbol is **Cl**; atomic number is **17**)

chlorophyll ['klɒrəfɪl] *noun* green pigment in plants, present mainly in the leaves

◊ **Chlorophyta** [klɒr'ɒfɪtə] *noun* green algae, the largest class of algae

◊ **chlorosis** [klɒ'rəʊsɪs] *noun* reduction of chlorophyll in plants, making the leaves turn yellow

COMMENT: chlorophyll absorbs light energy from the sun and supplies plants with the energy to fix CO_2 from the atmosphere. It is also used as a colouring (E140) in processed food

chocolate ['tʃɒklət] *noun* popular sweet food made from the cocoa bean (the beans are roasted, and then mixed with oils and sugar); **chocolate spot** = fungal disease of winter beans, occurring in two forms, *Botrytis cinerea* or the more severe *Botrytis fabae*. The disease appears as small round brown spots on leaves and stems. In bad attacks, the symptoms move to flowers and pods

choke [tʃəuk] **1** *noun* **(a)** stiff hairs inside the head of an artichoke; the hairs have to be removed before the heart can be eaten **(b)** disease of grasses appearing as a white fungus that develops out of the leaves and surrounds the young flower stalk. It reduces the seed crop **2** *verb* to kill an animal or plant by cutting off air or light; to stop movement; *the small plants were choked by weeds; the drainage channels were choked with water weed*

chop [tʃɒp] *verb* to cut up into pieces with a knife; **to chop down** = to cut down a tree, with an axe

◊ **chopper** [ˈtʃɒpə] *noun* machine for chopping; **straw chopper** = device attached to a combine harvester, which chops the straw into small pieces

Chorleywood bread process [tʃɔːlɪˈwʊd] method of making bread, developed by the British Baking Industries Research Association, in which the long fermentation period is eliminated by mixing the dough vigorously by mechanical means

chromosome [ˈkrəuməsəum] *noun* rod-shaped structure in the nucleus of a cell, formed of DNA which carries the genes

◊ **chromosomal** [krəuməˈsəuməl] *adjective* referring to chromosomes

COMMENT: each species has its own chromosomal make-up. The human cell has 46 chromosomes, 23 inherited from each parent. Cattle have 60 chromosomes, horses 64, and chickens 78. Plants have fewer chromosomes than animals: carrots have 18, maize 20, rice 24, etc. See also the comment at SEX

chronic [ˈkrɒnɪk] *adjective* (disease *or* condition) which lasts a long time; *chronic toxicity of the soil* compare ACUTE

chrysalis [ˈkrɪsəlɪs] *noun* stage in the development of a butterfly *or* moth, the hard case in which the pupa is protected

chrysanthemum [krɪˈzænθəməm] *noun* genus of composite plants, many of which

are cultivated for their flowers; the insecticide pyrethrum is derived from *Chrysanthemum roseum*

chufa [ˈtʃuːfə] *noun* a sedge *(Cyperus esculentus)* whose edible corms are used as a vegetable

churn [tʃɜːn] **1** *noun* large metal milk container **2** *verb* to shake a liquid violently to mix it

COMMENT: formerly milk was moved from farm to dairy in churns, but today most milk is collected by bulk tankers

CIAT = CENTRO INTERNACIONAL PARA AGRICULTURA TROPICALE the International Centre for Tropical Agriculture, set up at Palmira in Colombia in 1968; it researches into lowland tropical agriculture

CID = CATTLE IDENTIFICATION DOCUMENT document to be supplied with all male cattle sold at more than three months old

-cide [saɪd] *suffix* referring to killing; **herbicide** = substance which kills plants

cider [ˈsaɪdə] *noun* fermented drink made from apple juice

CIMMYT the International Centre for the Improvement of Maize and Wheat, established at El Batan, Mexico, in 1964

cinchona [sɪnˈkəunə] *noun* tropical tree, several species of which yield a bark containing quinine

cinnamon [ˈsɪnəmən] *noun* aromatic inner bark from a tropical tree, used as a spice *(Cinnamomum zeylanicum)*

circulation [sɜːkjuˈleɪʃn] *noun* movement of liquid or gas in a circle; **circulation of blood** = movement of blood round an animal's body; **circulation of carbon** = carbon atoms from carbon dioxide are incorporated into organic compounds in plants during photosynthesis. They are then oxidized into carbon dioxide again during respiration by the plants or by herbivores which eat them and by carnivores which eat the herbivores, so releasing carbon to go round the cycle again

citrus fruit [ˈsɪtrəs ˈfruːt] *noun* edible

fruits of evergreen citrus trees, grown throughout the tropics and subtropics; the most important are oranges, lemons, grapefruit and limes; citrus fruit have thick skin and are very acidic

city farm ['sɪti 'fɑːm] *noun* community project that utilizes wasteland in inner cities, or the urban fringe areas, to farm and garden with an ecological approach to the conservation management of land and resources

CIU = CROPS FOR INDUSTRIAL USE

Cl *chemical symbol for* chlorine

CLA = COUNTRY LANDOWNERS ASSOCIATION

clamp [klæmp] **1** *noun* method of storing root crops in the open; the crop is heaped into a pile and covered with straw and earth **2** *verb* to store crops or silage in a clamp

COMMENT: Silage is also kept in clamps, originally trenches dug into the ground into which the crop was tipped and then covered. Silage may be clamped as a mound often covered with polythene sheeting which is weighed down with old tyres or railway sleepers

clarts [klɑːts] *noun* dung attached to fleece of sheep

class [klɑːs] *see* WHEAT CLASS

clay [kleɪ] *noun* **(a)** weathered minerals which exhibit surface and cation exchange properties **(b)** particles in soils smaller than two microns (0.002mm) in diameter; **clay soils** = soils with more than 35% clay size material. Clay soils are sticky when wet and can hold more water than most other types of soil. They lie wet in the winter, and are liable to poaching; they are slow to warm in spring time. In long periods of dry weather, clay soils become hard and wide cracks may form

◊ **clayey** ['kleɪi] *adjective* formed of clay; *these plants do best in clayey soils*

◊ **claying** ['kleɪɪŋ] *noun* application of clay to sandy soils and black fen soils to improve their texture

◊ **claypan** ['kleɪpæn] *noun* hollow on the surface of clay land where rain collects

clean [kliːn] **1** *adjective* not dirty; **clean air**

= air which does not contain impurities *or* air which is free from pollution; **clean cattle** = cattle which have not been used for breeding; **clean crop** = measure of wheat or other cereal, not mixed with seeds of other plants; **clean land** = land which is free of weeds; **clean pasture** = pasture that has been left ungrazed for four to six weeks after contamination with parasitic worm larvae, and is considered free of the parasites **2** *verb* to make clean, by taking away dirt *or* impurities

cleansing ['klenzɪŋ] *see* AFTERBIRTH

clear [klɪə] *verb* to remove plants to prepare open land for cultivation; *they cleared hectares of jungle to make a new road to the capital; we are clearing rainforest at a faster rate than before*

◊ **clearance** ['klɪərəns] *noun* action of clearing land for cultivation

◊ **clearcut** ['klɪəkʌt] **1** *noun* cutting of all the trees in an area **2** *verb* to clear an area of forest by cutting all the trees; *see also* CUT

◊ **clearcutting** *or* **clearfelling** [klɪə'kʌtɪŋ *or* klɪə'felɪŋ] *noun* method of harvesting timber, where all the trees in an area are cut down at the same time

◊ **clearfell** [klɪə'fel] *verb* to clear an area of forest by felling all the trees; *the greatest threat to wildlife is the destruction of habitats by clearfelling the forest for paper pulp*

COMMENT: clearcutting is a way of managing a forest. Once the felled timber has been removed, the land is cleared of stumps and roots, and then sown with new tree seed

cleavers ['kliːvəz] *noun* a widespread weed *(Galium aparine)* affecting winter cereals, oilseed and early-sown spring crops (it has seed pods (burs) with tiny hooks which catch onto the coats of passing animals) (NOTE: other names are **bur** and **goosegrass)**

cleg [kleg] *noun* one of the horseflies

Cleveland Bay ['kliːvlənd 'beɪ] *noun* breed of light draught horse

clevis ['klevɪs] *noun* U-shaped iron attachment, used to couple an implement to a tractor towbar

click beetle ['klɪk 'biːtl] *noun* brown beetle *(Agriotes* species), whose larvae are wireworms which attack cereals by eating the plants just below the soil surface

climax ['klaɪmæks] *noun* final stable state in the development of an ecosystem; **climax community** = plant community which has been stable for many years and which is not likely to change unless the climate changes or there is human interference; **edaphic climax** = climax caused by the type of soil in a certain area

clingstone ['klɪŋstəun] *noun* varieties of peach, where the flesh is attached to the stone, as opposed to the freestone varieties

clip [klɪp] **1** *noun* total amount of wool obtained from a sheep or from a whole flock **2** *verb* to cut the wool from a sheep

◊ **clippers** ['klɪpəz] *noun* shears, used for clipping sheep

cloche [klɒʃ] *noun* covering of either glass or plastic, which can be carried from place to place and is used to protect seedbeds; crops protected by cloches can be sown or planted earlier than similar crops which have no protection; **tunnel cloche** = long continuous covering over rows of plants, usually made of plastic

clod [klɒd] *noun* large lump of soil which may be difficult to break down into tilth; clods will affect the quality of a seedbed

clone [kləun] **1** *noun* (i) group of cells derived from a single cell by asexual reproduction and so genetically exactly the same as the first cell; (ii) organism produced asexually (as by taking cuttings of a plant); *all Bramley's seedling apple trees are clones, since they are the result of grafting material from original specimens onto selected rootstocks* (NOTE: the word originally referred to a group of cells derived from the same source, but is now used in the singular to refer to a single organism) **2** *verb* to reproduce an individual organism by asexual means

◊ **cloning** ['kləunɪŋ] *noun* method of making an exact copy of a living organism by asexual reproduction, or possibly by genetic engineering

QUOTE livestock researchers in Adelaide have succeeded in producing the world's first triplet lambs cloned from a single egg. The technique, in which single cells are transplanted from a fertilized sheep's egg and transferred into the empty casing of other eggs for implantation into surrogate mother ewes, was developed by a team from the University of Adelaide
Australian

closed canopy ['kləuzd 'kænəpi] *noun* a canopy which either has achieved complete cover or intercepts 95% of visible light

closewool ['kləuswul] *noun* one of the breeds of sheep with short dense wool, such as the Devon closewool

Clostridium [klɒst'rɪdiəm] *noun* type of bacterium

◊ **clostridial** [klɒst'rɪdiəl] *adjective* referring to Clostridium; **clostridial disease** = disease (such as pasteurellosis) caused by Clostridium

COMMENT: species of Clostridium cause botulism, tetanus and gas gangrene, but they also serve to increase the nitrogen content of soil

clot [klɒt] *verb* to coagulate; **clotted cream** = cream which has been heated and so becomes more solid

clove [kləuv] *noun* **(a)** dried flower bud of a tree *(Eugenia caryophyllata)* used for flavouring **(b) clove of garlic** = small bulb in a cluster of garlic

clover ['kləuvə] *noun* large genus *(Trifolium)* of leguminous plants, with trefoil leaves and small flowers

COMMENT: clovers are essential plants for the longer ley and permanent pasture. They are nitrogen fixing plants, and with their creeping habit of growth they knit the sward together and help keep out other weeds. The clovers of agricultural importance are the red and white clovers

cloxacillin [klɒksə'sɪlɪn] *noun* type of antibiotic

club root ['klʌb 'ruːt] *noun* fungal disease *(Plasmodiophora brassicae)* affecting brassicas, causing swelling and distortion of the roots and stunted growth (NOTE: also called **finger and toe)**

Clun forest ['klʌn 'fɒrɪst] *noun* hardy grass hill breed of sheep; it has fine dense fleece, a dark brown face and a permanent topknot which extends out over the forehead

cluster ['klʌstə] *noun* the four cup attachments of a milking machine, attached to the teats of the cow's udder

clutch [klʌtʃ] *noun* set of eggs laid by a bird

Clydesdale ['klaɪdzdeɪl] *noun* breed of heavy draught horse, originating in Scotland; brown or black in colour, with a mass of white 'feathers' at the feet

Co *symbol for* cobalt

coarse [kɔːs] *adjective* (particle *or* grain of sand) which is larger than others; *coarse sand fell to the bottom of the liquid as sediment, while the fine grains remained suspended;* **coarse grains** = cereal crops (such as maize, barley, millet, oats, rye and sorghum) which are less fine than wheat or rice

coat [kəut] *noun* the hair on the body of an animal; *some breeds of goat have very thick coats*

COMMENT: most animals have a thicker topcoat of coarse hair, which gives protection against the weather and a softer undercoat of fine wool which keeps the animal's body warm

cob [kɒb] *noun* **(a)** sturdy short-legged riding horse **(b) corn cob** = seed head of maize **(c)** mixture of clay, gravel and straw used as a building material **(d) cob nut** = large hazel nut

cobalt ['kəubɔːlt] *noun* metallic element, which forms the centre piece in the molecular structure of vitamin B_{12}. Sheep grazing on land whose soil is deficient in cobalt may suffer from a wasting disease known as 'pining' (NOTE: symbol is **Co**; atomic number is **27**)

coccidiosis [kɒksɪdɪ'əusɪs] *noun* parasitic disease of livestock and poultry affecting the intestines

◊ **coccidioidomycosis** [kɒksɪdɪəuaɪdəumaɪ'kəusɪs] *noun* lung disease caused by inhaling spores of the fungus *Coccidioides immitis*

coccus ['kɒkəs] *noun* ball-shaped bacterium (NOTE: plural is **cocci**)

Cochin ['kɒtʃɪn] *noun* breed of domestic fowl originating in China

cochineal [kɒtʃɪ'niːl] *noun* red colouring matter obtained from the dried body of an insect, the female concilla *(Coccus cacti)* found in Mexico, Central America and the West Indies

cock [kɒk] *noun* (i) male bird; (ii) for poultry, a male bird over 18 months old

◊ **cockerel** ['kɒkrəl] *noun* young male chicken, up to 18 months old

◊ **cocksfoot** ['kɒksfut] *noun* perennial grass *(Dactylis glomerata)*. A high-yielding, deep-rooting grass, which is resistant to drought and sometimes used in pasture

cocoa ['kəukəu] *noun* product of the *Theobroma cacao* tree

COMMENT: cocoa is obtained from the beans, which are seeds contained in a red or green fleshy fruit. The beans contain a fat (cocoa butter), which is removed in preparing cocoa for drinking. Cocoa beans are the raw material of chocolate, and extra fat and sugar are added in its preparation. The main producers of cocoa are Cote d'Ivoire, Brazil and Ghana

coconut ['kəukənʌt] *noun* tropical palm tree *(Cocos nucifera)* with a large hard fruit containing a white edible pulp; the outer husk is used to make a rough fibre called coir; **coconut oil** = oil extracted from dried coconut flesh or copra

COMMENT: the dried flesh of the coconut is called 'copra'; it is the main form in which coconuts are traded on the world markets; it is used for the extraction of coconut oil, which is used in making margarine, soap and other products. The Philippines, Malaysia and Mexico are all important producers

cocoyam ['kəukəujæm] *see* TARO

Codex ['kəudeks] a United nations food standards body run by FAO and WHO to develop international food safety and quality standards

codlin *or* **codling** ['kɒdlɪn *or* 'kɒdlɪŋ] *noun* an apple with a long tapering shape

◊ **codling moth** ['kɒdlɪŋ 'mɒθ] *noun* *Cydia pomonella,* a serious pest; its larvae burrow into apple fruit

coffee ['kɒfiː] *noun* a shrub *(Coffea)* which bears fruit (coffee cherries) each of which contains two beans, which are roasted and ground to make the drink; **coffee berry borer** = small beetle which lays its eggs inside green coffee berries; **coffee rust** = fungus disease which attacks coffee plants; **green coffee** = fresh coffee beans, before they are roasted

COMMENT: the two main varieties of coffee are Arabica and Robusta. The Arabica shrub, *(Coffea arabica)* was originally grown in the southern parts of the highlands of Ethiopia, and was later introduced into south-western Arabia. The Arabica plant only grows well on altitudes of 1,000m and above. It represents 75% of the world's total coffee production. Arabica coffee beans are generally considered to produce a higher quality drink than those obtained from the Robusta coffee plant *(Coffea canephora)* which originated in West Africa. Robusta coffee has a stronger and more bitter taste than Arabica. The most important area for growing coffee is South America, especially Bolivia, Brazil and Colombia, though it is also grown in Kenya and Indonesia

coir [kɔɪə] *noun* rough fibre from the outer husk of coconuts, used as a material for stuffing chairs and mattresses, and also to make floor matting

cola ['kəʊlə] *noun* tree *(Cola acuminata)* which is indigenous to West Africa, but which is also grown in the West Indies and South America. The nut-like fruit contain caffeine, and can be chewed or used to make 'cola' drinks

cold [kəʊld] *adjective* **cold frame** = protective covering, usually of glass mounted on a wooden frame, used to give protection to young plants and seedlings; **cold shortening** = chilling meat too quickly after slaughter, which makes it tough; **cold storage** = keeping perishable produce in a refrigerated room or container, before moving it to market or to a retailer. The low temperature inhibits bacterial and fungal activity

Colbred ['kəʊlbred] *noun* breed of sheep of medium size with white face. When used in cross-breeding, they are capable of transmitting high fertility and high milking capacity

coleoptile [kɒlɪ'ɒptaɪl] *noun* a cylindrical sheath of tissue that encloses and protects young shoots of grasses and cereals during growth to the soil surface

colic ['kɒlɪk] *noun* pain in any part of the intestinal tract, especially a symptom of abdominal pain in horses

coliform ['kɒlɪfɔːm] *adjective* (bacteria) which are similar to *Escherichia coli* and are commonly found in animal intestines; if *Escherichia coli* is found in water, it indicates that the water has been polluted by faeces, although it may not cause any illness

collar ['kɒlə] *noun* leather-covered roll put round a horse's neck, to carry the weight of a plough or cart which the horse is pulling; **collar work** = hard work done by a horse as it pulls a plough

collective farming [kə'lektɪv 'fɑːmɪŋ] *noun* type of farm organization which originated in the USSR. In 1994 an executive order was signed making possible the privatization of the 27,000 collectives in Russia. There are now at least 100,000 small farms in Russia

COMMENT: formerly, all land was owned by the State, and amalgamated into large collective farms which were either leased permanently to shareholder farmworkers who ran the holding as a single unit (kolkhoz), or run by a manager, who directed the work of labourers (sovkhoz)

collie ['kɒli] *noun* Scottish breed of sheepdog

colloid ['kɒlɔɪd] *noun* substance with very small particles, which form a stable suspension in a liquid

◊ **colloidal** [kə'lɔɪdl] *adjective* (very fine particle) which does not settle but remains in suspension in a liquid; clay particles and humus in the soil are colloidal

◊ **colloidally** [kə'lɔɪdəli] *adverb* **colloidally dispersed particles** = particles which remain in suspension in a liquid

colluvium [kə'luːvɪəm] *noun* material transported across and deposited on slopes as a result of erosion or mass movement

colon ['kəʊlɒn] *noun* the large intestine, running from the caecum to the rectum

colony ['kɒləni] *noun* group of animals *or* plants *or* microorganisms living together in a certain place; *a colony of ants*

◊ **colonial** [kə'ləʊnɪəl] *adjective* referring to a colony; **colonial animals** = animals (such as ants) which usually live in colonies

◊ **colonization** [kɒlənaɪ'zeɪʃn] *noun* act of colonizing a place; *islands are particularly subject to colonization by* *species of plants or animals introduced by man*

◊ **colonize** ['kɒlənaɪz] *verb* to begin to live in a place (as a group); *derelict city sites rapidly become colonized by plants; rats have colonized the river banks*

◊ **colonizer** ['kɒlənaɪzə] *noun* animal *or* plant which colonizes an area

Colorado beetle [kɒlə'rɑːdəʊ 'biːtl] *noun* beetle *(Leptinotarsa decemlineata)* with black and yellow stripes which eats and destroys potato plants

colostrum [kə'lɒstrəm] *noun* fluid secreted by an animal's mammary glands at the birth of young and before the true milk starts to flow; it is very rich in proteins and contains important vitamins and antibodies

QUOTE always make sure that every kid has suckled a good quantity of colostrum. This will protect it from infection and prevent hypothermia
Smallholder

colt [kəʊlt] *noun* young male horse, less than four years old (in the case of thoroughbreds, less than five years old)

◊ **coltsfoot** ['kəʊltsfʊt] *noun* perennial weed *(Tussilago farfara)*

colza ['kɒlzə] *see* RAPE **colza oil** = RAPE OIL

COMA = COMMITTEE ON MEDICAL ASPECTS OF FOOD POLICY

comb [kəʊm] *noun* the red fleshy crest on a fowl

combine 1 ['kɒmbaɪn] *noun* = COMBINE HARVESTER **2** [kəm'baɪn] *verb* to harvest using a combine harvester; *the men have been combining all day during the fine weather;* **combining peas** = peas grown on a large scale, which are harvested with a combine harvester

combine drill ['kɒmbaɪn 'drɪl] *noun* drill which sows grain and fertilizer at the same time; some drills have separate tubes for the seed and fertilizer, others have one tube for both. It is important to clean combine drills after use, as the corrosive action of the fertilizer can damage the tubes

combine harvester ['kɒmbaɪn 'hɑːvɪstə] *noun* machine used to harvest a vast range of crops: cereals, grass, clover, peas, oilseed rape, and other arable crops. The combine cuts the crop, passes it to the threshing mechanism, then sorts the grain or seed from the straw or chaff. The straw is left in a swath behind the combine, and the chaff is blown out of the back. The grain is lifted to a hopper from which it is unloaded into trailers. Most combine harvesters are self-propelled, with a cab for the driver, power-steering and monitoring systems for the key components. Special attachments used with combines include straw spreaders, straw choppers, pick-up attachments for grass and clover crops, and maize pickers

combing wool ['kəʊmɪŋ 'wʊl] *noun* long-stapled wool, suitable for combing and making into worsted

comfrey ['kʌmfri] *noun* herb of the genus *Boraginaceae*. A medicinal herb, also used in salads. Russian comfrey is grown as a fodder crop. Comfrey is also used for composting

Comice ['kɒmiːs] *see* DOYENNE

commensal [kə'mensəl] *noun & adjective* (plant *or* animal) which lives on another plant *or* animal but does not harm it or influence it in any way, though one of the two may benefit from the other (NOTE: if it causes harm it is a **parasite;** if the two organisms are dependent on each other they exist in **symbiosis)**

◊ **commensalism** [kə'mensəlɪzm] *noun* existing together as commensals

commercial [kə'mɜːʃl] *adjective* referring to business; produced to be sold for profit; **commercial grain farming** = highly-mechanized agricultural system in which large areas of mid-latitude grasslands are given over to cereal cultivation; **commercial grazing** = RANCHING; **commercial market** = market for cattle or sheep for meat, not for breeding (as opposed to the pedigree market); **commercial seed** = seed sold as being true to kind, but not necessarily pure

comminute ['kɒmɪnjuːt] *verb* to grind meat into very small pieces (used with odd pieces of meat removed from the carcass before processing them to make sausages and other processed meat products)

Commission of the European Union *noun* the executive body of the European Union

commodity [kə'mɒdɪti] *noun* substance sold in very large quantities, such as raw materials or foodstuffs such as corn, rice, butter; **primary** *or* **basic commodities** = farm produce grown and sold in large quantities; **commodity exchange** = place where commodities are bought and sold; **commodity futures** = trading in commodities for delivery at a later date (often, the produce will not yet have been grown or harvested); **commodity mountain** = surplus of a certain agricultural product produced in the EU, for example the 'butter mountain'

common ['kɒmən] **1** *noun* land on which anyone can walk, and may have the right to keep animals, pick wood, etc. (NOTE: now usually used in place names such as **Clapham Common**) **2** *adjective* **(a)** referring to *or* belonging to several different people or to everyone; **common land** = land on which anyone can walk and may have the right to keep animals, pick wood, etc.; **common prices** = the prices obtained by all EU farmers for a wide range of their products, including beef, cereals, milk products and sugar; EU regulations involve control on imports and intervention buying. These prices are reviewed each year **(b)** ordinary; frequently found; **common fumitory** = widespread weed *(Fumaria officinalis)* also known as 'beggary'; **common hemp nettle** = weed *(Galeopsis tetrahit)* found in spring cereals and vegetables (also called 'day-nettle', 'glidewort')

Common Agricultural Policy (CAP) *noun* agreement between members of the EU to protect farmers by paying subsidies to fix prices of farm produce

COMMENT: the European Union has set up a common system of agricultural price supports and grants. The system attempts to encourage stable market conditions for agricultural produce, to ensure a fair return for farmers and reasonable market prices for the consumer, and finally to increase yields and productivity on farms in the Union. A system of common prices for the main farm products has been established with intervention buying as the main means of market support. The first major reforms in 30 years were carried out in 1992. The objectives were to control surpluses and to reduce support costs to the taxpayer and to comply with the demands of GATT. The reforms included arable set-aside, suckler cow quotas, ewe quotas, price reductions on oilseeds, peas, beans, cereals and beet. Cereal prices will be cut by 29% with income compensation available and linked to an obligation to set aside of 15% of arable land. In 1995 it was estimated that the CAP spent about £30 billion a year on farm subsidies throughout the EU

community [kə'mjuːnɪti] *noun* group of different organisms which live together in an area, and which are usually dependent on each other for existence; **climax community** = plant community which has been stable for many years and which is not likely to change unless the climate changes or there is human interference

compact [kəm'pækt] *verb* to make the ground hard and difficult to work, as by driving over it with heavy farm machinery

◊ **compacting** *or* **compaction** [kəm'pæktɪŋ *or* kəm'pækʃn] *noun (of soil)* action of becoming hard and compacted

companion plants [kəm'pænjən 'plɑːnts] *noun* (i) plants which grow best with other plants; (ii) plants which are grown with others by horticulturists because they encourage growth *or* because they reduce pest infestation

COMMENT: some plants grow better when planted near others. Beans and peas help root plants such as carrots and beetroot. Most herbs (except fennel) are helpful to other plants. Marigolds help reduce aphids if they are planted near plants such as broad beans or roses which are subject to aphid infestation. The strong smell of onions is disliked by the carrot fly, so planting onions near carrots makes sense. On the other hand, most other plants (and especially peas and beans) dislike onions and will not grow well near them

compensatory growth [kɒmpən'seɪtəri 'grəʊθ] *noun* growth that occurs after a period of under-feeding when the animal regains lost weight

complementary feeders [kɒmplɪ'mentəri 'fiːdez] *noun* animals which feed in areas which complement other animals (i.e. they do not compete with them); so goats, which browse, complement sheep which graze

complete diet [kəm'pliːt 'daɪət] *noun* winter feed for livestock involving mixing of concentrates with forage and allowing the animals free feeding of the mix

Compositae [kəm'pɒzɪtaɪ] *noun* common and very large family of plants, where the flowers are made of many florets arranged in heads (as in dandelions)

composition [kɒmpə'zɪʃn] *noun* **milk composition** = the percentages of protein, lactose, fat, minerals and water which make up milk; the composition varies according to the breed of cow, but average percentages are: protein (3.4%), milk sugar (4.75%), fat (3.75%), minerals (0.75%), water (87.35%): **milk compositional quality scheme** = a scheme organized by Milk Marque, which categorizes milk and pays farmers according to the quality of milk produced

compost ['kɒmpɒst] **1** *noun* **(a)** vegetation decomposed under aerobic conditions; **compost activator** = chemical added to a compost heap to speed up the decomposition of decaying plant matter; **compost heap** = pile of rotting vegetation in a garden, which, when fully rotted, can be spread on the soil as a fertilizer or mulch **(b)** prepared soil or peat mixture in which plants are grown in horticulture; **mushroom compost** = special growing medium for commercial production of mushrooms **2** *verb* to make waste into compost

◊ **composting** ['kɒmpɒstɪŋ] *noun* action of encouraging the breaking down of organic waste, as in a compost heap

compound ['kɒmpaʊnd] *noun* substance formed of two or more chemical elements

COMMENT: compounds are stable (i.e. the proportions of the elements in them are always the same) but can be split into their basic elements by chemical reactions

◊ **compound feed** ['kɒmpaʊnd 'fiːd] *noun* animal feed made up of several different ingredients, including vitamins and minerals, providing a balanced diet; usually fed to animals in the form of compressed pellets

◊ **compounder** [kɒm'paʊndə] *noun* company which produces compound feed

◊ **compound fertilizer** ['kɒmpaʊnd 'fɜːtɪlaɪzə] *noun* fertilizer that supplies two or more of the nutrients N, P and K, available in the form of granules which are easy to store and handle. Trace elements, such as boron and iron, may be added

(NOTE: in US English called **mixed fertilizer**)

compulsory [kəm'pʌlsəri] *adjective* which is forced or ordered by an authority; *the compulsory slaughter of infected animals*

concave [kɒn'keɪv] *noun* part of a combine harvester, a curved dish which catches the grain after it has been threshed

concentrate ['kɒnsəntreɪt] **1** *noun* **(a)** strength of a solution; quantity of a substance in a certain volume **(b)** strong solution which is to be diluted **(c)** **concentrates** = animal feedingstuffs with a high food value relative to their bulk. Concentrates are rich in protein and carbohydrates; they also contain trace elements, minerals and vitamins and are complete foods. Concentrates are available through commercial suppliers or may be made up on the farm **2** *verb* to reduce a solution and increase its strength by evaporation

◊ **concentration** [kɒnsən'treɪʃn] *noun* amount of a substance in a certain volume *or* in a solution

QUOTE concentrates should not be overfed, for high protein levels are thought to aggravate milk fever
Farmers Weekly

condition [kən'dɪʃn] *noun* state (of health *or* of cleanliness) of an animal; *(in breeding)* the amounts of muscle and fat present in an animal; **condition scoring** = a method of assessing the state of body condition of animals; scores range from 0-5 for cattle and 1-9 for sows. Low condition scores indicate thinness, and high scores fatness. A score of about 3 is ideal

◊ **conditioned reflex** [kən'dɪʃnd 'riːfleks] *noun* automatic reaction by an animal to a stimulus; a normal reaction to a normal stimulus learned from past experience

◊ **conditioner** [kən'dɪʃənə] *noun* substance used to improve the condition of the soil

◊ **conditioning** [kən'dɪʃənɪŋ] *noun* **(a)** the preparation of crops for harvesting **(b)** making meat more tender by keeping it for some time at a low temperature **(c)** preparation of grain for milling by adding water to it, so as to ensure that the grain has the correct moisture content

cone [kəʊn] *noun* **(a)** fruit of a conifer, a hard seedcase, which splits into many small branches to release the seed **(b)** fruit of the female hop plant, which is separated from leaves and other debris before being dried in an oasthouse

Confederation of European Maize Producers organization representing the interests of European farmers who produce maize

Conference ['kɒnfərəns] *noun* popular variety of dessert pear; it has a long shape and keeps very well

conformation [kɒnfɔː'meɪʃn] *noun* the general shape of an animal or bird; *carcass conformation is very important when buying cattle at auction*

COMMENT: conformation is important in the Carcass Classification System. There are five conformation classes, called EUROP: E = excellent; U (= good); R (= average); O (= below average); P (= poor)

QUOTE commercial breeders have greatly improved growth rate and conformation of the modern turkey
Poultry Service Symposium 18

congenital [kən'dʒenɪtl] *adjective* which exists at or before birth; **congenital disorder** = disorder which is present at birth

COMMENT: an animal may be abnormal at birth because of a genetic defect, such as misshapen heads of calves; other congenital disorders such as swayback in lambs, may be caused by deficiencies in the mother (in the case of swayback, maternal copper deficiency)

conifer ['kɒnɪfə] *noun* evergreen tree with long thin leaves (called needles), which produces fruit in the form of cones; *afforestation with conifers has had a noticeable effect on the landscape of the upland region*

◊ **coniferous** [kə'nɪfərəs] *adjective* (tree) which has cones

COMMENT: Conifers are members of the order Coniferales, and include pines, firs, spruce, etc. They are natives of the cooler temperate regions, are softwoods and grow very fast. They exude resin, and this, together with the hardness of the needles makes it impossible for plants to grow on the forest floor beneath them. If conifers are planted close together they will grow straight and tall, making them ideal for cheap construction material. Because of this, they are frequently used in afforestation plantations in places where they would not have grown naturally, and where, unfortunately, they hide the landscape and reduce wildlife

conillon [kɒ'niːjɒn] *noun* Robusta coffee grown in Brazil

connective tissue [kə'nektɪv 'tɪʃuː] *noun* tissue which forms the main part of bones, cartilage, tendons and ligaments in which a large proportion of fibrous material surrounds the tissue cells

conserve [kən'sɜːv] *verb* to keep *or* not to waste

◊ **conservancy** [kən'sɜːvənsi] *noun* official body which protects a part of the environment, such as a river; **Nature Conservancy Council** = official body in the UK responsible for conservation of fauna and flora

◊ **conservation** [kɒnsə'veɪʃn] *noun* **(a)** keeping *or* not wasting; **conservation of energy** = making consumption of energy more efficient; preventing loss *or* waste of energy (such as the loss of heat from buildings); **conservation of soil** *or* **soil conservation** = preventing soil from being eroded *or* overcultivated (by irrigation, mulching, etc.) **(b)** maintenance of environmental quality *or* resources by the use of ecological knowledge and principles; **nature conservation** = preserving nature as it is at present, and taking steps to make sure that it is not polluted *or* disturbed

◊ **conservationist** [kɒnsə'veɪʃnɪst] *noun* person who promotes the preservation of the countryside and the careful management of natural resources

◊ **Conservation Reserve Program (CRP)** *US* federal programme which pays farmers to let land lie fallow

consolidation [kɒnsɒlɪ'deɪʃn] *noun* **farm consolidation** *or* **land consolidation** = joining small plots of land together to form larger farms or bringing scattered units together to form large fields (consolidation is promoted by most governments to overcome the inefficiencies associated with fragmentation of land into small scattered holdings)

consume [kən'sjuːm] *verb* to use up *or* to

burn (fuel) *or* to eat (foodstuffs); *the population consumes ten tonnes of foodstuffs per week; the new pump consumes only half the fuel which a normal pump would use*

◊ **consumer** [kən'sjuːmə] *noun* person *or* company which buys and uses goods and services; **consumer council** = group representing the interests of consumers; **consumer panel** = group of consumers who report on products they have used so that the manufacturers can improve them or use what the panel says about them in advertising; **consumer protection** = protecting consumers against unfair *or* illegal traders; **consumer research** = research into why consumers buy goods and what goods they really want to buy

◊ **consumerism** [kən'sjuːmərɪzm] *noun* movement for the protection of the rights of consumers; **green consumerism** = movement to encourage people to buy food and other products which are environmentally good (such as organic foods, lead-free petrol, etc.)

consumption [kən'sʌmpʃn] *noun* using a fuel; eating or drinking food *or* liquid; *world sugar consumption has increased at an average of 2m tonnes per year; 3% of all food samples were found to be unfit for human consumption through contamination by lead;* **home consumption** *or* **domestic consumption** = use of something in the home

contact ['kɒntækt] *noun* touching someone *or* something; **contact animal** = animal which has had contact with a diseased animal and which may need to be isolated; **contact herbicide** *or* **contact weedkiller** = substance (such as paraquat) which kills the plants that it touches, used to control annual weeds; **contact insecticide** = substance (such as DDT) which kills insects which touch it, used against insects such as mosquitoes

contagious [kən'teɪdʒəs] *adjective* (disease) which can be transmitted by touching an infected animal *or* by touching objects which an infected animal has touched; **contagious abortion** = Brucellosis, an infectious disease, which is usually associated with cattle where it results in reduced milk yields, infertility and abortion

contaminate [kən'tæmɪneɪt] *verb* to make something impure by adding

something to it; *the cans of vegetables have been contaminated by oil; 7% of the soil is contaminated by heavy metals*

◊ **contaminant** [kən'tæmɪnənt] *noun* substance which contaminates

◊ **contamination** [kəntæmɪ'neɪʃn] *noun* making impure; *contamination of the water supply by runoff*

contour ['kɒntʊə] *noun* line drawn on a map to show ground of the same height above sea-level; **contour farming** = method of cultivating sloping land, where the land is ploughed along a level, rather than down the slope, so reducing erosion; **contour ridging** = ploughing across the side of a hill so as to create ridges along the contours of the soil which will hold water and prevent erosion; **contour strip cropping** = planting different crops in bands along the contours of sloping land so as to prevent soil erosion

COMMENT: in contour farming, the ridges of earth act as barriers to prevent soil being washed away and the furrows retain the rainwater

contract work ['kɒntrækt 'wɜːk] *noun* work carried out by specialist firms on a contract, which involves payment for work carried out, such as the provision of a drainage system or combining a crop; *contract rearing of calves avoids the use of land and labour, but needs capital and a competent rearer*

◊ **contractor** [kən'træktə] *noun* company or person who carries out contract work for a farmer

control [kən'trəʊl] **1** *noun* **(a)** restraining *or* keeping in order; **control area** = area were controls are operating to prevent the spread of a disease within the area, usually a larger area than the infected area. A movement licence is needed if livestock are to be moved out of a control area **(b)** *(in experiments)* sample used as a comparison with the one being tested; **control group** = group of substances which are not being tested, but whose test data is used as a comparison **2** *verb* to keep in order; *the veterinary service is trying to control the epidemic; they were unable to control the spread of the pest;* **controlled atmosphere packaging** = packaging of foods in airtight containers in which the air has been treated by the addition of other gases (this allows a longer shelf-life); **controlled grazing** = system of grazing in which the number of livestock is linked to the pasture available, with moveable fences being erected to restrict the area being grazed

convenience foods [kən'vi:niəns 'fu:dz] *noun* food which has been prepared so that it is ready to be served, after simply reheating it

convert [kən'vɜ:t] *verb* to change to something different

◊ **conversion** [kən'vɜ:ʃn] *noun* act of converting; *the conversion of oxygen into ozone*

◊ **converter** [kən'vɜ:tə] *noun* device *or* process which converts; **methane converter** = process which takes the gas produced by slurry from livestock and processes it into a usable form as gas

cook chill *or* **cook freeze** [kʊk 'tʃɪl *or* kʊk 'fri:z] *noun* methods of preparing food for preserving, where the food is cooked to a certain temperature and then chilled or frozen

coomb [ku:m] *noun* measure of cereals, equalling one sack or four bushels

coop [ku:p] *noun* cage for poultry

COPA *(EU)* = COMMITTEE OF PROFESSIONAL AGRICULTURAL ORGANIZATIONS

copper ['kɒpə] *noun* metallic trace element, essential to biological life; in plants, copper compounds act as catalysts in helping respiration and helping the plant to make use of available iron in the soil; **copper sulphate (CuSO$_4$)** = crystalline salt used in the preparation of Bordeaux mixture; also used in the treatment of copper deficiency in animals (NOTE: the chemical symbol is **Cu**; atomic number is **29**)

◊ **copper deficiency** ['kɒpə dɪ'fɪʃənsi] *noun* lack of copper in an animal's diet, sometimes caused by poisoning with molybdenum

COMMENT: symptoms of copper deficiency vary, but can include lack of growth and change of colour, where black animals turn red or grey. In severe cases, bones can fracture, particularly the shoulder blade. Diarrhoea can also occur, as well as anaemia. Copper deficiency in

ewes can cause swayback in lambs. The condition is treated with injections of copper sulphate

coppice ['kɒpɪs] **1** *noun* area of trees which have been cut down to the ground to allow shoots to grow which are then harvested **2** *verb* to cut trees down to near the ground to produce strong straight shoots which are then harvested; *coppicing is a traditional method of wood management; coppiced wood can be dried for use in wood-burning stoves* compare POLLARD

COMMENT: the best trees for coppicing are those which naturally send up several tall straight stems from a bole, such as hazel and sweet chestnut. In coppice management, the normal cycle is about five to ten years of growth, after which the stems are cut back. Thick stems are dried and used as fuel, or for making charcoal. Thin stems are used for fencing. Up to 1m hectares of surplus arable land could be used for coppice woodland by the end of the century. Cash aid under the set-aside scheme could be used for short-rotation energy coppicing

copra ['kɒprə] *noun* dried pulp of a coconut, from which coconut oil is extracted by pressing

copse [kɒps] *noun* small wood; *see also* COPPICE

cordon ['kɔ:dən] *noun* trained fruit tree, whose growth is restricted to the main stem by pruning; *compare* ESPALIER

cordwood ['kɔ:dwʊd] *noun* pieces of cut tree trunks, all of the same length, ready for transporting

co-responsibility levy [kəʊrɪspɒnsɪ'bɪlɪti 'levi] *noun* levy on overproduction introduced in the EU in 1987; the levy shared the cost of disposal of surpluses between the community and the producers

cork [kɔ:k] *noun* layer of dead cells under the bark of a tree, especially the thick layer of this in the cork oak

COMMENT: cork is harvested by cutting large sections of bark off a cork oak tree, while still leaving enough bark on the tree to ensure that it will continue to grow

corm [kɔ:m] *noun* bottom part of the stem

of a plant which, like a bulb, can be preserved and from which the plant sprouts again in the spring

| COMMENT: crocuses, gladioli and cyclamens have corms, not bulbs

corn [kɔːn] *noun* **(a)** (i) any cereal crop, but especially wheat in England; (ii) in Scotland, also used to refer to oats; **corn cockle** = poisonous weed *(Agrostemma githago)* with a tall stem and purple flowers; **corn marigold** = common weed *(Chrysanthemum segetum)* (NOTE: also known as **golden cornflower, yellow oxeye) corn poppy** = common weed *(Papaver rhoeas)* affecting cereals (NOTE: also known as **cock's comb, field poppy) corn spurrey** = common weed *(Spergula arvensis)* with matted growth which makes it difficult to eradicate in row crops (NOTE: also known as **farmer's ruin, sand grass, yarr) (b)** *(mainly US)* maize; **corn cob** = seed head of maize; **corn-on-the-cob** = head of maize, cooked and eaten as a vegetable; **corn gluten** = high protein animal feed obtained after corn starch has been extracted from maize; **corn oil** = vegetable oil obtained from maize grains, used for cooking and as a salad oil; **corn starch** = cornflour; **flour corn** = variety of maize with large soft grains and friable endosperm, making it easy to grind to flour

◊ **Corn Belt** ['kɔːn 'belt] *noun* region of the mid-western United States, covering the states of Illinois, Indiana, Iowa, Minnesota, Nebraska and Ohio, where maize is grown on a vast scale

◊ **corned beef** ['kɔːnd 'biːf] *noun* beef which has been cured in brine and is preserved in cans

◊ **cornflour** *or* *US* **corn starch** ['kɔːnflauə *or* 'kɔːn 'staːtʃ] *noun* flour extracted from maize grain, used in cooking (it contains a high proportion of starch, and is used for thickening sauces)

◊ **cornflower** ['kɔːnflauə] *noun* common weed *(Centaurea cyanus)* with tall stems and bright blue flowers

corolla [kə'rɒlə] *noun* ring of brightly coloured petals on a flower, its main function being to attract pollinating insects

corpuscle ['kɔːpʌsl] *noun* small round blood cell; **red corpuscle** = red blood cell which contains haemoglobin and carries oxygen to the tissues; **white corpuscle** =

blood cell which does not contain haemoglobin

corral [kə'raːl] **1** *noun* pen for horses or cattle **2** *verb* to put horses or cattle in pens

Corriedale ['kɒrɪdeɪl] *noun* New Zealand breed of sheep (originally from longwool rams and Merino ewes); now bred in Australia both for meat and its thick 27-micron wool

Corsican pine ['kɔːsɪkən 'paɪn] *noun* fast growing conifer *(Pinus nigra)*

cos [kɒs] *noun* type of lettuce with long darker green leaves

COSHH = CONTROL OF SUBSTANCES HAZARDOUS TO HEALTH

COMMENT: farmwork may involve exposure to many substances which can be hazardous to health. Safety in the use of pesticides extends not only to protecting the user but also the need to protect the environment

cosset lamb ['kɒsɪt 'læm] *noun* lamb which has been reared by hand

cote [kəʊt] *noun* **dove cote** = small shelter for doves

Cotentin ['kɒtəntæn] *noun* French breed of sheep from the Cotentin peninsula of Normandy

Cotswold ['kɒtswəʊld] *noun* breed of sheep from the Cotswold hills, now becoming rare

cottage ['kɒtɪdʒ] *noun* small house in the country; **cottage garden** = flower garden containing old-fashioned flowers; **tied cottage** = house belonging to a farm which is let to a farm worker, but only while he is employed by the farm; **cottage piggery** = pig housing with low roofs and an open yard

cotton ['kɒtən] *noun* white downy fibrous substance surrounding the seeds of the cotton plant, a subtropical plant *(Gossypium sp.)*. Cotton is the world's most important textile fibre; **cotton gin** = machine which separates the seeds from the cotton fibres; **cotton grasses** = types of grass found in boggy areas, only of limited value for grazing

◊ **cottonseed** ['kɒtənsiːd] *noun* seed of

the cotton plant, one of the world's most important sources of oil; **cottonseed cake** *or* **cottonseed meal** = residue of cottonseed after the extraction of oil; it is used as a feeding stuff

COMMENT: cotton is widely grown in tropical and sub-tropical areas, including China, India, Pakistan, Paraguay and the southern states of the USA; it is the main crop of Egypt. It is sold packed in standard bales

cottonwood ['kɒtənwʊd] *noun* North American hardwood tree

cotyledon [kɒtɪ'liːdən] *noun* first leaf of a plant as a seed sprouts

COMMENT: cotyledons are thicker than normal leaves, and contain food for the growing plant. Plants are divided into two groups: those which produce a single cotyledon (monocotyledons) and those which produce two cotyledons (dicotyledons)

couch grass ['kuːtʃ 'grɑːs] *noun* kind of grass *(Agropyron repens)* with long creeping rhizomes, which is difficult to eradicate from cultivate crops; also called 'scutch' or 'twitch'

coulter ['kəʊltə] *noun* part of the plough which goes into the soil and makes the vertical cut; **drill coulter** = coulter which makes a drill for sowing seed; common types of drill coulters include the Suffolk, the Hoe and the disc

COMMENT: there are several types of coulter: the disc coulter cuts the side of the furrow about to be turned; the knife coulter serves the same purpose, but is now little used; the skim coulter turns a small slice off the corner of the furrow about to be turned and throws it into the bottom of the one before; it is attached to the beam behind the disc coulters

Council for the Protection of Rural England (CPRE) an organization which campaigns for the conservation of the English countryside

country ['kʌntri] *noun* area of land which is not a town or city; *we live in the country; country areas are less polluted than urban areas;* **country code** = set of guidelines concerning the way the public should behave in the country, such as shutting farm gates and removing litter; **country**

park = area of land in the country with special facilities for the public

◊ **Country Landowners Association** organization representing the interests of landowners

countryside ['kʌntrisaɪd] *noun* the land excluding towns and cities; **Countryside Commission** = official organization established in the UK in 1968, which supervises the planning of the countryside and how it can be used for the enjoyment of the population. It is particularly concerned with National Parks and Areas of Outstanding Natural Beauty, together with rights of way (such as footpaths and bridle paths) in the country

◊ **Countryside Premium Scheme** a pilot scheme administered by the Countryside Commission which involves paying farmers £200 per hectare to leave land fallow. Further grants will be made to put the land to a conservation use, for example, £45 for providing habitats for ground-nesting birds; £120 for creating meadowland for wildlife. The scheme is to operate in seven Eastern counties for a period of five years

◊ **Country Stewardship Scheme** scheme launched in 1991 which aims at encouraging farmers and landowners to undertake positive management of the following landscape types: chalk and limestone grasslands, heathlands, river valley landscapes, the coast and uplands

coupe [kuːp] *noun* area of a forest in which trees have been cut

couple ['kʌpl] *verb* to attach an implement, such as a harrow, to a tractor

◊ **coupling** ['kʌplɪŋ] *noun* attachment which couples an implement to a tractor

courgette [kɔː'ʃet] *noun* marrow fruit at a very immature stage in its development, cut when between 10 and 20 cm long; it may be green or yellow in colour; also called 'zucchini' in Italy

course [kɔːs] *noun* length of time in a rotation, when the land is growing a particular crop; *the Norfolk four-course rotation has turnips, followed by spring barley, red clover and winter wheat; so each crop will only be grown on the same land in one year out of four*

cover ['kʌvə] **1** *noun* (i) something which is

put over something to close it *or* protect it; (ii) amount of soil surface covered with plants; (iii) plants grown to cover the surface of the soil; *grass cover will provide some protection against soil erosion;* **ground cover** = plants which cover the soil **2** *verb* to copulate with; *a bull covers a cow*

◊ **cover crop** ['kʌvə'krɒp] *noun* **(a)** crop which is sown to cover the soil and prevent it from drying out and being eroded; when the cover crop has served its purpose, it is usually ploughed in, hence leguminous plants which are able to enrich the soil are often used as cover crops **(b)** crop which is grown to give protection to another crop which is sown with it; *see also* NURSE CROP **(c)** crop which is sown to give cover for game birds

◊ **covered smut** ['kʌvəd 'smʌt] *noun* fungal disease *(Ustilago hordei)* affecting oats and barley

cow [kaʊ] *noun* female bovine animal; **cow beef** = beef from dairy cows no longer needed for milk production; **cow bell** = bell worn round neck of a cow to make it easier for the farmer to locate the animal

◊ **cowboy** ['kaʊbɔɪ] *noun US* man who looks after cattle on a ranch

◊ **cowman** ['kaʊmæn] *noun* person in charge of a dairy herd (NOTE: plural is **cowmen**)

◊ **cowpea** ['kaʊpiː] *noun* legume *(Vigna unguiculata)* grown throughout the subtropics and tropics as a pulse and green vegetable; it is grown for fodder, as a vegetable and as green manure

◊ **cowpox** ['kaʊpɒks] *noun* infectious viral disease of cattle, which can be transmitted to humans. It is used as part of the vaccine against smallpox

◊ **cowshed** ['kaʊʃed] *noun* building for cows

Cox's orange pippin ['kɒksɪs ɒrɪndʒ 'pɪpɪn] *noun* popular variety of dessert apple. The most important commercially grown apple in the UK

CPRE = COUNCIL FOR THE PROTECTION OF RURAL ENGLAND

cramp [kræmp] *noun* spasm of the muscles where the muscle may remain contracted for some time; **layer's cramp** = condition found in pullets after the first few weeks of their laying life: the bird appears weak, but the trouble usually disappears after a few days

cranefly ['kreɪnflaɪ] *noun* a common pest *(Tipula)*. The larvae are leatherjackets, which affect cereal crops, feeding on the crops in spring, eating away the roots and stems (NOTE: the fly is also known as the **daddy-long-legs**)

cratch [krætʃ] *noun* rack used for feeding livestock out of doors

crate [kreɪt] *noun* large strong box; **farrowing crate** = special box in which sows are housed when giving birth to piglets

craving ['kreɪvɪŋ] *noun* strong desire for something, such as a craving for salt

crawler tractor ['krɔːlə 'træktə] *noun* large powerful caterpillar tractor, used for heavy work

crazy chick disease ['kreɪzi 'tʃɪk dɪ'ziːz] *noun* disease of chicks associated with a diet which is too rich in fats or deficiency of vitamin E; the symptoms include falling over and paralysis

cream ['kriːm] *noun* oily part of milk, containing fats, which gathers on the top of standing milk; cream is used to make butter, or can be sold to consumers as liquid (double or single), clotted or sterilized

◊ **creamery** ['kriːməri] *noun* factory where butter and other products are made from milk

creep [kriːp] *noun* small entrance through which young animals can pass; **creep feed** = feed given to small animals during creep feeding; **creep feeding** = allowing a young animal (such as a calf) access to concentrates through a small entrance, while the adult cow is unable to reach the feed; **creep grazing** = type of rotational grazing using creep gates, which allow the lambs access to the pasture before the ewes; *see also* FORWARD CREEP GRAZING

creosote ['kriːəsəʊt] *noun* liquid derived from coal tar, used as a pesticide or to paint timber to preserve it from rot *or* to prevent attacks by insects

Crescent [kresənt] *see* FERTILE CRESCENT

cress [kres] *noun* plant *(Lepidium sativum)* used as a salad vegetable

crest [krest] *noun* comb on the head of a cock bird

◊ **crested dogstail** ['krestɪd 'dɒgzteɪl] *noun* perennial grass *(Cynosurus cristatus)* which is not very palatable because of its wiry inflorescences, and is used in seed mixtures for lawns

crib [krɪb] *noun* holder for fodder, with sides made of wooden bars

crimp [krɪmp] *verb* to condition fresh cut grass, by nipping the stems and releasing the sap

◊ **crimper** ['krɪmpə] *noun* machine for crimping grass, similar to roller crushers

Criollo [krɪ'əʊləʊ] *noun* breed of improved Spanish longhorn cattle found in South America; they are used for milk production

crispbread ['krɪspbred] *noun* dry biscuits made from rye

◊ **crisphead** ['krɪsphed] *noun* variety of lettuce with stiff leaves

croft [krɒft] *noun* small farm in the Highlands and Islands of Scotland

◊ **crofter** ['krɒftə] *noun* joint tenant of a divided farm in Scotland

◊ **crofting** ['krɒftɪŋ] *noun* system of farming in Scotland, where the arable land of small farms, which was previously held in common, was divided among the joint tenants into separate crofts, while the pasture remains in common

crook [krʊk] *noun* long-handled staff with a hooked end, used by shepherds to catch sheep

crop [krɒp] **1** *noun* **(a)** plants grown for food; *crop breeders depend on wild plants to develop new and stronger strains;* **crop circles** = usually circular patterns occurring in cereal stands, where crops have been flattened (attributed to local winds, UFOs and pranksters); **crop growth rate** = the rate of increase in dry weight per unit area of all or part of a sward; **crop husbandry** = practice of growing and harvesting crops; **crop standing** = herbage growing in a field before harvest; **crop rotation** *or* **rotation of crops** = system of cultivation where two or more crops are grown in a field in a fixed sequence, covering three consecutive growing seasons; this is done to prevent the nutrients in the soil being totally used up (each plant uses different nutrients, and the land is allowed to lie fallow for the fourth year). Rotation also helps to limit damage to crops from pests and diseases; **crop year** = period of twelve months calculated as the time from the sowing and harvesting of one crop until the next sowing season **(b)** amount of produce from a plant *or* from a field; *the tree has produced a heavy crop of apples; the first crop was a failure; the potato crop was bad this year* **(c)** bag-shaped part of a bird's throat where food is stored before digestion **2** *verb (of plants)* to produce fruit; *a strain of rice which crops heavily*

◊ **cropland** ['krɒplænd] *noun* agricultural land which is used for growing crops

◊ **cropper** ['krɒpə] *noun* **heavy cropper** = tree *or* plant which produces a large crop of fruit

cross [krɒs] **1** *noun* breed of plants *or* animals from two different breeds or varieties; individual plant *or* animal bred from two different breeds; *a cross between two strains of cattle* **2** *verb* to breed a plant *or* animal from two different breeds; *they crossed two strains of rice to produce a new strain which is highly resistant to disease*

◊ **crossbar** ['krɒsbɑː] *noun* bar which goes across something to make it more solid, as the crossbar of a rake

◊ **crossbred** ['krɒsbred] *adjective & noun* (animal) which is the result of breeding from two breeds; *a herd of crossbred sheep or a herd of crossbreds*

◊ **cross-breed** ['krɒsbriːd] **1** *noun* animal which is bred from two different pure breeds **2** *verb* to breed new breeds of animals, by mating animals of different pure breeds

◊ **cross-breeding** ['krɒsbriːdɪŋ] *noun* mating of animals of different breeds, both being purebred, in order to combine the best characteristics of the two breeds

◊ **cross-fertilization** [krɒsfɜːtəlaɪ'zeɪʃn] *noun* fertilizing one individual plant from another of the same species

◊ **cross-infection** [krɒsɪn'fekʃn] *noun* infection of other animals in a herd or flock from an infected animal

◊ **cross-pollination** [krɒspɒlɪ'neɪʃn] *noun* pollinating a plant with pollen from another plant of the same species (the pollen goes from the anther of one plant to the stigma of another)

COMMENT: all of these processes avoid inbreeding, which may weaken the species. Some plants are self-fertile (i.e. they are able to fertilize themselves) and do not need pollinators, but most benefit from cross-fertilization and cross-pollination

croup [kru:p] *noun* part of the back of a horse, near the tail

crown [kraʊn] *noun* (i) top part of a tree or plant, where the main growing point is; (ii) the perennial rootstock of some plants; *the disease first affects the lower branches, leaving the crowns still growing;* **crown graft** = type of graft where a branch of a tree is cut across at right angles, slits made in the bark around the edge of the stump, and shoots inserted into the slits; **crown rust** = fungal disease affecting oats, causing the grain to shrivel

CRP = CONSERVATION RESERVE PROGRAM

Cruciferae [kru:'sɪfəri:] *noun* family of common plants whose flowers have four petals: the family includes forage crops such as turnips and swedes, and vegetables such as cabbages and cauliflowers

crude fibre [kru:d 'faɪbə] *noun* term used in analysing foodstuffs, as a measure of digestibility. Fibre is necessary for good digestion, and lack of it can lead to diseases in the intestines

◊ **crude protein** [kru:d 'prəʊti:n] *noun* an approximate measure of the protein content of foods

cruels [kru:lz] *noun* popular name for ACTINOBACILLOSIS

crumb [krʌm] *noun* (a) soft inside part of baked bread, surrounded by the harder crust (b) arrangement of soil particles in a group: *see also* PED

◊ **crumbly** ['krʌmbli] *adjective* which falls apart into particles; *a crumbly soil*

crupper ['krʌpə] *noun* a strap fixed to the back of a saddle and looped under the horse's tail

crush ['krʌʃ] **1** *noun* steel or wood appliance, like a strong stall, used to hold livestock when administering injections or when the animal is being inspected by a veterinary surgeon **2** *verb* to press with a

heavy weight (as in crushing seeds to extract oil); **crush margin** = difference in price between the unprocessed seed and the product extracted after crushing; **crushing mill** = machine used to flatten grain before feeding it to livestock; **crushing subsidy** = payment made in the EU to oil producers to compensate for the difference between vegetable oil prices in the EU and those outside

◊ **crusher** ['krʌʃə] *noun* company *or* factory which specializes in crushing seed to extract oil

crust [krʌst] *noun* **(a)** hard layer which forms on the surface of something, such as the crust of salts formed on soil surfaces after evaporation **(b)** hard baked surface on bread

crutch [krʌtʃ] *noun* forked pole, used when dipping sheep, to push the animal's shoulders down into the liquid

cryophilous [kraɪ'ɒfɪləs] *adjective* (plant) which needs a spell of cold weather to grow properly

COMMENT: cryophilous crops need a certain amount of cold weather in order to provide flowers later in the growing period. If such crops do not undergo this cold period, their growth remains vegetative, or they only form abortive flowers with no seeds. Cereal crops, such as wheat, barley and oats; peas; root crops like sugar beet and potatoes, are all cryophilous

cryophyte ['kraɪəfaɪt] *noun* plant which lives in cold conditions (as for example in snow)

cryptosporidiosis [krɪpəʊspərɪdɪ'əʊsɪs] *noun* disease of humans, caused by bacteria found in animals and in contaminated water

crystallized fruit ['krɪstəlaɪzd 'fru:t] *noun* fruit which has been preserved by soaking in a strong sugar solution (used particularly in Middle Eastern countries)

Cs *symbol for* caesium

CSCT = CENTRAL SCOTLAND COUNTRYSIDE TRUST

CSF = CLASSICAL SWINE FEVER

CSTS = CRAFT SKILLS TRAINING SKILLS

Cu *symbol for* copper

cube [kju:b] **1** *noun* small square block or pellet **2** *verb* to press animal feed into small square pellets

◊ **cubed** [kju:bd] *adjective* in small blocks; **cubed concentrates** = concentrates for livestock in the form of small cubes

◊ **cuber** ['kju:bə] *noun* machine used for making cubes or pellets from meal; meal mixed with molasses is forced through small holes and cut into various lengths

cubicle ['kju:bɪkl] *noun* a compartment, similar to a stall, for housing a single cow or bull; the floor of each cubicle is covered with straw or sawdust for bedding. Cubicles are usually arranged in rows backed by a dunging passage

QUOTE cubicle buildings for dairy cows do not last for ever. Some buildings have a lifespan of 40 years or more, but others put up over the last two decades are now coming to the end of their useful life
Farmers Weekly

QUOTE minimum dimensions for a cubicle are 2.1m x 1.2m, with a front to rear fall of 100mm
Practical Farmer

QUOTE sawdust is used on the cubicle beds, which have an overall length of 2.15m and are 1.18m wide to accommodate Holstein Friesian cows
Farm Building Progress

cucumber ['kju:kʌmbə] *noun* creeping plant *(Cucumis sativus)* with long green fleshy fruit, used as a salad vegetable

COMMENT: cucumbers are native to India, and are used as a cooked vegetable in oriental cooking; they are a major glasshouse crop

Cucurbitaceae [kʊkɜ:bɪ'teɪsɪi] *noun* Latin name for vine crops, the family of plants including melons, marrows and gourds (NOTE: also called **the cucurbits**)

cud [kʌd] *noun* food that ruminating animals bring back from the first stomach into the mouth to be chewed again

cull [kʌl] **1** *noun* **(a)** killing a certain number of living animals to keep the population under control or to remove excess animals from a herd or flock; **deer**

cull *or* **dairy cow cull** = killing a certain number of deer *or* dairy cows **(b)** animal separated from the herd or flock and killed, usually because it is old or of poor quality; **cull cows** *or* **cull ewes** *or* **cull sows** = cows, ewes or sows which are removed from the herd or flock and sold for slaughter **2** *verb* to kill a certain number of animals in order to keep the population under control, or to remove old or poor quality animals from a herd or flock

COMMENT: in the management of large wild animals without predators, such as herds of deer in Europe, it is usual to kill some mature animals each year to prevent a large population forming and overgrazing the pasture. Without culling, the population would seriously damage their environment and in the end die back from starvation. In management of dairy cattle, animals are culled from herds to eradicate disease

QUOTE introducing superior quality genotypes into the population can only improve the entire breed if inferior ones are culled
Appropriate Technology

culm [kʌlm] *noun* stem of a grass which produces flowers; **malt culms** = roots and shoots of partly germinated malting barley; a by-product of the malting process, the culms are used as a feedingstuff for livestock

cultivate ['kʌltɪveɪt] *verb* **(a)** to dig and manure the soil ready for growing crops; *the fields are cultivated in the spring, ready for sowing corn;* **cultivated land** = land which has been dug or manured for growing crops **(b)** to grow (a crop); *they cultivate rice in paddy fields*

◊ **cultivar** ['kʌltɪvɑ:] *noun* variety of a plant which has been developed under cultivation and which does not occur naturally in the wild, but which is a distinct subspecies

◊ **cultivation** [kʌltɪ'veɪʃn] *noun* action of cultivating land; **land under cultivation** = land which is being cultivated *or* which has crops growing on it; *the area under rice cultivation has grown steadily in the past 40 - 50 years;* **shifting cultivation** = form of cultivation practised in some tropical countries, where land is cultivated until it is exhausted and then left as the farmers move on to another area

◊ **cultivator** ['kʌltɪveɪtə] *noun* **(a)** person

who cultivates land **(b)** implement used for seedbed preparation, especially after ploughing

COMMENT: a cultivator has a frame with a number of tines which break up and stir the soil as the implement is pulled across the surface. There are several types of tine, both rigid and spring-loaded. Cultivators can also be used for cleaning stubble and general weed control; the tines can be grouped together so that they pass easily between the rows of growing plants

culture ['kʌltʃə] **1** *noun* bacteria *or* tissues grown in a laboratory; **culture medium** *or* **agar** = liquid *or* gel used to grow bacteria *or* tissue; **stock culture** = basic culture of bacteria from which other cultures can be taken **2** *verb* to grow bacteria in a culture medium; **cultured milk products** = products (such as yoghurt) made from milk which has been exposed to harmless bacteria

culvert ['kʌlvət] *noun* covered drain for waste water

cumin ['kjuːmɪn] *noun* Mediterranean aromatic herb (*Cuminum cyminum*) used for flavouring

curb [kɜːb] *noun* strap passing under the lower jaw of a horse, used as a check

curd [kɜːd] *noun* **(a)** coagulated substance formed by action of acid on milk, used to make into cheese **(b) curds** = tight flower heads formed on brassicas, such as cauliflowers and broccoli

cure [kjʊə] *verb* to preserve meat by salting or smoking

COMMENT: meat is cured by keeping in brine for some time; both salting and smoking have a dehydrating effect on the meat, preventing the reproduction and growth of microorganisms harmful to man

currant ['kʌrənt] *noun* small dried seedless black grape; **blackcurrant** = soft fruit (*Ribes*) grown for its small black berries, used also in making soft drinks; **red currant** = soft fruit, with bright red berries, growing on bushes, and used for making jam

curry ['kʌri] **1** *noun* Indian spices, or food prepared with spices **2** *verb* to rub down or dress a horse

◊ **currycomb** ['kʌrɪkəʊm] *noun* special comb used for grooming a horse

custard apple ['kʌstəd 'æpl] *noun* fruit of the *Annona* genus, grown in the West Indies for its sweet pulpy fruit

cut [kʌt] **1** *noun* **(a)** act of cutting hay or other plants; *it is necessary to get enough silage from three cuts to see the herd through the winter* **(b)** act of cutting down trees; **partial cut** = method of foresting where only some trees are cut, leaving others standing to seed the area which has been left clear by cutting; *see also* CLEARCUT **2** *verb* to remove trees with a saw or axe

cutch [kʌtʃ] *noun* dual-purpose breed of goat found in South India; the animals are black, brown or white, or mixtures of these colours. It is a prolific breeder

cuticle ['kjuːtɪkl] *noun* a continuous waxy layer that covers the aerial parts of a plant to prevent excessive water loss

cutter ['kʌtə] *noun* pig finished for both the fresh meat and the processing markets at weights similar to bacon pigs, i.e. 80-90kg live weight

◊ **cutter bar** ['kʌtə 'bɑː] *noun* device on a mower or combine harvester, formed of a number of metal fingers which support the knife; **cutter bar mower** = machine used to cut grass and other upright crops; the knife cutter bar mower has mostly been replaced by the rotary mower

cutting ['kʌtɪŋ] *noun* small piece of a plant used for propagation; **leaf cutting** *or* **root cutting** *or* **stem cutting** = piece of a leaf *or* root *or* stem cut from a living plant and put in soil where it will sprout

COMMENT: taking cuttings is a frequently used method of propagation which makes sure that the new plant is an exact clone of the one from which the cutting was taken. Leaf cuttings are used to propagate ornamental plants such as begonias (small pieces of leaf with a rib attached will root if laid on the surface of the compost). Root cuttings are used to propagate some perennials

cutworm ['kʌtwɜːm] *noun* caterpillar of the turnip moth and the garden dart moth, which attacks plants such as turnips, swedes and potatoes by eating their roots and stems

CVL = CENTRAL VETERINARY LABORATORY

cwt = HUNDREDWEIGHT

cyanocobalamin [saɪənəʊkəʊˈbæləmiːn] *noun* vitamin B^{12}

cycle [saɪkl] *noun* series of events which recur regularly; *industrial waste upsets the natural nutrient cycle;* **carbon cycle** *or* **circulation of carbon** = carbon atoms from carbon dioxide are incorporated into organic compounds in plants during photosynthesis. They are then oxidized into carbon dioxide again during respiration by the plants or by herbivores which eat them and by carnivores which eat the herbivores, so releasing carbon to go round the cycle again; **hydrological cycle** = cycle of events when water in the sea evaporates, forms clouds which deposit rain as they pass over land, the rain then drains into rivers which return the water to the sea; **nitrogen cycle** = process by which nitrogen enters living organisms (the nitrogen is absorbed into green plants in the form of nitrates, the plants are then eaten by animals, and the nitrates are returned to the ecosystem through the animal's excreta or when an animal or a plant dies)

◊ **cyclical** [ˈsɪklɪkl] *adjective* referring to cycles

cypress [ˈsaɪprəs] *noun* tree *(Cupressus sempervirens)* of the Mediterranean region; its wood is used for furniture

cyst nematodes [ˈsɪst ˈnemətəʊdz] *noun* dark brown lemon-shaped cysts, which live and breed in the roots of cereals, mainly oats; the crops will show patches of stunted yellowish-green plants

cytoplasm [ˈsaɪtəʊplæzm] *noun* substance inside the membrane of a cell, surrounding the nucleus

Dd

Vitamin D ['vɪtəmɪn 'diː] *noun* vitamin which is soluble in fat and found in butter, eggs and fish (especially cod liver oil); it is also produced by the skin when exposed to sunlight

COMMENT: Vitamin D helps in the formation of bones and lack of it causes rickets in children

DA = DISADVANTAGED AREA

daddy-long-legs [dædɪ 'lɒŋ legz] *noun* popular name for the cranefly

DAFS = DEPARTMENT OF AGRICULTURE AND FISHERIES, SCOTLAND

dag [dæg] **1** *noun* tuft of dirty wool round the tail of a sheep **2** *verb* to remove dirty wool from the hindquarters of a sheep

dairy ['deəri] *noun* **(a)** building used for cooling milk at the farm, before it is taken to a commercial factory **(b)** company which receives milk from farms and bottles it and distributes it to the consumer; company which produces cream, butter, cheese and other milk products **(c) dairy cows** = cows and heifers kept for milk production and for rearing calves to replace older cows in a dairy herd; **dairy farm** = farm which is principally engaged in milk production; **dairy farming** = rearing cows for milk production; **dairy fax** = viewdata service operated by the former Milk Marketing Board in England and Wales; it gave information about milk quality results and other general information of interest to dairy farmers on TV; **dairy followers** = young dairy cattle, intended to replace older cows in due course; **dairy herd** = herd of dairy cows; **dairy products** = foods prepared from milk, such as butter, cream, cheese, yoghurt, etc.

◊ **dairying** ['deəriɪŋ] *noun* agricultural system which involves the production of milk and other dairy products from cows kept on special farms

COMMENT: in 1994 over 29,000 dairy farmers in England and Wales produced around 11 billion litres of milk worth some £2.3 billion at the farm gate or nearly 17% of the total value of United Kingdom agricultural output

◊ **dairyman** ['deərɪmən] *noun* person who works with dairy cattle; person employed in a commercial dairy

◊ **Dairy Shorthorn** ['deəri 'ʃɔːthɔːn] *noun* dual-purpose breed of cattle; the colour may be red, white or red and white

DAISY = DAIRY INFORMATION SYSTEM

dal *or* **dhal** [dɑːl] *noun* Indian term for pulses, such as lentils; also used for curries and soups prepared from these pulses

Dalesbred ['deɪlzbred] *noun* local sheep of the Swaledale type; it has a white spot on either side of a black face, with a grey muzzle; it provides a long coarse fleece

dam [dæm] *noun* **(a)** female parent of an animal **(b)** construction built to block a river

COMMENT: dams are constructed either to channel the flow of water into hydroelectric power stations or to regulate the water supply to irrigation schemes. In tropical areas, dams encourage the spread of bacteria, insects and parasites, leading to an increase in diseases such as bilharziasis; they may increase salinity in watercourses. A dam also holds back silt which would normally be carried down the river and be deposited as fertile soil in the plain below

Damascus [də'mæskəs] *noun* breed of dairy goat found in Syria, Lebanon and Cyprus; it has a red and white coat with long lop ears

Damietta [dæmɪ'etə] *noun* type of Egyptian cattle; *see* EGYPTIAN

damp off ['dæmp 'ɒf] *verb (of seedlings)* to

die from a fungus which spreads in warm damp conditions and attacks the roots and lower stems

COMMENT: damping off is a common cause of loss of seedlings in greenhouses

damson ['dæmzən] *noun* small dark purple plum *(Prunus damascena)*

dandelion ['dændɪlaɪən] *noun* yellow weed *(Taraxacum officinale)* found in grassland; also sometimes eaten as salad

DANI = DEPARTMENT OF AGRICULTURE NORTHERN IRELAND

Danish red ['deɪnɪʃ 'red] *noun* dual-purpose breed of cattle, originating in Jutland, Denmark

danthonia [dæn'θəʊnɪə] *noun* tufted pasture grass found in Australia and New Zealand

dapple [dæpl] *noun* rounded patches of colour, especially on a horse

darnel ['dɑːnəl] *noun* common weed *(Lolium temulentum)* which affects cereals and is poisonous to animals

Dartmoor ['dɑːtmɔː] *noun* breed of large moorland sheep, white-faced with black spots, and a long curly fleece

date [deɪt] *noun* fruit of the date palm, a staple food of many people in the Middle East; **date palm** = a tall subtropical tree *(Phoenix dactylifera)* with very large frond leaves; its fruit are dates

COMMENT: the biggest producers of dates are Saudi Arabia, Iraq and Algeria, though most Middle Eastern countries produce small quantities

Daucus ['dɑʊkəs] *Latin name* the family of plants which includes the carrot

day-nettle ['deɪ 'netl] *noun* another name for the common hemp nettle

day-old chick ['deɪ əʊld 'tʃɪk] *noun* chick up to 24 hours old, sent from a breeder or hatchery to a buyer

DDT [diːdiː'tiː] = DICHLORO-DIPHENYLTRICHLORO-ETHANE **DDT-resistant insect** = insect which has build up a resistance to DDT

COMMENT: DDT is an organochlorine insecticide, which remains for a long time as a deposit in animal organisms. It accumulates as a breakdown product in fatty material and gradually builds up in the food chain as smaller animals are eaten by larger ones, with the result that raptors (such as eagles and hawks) can be killed by the DDT which builds up in their systems from the bodies of smaller animals which they have eaten

dead heading ['ded 'hedɪŋ] *noun* cutting the dead flower heads from a plant, so as to prevent the formation of seeds

◊ **dead-in-shell** ['ded ɪn 'ʃel] *adjective* referring to chicks which die in the egg, because they cannot break out, or can only break part of the way out of the shell

◊ **deadly nightshade** ['dedli 'naɪtʃeɪd] *noun* poisonous plant *(Atropa belladonna)* sometimes eaten by animals

◊ **deadnettle** ['dednetl] *see* RED DEADNETTLE

◊ **dead stock** ['ded 'stɒk] *noun* a comprehensive term for all implements, tools, appliances and machines used on a farm; it can also be used to include seed, fertilizer and feedingstuffs

◊ **deadweight** ['dedweɪt] *noun* weight of a dressed carcass; **Deadweight Certification Centre** = slaughterhouse which has been given approval by the Meat and Livestock Commission, where cattle and sheep carcasses are presented for certification as being eligible for premium payment

decay [dɪ'keɪ] **1** *noun* rotting of dead organic matter; *wood is treated to resist decay* **2** *verb (of organic matter)* to rot

deciduous [de'sɪdjuəs] *adjective* **(a)** (woody plant) which loses its leaves in the autumn and produces new leaves from buds the following spring **(b) deciduous teeth** = the first teeth of an animal, in humans called milk teeth, which are gradually replaced by permanent teeth

decompose [diːkəm'pəʊz] *verb (of organic material)* to break down into simpler chemical compounds

◊ **decomposer** [diːkəm'pəʊzə] *noun* organism (such as a fungus) which obtains its energy from decomposing matter in the ecosystem; animal (such as the earthworm) which helps the decomposition of organic matter

◊ **decomposition** [di:kɒmpə'zɪʃn] *noun* action of decomposing organic matter

decortication [di:kɔ:tɪ'keɪʃn] *noun* removing husks from seeds

decumbent [dɪ'kʌmbənt] *adjective* (plant stems) which lie on the surface of the soil for part of their length, but turn upwards at the end

deep freezing [di:p'fri:zɪŋ] *noun* storing of produce in temperatures below freezing point; many crops, such as peas and beans, are grown specifically for the freezer market

deep litter ['di:p'lɪtə] *noun* system of using straw, wood shavings, peat moss or sawdust for bedding cattle or poultry

COMMENT: for poultry an inch of well-composted horse manure is laid down first; on this wood shavings, peat moss, or cut straw are placed. The litter is changed after each crop of birds. Deep litter also has value as a manure. For cattle, straw, shavings and sawdust form a deep litter; warmth is given off as faeces in the litter ferment, and additions of fresh litter can be made on top of the old

deep ploughing ['di:p 'plauɪŋ] *noun* ploughing very deep into the soil (used when reclaiming previously virgin land for agricultural purposes)

QUOTE some farmers are opposed to deep ploughing because they feel they will bury the high nutrient and organic matter concentration in the top soil layer. However, experience shows that deep ploughing in suitable soils has brought about improvements in soil condition
Practical Farmer

deep-rooted *or* **deep-rooting** [di:p'ru:tɪd *or* di:p'ru:tɪŋ] *adjective* (plant) with long roots which go deep into the soil (as opposed to a surface-rooting plant)

deer [di:ə] *noun* ruminant animal; the males have distinctive antlers; the meat of deer is venison; **deer farming** = commercial farming of deer to be sold as venison; **deer forest** = extensive tract of upland, usually treeless, but managed by keepers to provide deer-stalking; **farmed deer** = deer which are raised on deer farms on land enclosed by deer-proof barriers, for meat or skins or other by-products, or as breeding stock

◊ **deer-stalking** ['di:ə 'stɔ:kɪŋ] *noun* hunting of deer in the wild

COMMENT: there are three wild species in the UK: the fallow deer *(Dama dama)* the roe deer *(Capreolus capreolus)* and the red deer *(Cervus elaphus)*, which is also raised commercially. Deer are hardy animals, and are well adapted to severe winters; they can suffer from tuberculosis, and the British government has introduced a compulsory slaughter scheme for animals suffering from the disease. in 1993 in England and Wales, the total number of farmed deer holdings was 227 and the number of farmed deer was 40,616. The west of Scotland is the most important single area for deer in the UK; there are over 140 deer farms in the Highlands and Islands of Scotland

deficiency [dɪ'fɪʃənsi] *noun* lack of something essential; **deficiency disease** = disease of plants *or* animals caused by the lack of an essential nutrient, such as a vitamin, protein, mineral element or amino acid: heart rot and stem rot are two common plant diseases of this type; **deficiency payment** = payment made to a producer, where the price for a commodity at the market does not reach a preset guaranteed price

◊ **deficient** [dɪ'fɪʃənt] *adjective* lacking something essential; *the soil is deficient in important nutrients; scrub plants are well adapted to this moisture-deficient habitat*

definite inflorescence ['defɪnɪt ɪnflɔ:'resəns] *noun* inflorescence where the main stem ends in a flower and stops growing when the flower is produced

definitive host [de'fɪnɪtɪv 'həust] *noun* host on which a parasite settles permanently, lives and produces its eggs

deflector plate [dɪ'flektə 'pleɪt] *noun* attachment in a slurry spreader which spreads the slurry over a wide area

deflocculation [di:flɒkjʊ'leɪʃn] *noun* state where clay particles repel each other instead of sticking together; *deflocculation may occur, when clays are worked in a wet condition or if the soil becomes saline* (NOTE: the opposite is **flocculation**)

defoliate [di:'fəʊlieɪt] *verb* to remove leaves from plants

◊ **defoliation** [di:fəʊli'eɪʃn] *noun* removing leaves from plants

◊ **defoliant** [di:'fəʊliənt] *noun* type of herbicide which makes the leaves fall off plants

deforest [di:'fɒrɪst] *verb* to cut down forest trees for commercial purposes *or* to make arable land; *timber companies have helped to deforest the tropical regions; about 40,000 square miles are deforested each year*

◊ **deforestation** [di:fɒrɪs'teɪʃn] *noun* cutting down forest trees to make arable land

> QUOTE the main reason for the drought and its severity is deforestation like that which has taken place in Ethiopia and Brazil. The destruction of the trees destroys the land's watersheds. The lack of trees also allows the fertile topsoil to blow away
> *Appropriate Technology*

> QUOTE deforestation is the result of the deliberate conversion of forested lands to other economic uses and the collection of fuelwood. Stripping the land of its tree cover exposes thin tropical soils to rapid surface erosion
> *Forestry Chronicle*

degrade [dɪ'greɪd] *verb* (i) to reduce the quality of something; (ii) to make a chemical compound decompose into its elements; *the land has been degraded through overgrazing; ozone may make nutrient leaching worse by degrading the water-resistant coating on pine needles*

◊ **degradable** [dɪ'greɪdəbl] *adjective* (substance) which can be degraded; *compare* BIODEGRADABLE

◊ **degradation** [degrə'deɪʃn] *noun* (i) making worse; becoming worse; (ii) decomposition of a compound into its elements: *chemical degradation of the land can be caused by over-use of fertilizers and by pollutants from industrial processes;* **degradation of air** = making clean air become polluted

dehair [di:'heə] *verb* to remove hard hairs from fine goat fibres such as angora

dehisce [di:'hɪs] *verb* (of ripe seed pod or fruit) to burst open to release seeds

◊ **dehiscence** [di:'hɪsns] *noun* sudden bursting of a seed pod when it is ripe allowing the seeds to scatter

◊ **dehiscent** [di:'hɪsnt] *adjective* (seed pod) which bursts open to allow the seeds to scatter

dehorn [di:'hɔ:n] *verb* to remove the horns of an animal, done by disbudding when the animal is young (note that by selective breeding, many breeds are now hornless)

dehusk [di:'hʌsk] *verb* to remove the husk from seeds such as corn

dehydrate [di:haɪ'dreɪt] *verb* to remove water from something, to preserve it; **dehydrated milk** = milk which has been dried and reduced to a powder

◊ **dehydration** [di:haɪ'dreɪʃn] *noun* removing water from something to preserve it

> COMMENT: food can be dehydrated by drying in the sun (as in the case of dried fruit), or by passing through various industrial processes, such as freeze-drying

deintensified farming [di:ɪn'tensɪfaɪd 'fɑ:mɪŋ] *noun* farming which was formerly intensive, using chemical fertilizers to increase production, but has now become extensive; overproduction of cereals has led to less intensive farming; *see also* SET-ASIDE

denature [di:'neɪtʃə] *verb* (i) to make something change its nature; (ii) to convert a protein into an amino acid; **denatured wheat** = wheat which has been stained to make it unusable for human consumption

◊ **denaturing** [di:'neɪtʃərɪŋ] *noun* process of staining wheat grain with a dye, so as to make it unusable for human consumption; such grain may be used as animal feed

dendrochronology [dendrəʊkrɒ'nɒlədʒi] *noun* finding the age of wood by the study of tree rings

denitrification [di:naɪtrɪfɪ'keɪʃn] *noun* conversion by bacteria of nitrates into gaseous nitrogen products which are then lost into the atmosphere

dental ['dentəl] *adjective* referring to teeth; **dental formula** = a formula indicating the number of each type of teeth for a certain animal species

◊ **dentition** [den'tɪʃn] *noun* arrangement of teeth in an animal's mouth; an examination of an animal's teeth may help in estimating its age

denude [dɪ'njuːd] *verb* to make bare; *the timber companies have denuded the mountains*

◊ **denudation** [dɪnjuː'deɪʃn] *noun* making land or rock bare by cutting down trees, so contributing to erosion

Department of Agriculture department of the US federal government, which deals with agricultural matters

deposition [depə'zɪʃn] *noun* action of depositing; **dry deposition** = falling of dry particles from polluted air (they form a harmful deposit on surfaces such as buildings or leaves of trees)

depress [dɪ'pres] *verb* to make a price lower; *overproduction of some items in the EU may depress the price level in the open market*

Derbyshire Gritstone ['dɑːbɪʃə 'grɪtstəʊn] *noun* blackfaced, hornless hardy breed of sheep, which produces a soft fleece of high quality (NOTE: the name comes from a typical millstone grit in the Peak District of Derbyshire)

derelict ['derəlɪkt] *adjective* (land) which is not used for anything

derris ['derɪs] *noun* powdered insecticide extracted from the root of a tropical plant, used against fleas, lice and aphids

desalinate [diː'sælɪneɪt] *verb* to remove salt from water

◊ **desalination** [diːsælɪ'neɪʃn] *noun* removing salt from a substance such as sea water; **desalination plant** = factory which takes sea water and removes the salt to produce fresh drinking water

COMMENT: desalination can refer simply to the removal of salts from soil by the action of rain. It is most commonly used to mean the process of removing salt from sea water to make it fit to drink. Desalination plants work by distillation, dialysis or by freeze drying. The process is very expensive, and is only cost-effective in desert countries where the supply of fresh water is minimal

desert ['dezət] *noun* area of land with very little rainfall, arid soil and little or no vegetation; the soil is unproductive in terms of food production, being unable to support plant life with the exception of

sparse vegetation; **desert soil** = the soil in a desert (it is normally sandy with little organic matter)

◊ **desertification** [dɪzɜːtɪfɪ'keɪʃn] *noun* process of becoming a desert; *changes in the amount of sunlight reflected by different vegetation may contribute to desertification; increased tilling of the soil, together with long periods of drought have brought about the desertification of the area*

◊ **desertify** [dɪ'zɜːtɪfaɪ] *verb* to make land into a desert; *it is predicted that half the country will be desertified by the end of the century*

COMMENT: a desert will be formed in areas where rainfall is very low (less than 25 centimetres per annum) whether the region is hot or not. About 30% of all the land surface of the earth is desert or in the process of becoming desert. The spread of desert-type conditions in arid and semi-arid regions is caused not only by climatic conditions, but also by human pressures. So overgrazing of pasture and the clearing of forest for fuel and for cultivation, both lead to the loss of organic material, a reduction in rainfall by evaporation and to soil erosion. Reclaiming desert is a very expensive operation and usually not very successful

desiccate ['desɪkeɪt] *verb* to dry out; *cattle travel across the desiccated pastures in search of fodder*

◊ **desiccant** ['desɪkənt] *noun* (i) substance which dries; (ii) type of herbicide which makes leaves wither and die before harvesting

◊ **desiccation** [desɪ'keɪʃn] *noun* act of drying out (the soil); *the greenhouse effect may lead to climatic changes such as the vast desiccation of Africa*

dessert fruit [dɪ'zɜːt 'fruːt] *noun* fruit which are sweet and can be eaten raw, as opposed to being cooked; *a variety of dessert gooseberry*

determination [dɪtɜːmɪ'neɪʃn] *noun* identifying what something is; **sex determination** = way in which the sex of an individual organism is determined by the number of chromosomes which make up its cell structure

Devon ['devən] *noun* breed of fine-boned dual-purpose cattle; North and South Devons are dark red, and belong to a type of red cattle bred for centuries in England; they thrive on pasture which would not be sufficient for larger breeds, and provide both meat and milk (NOTE: they are commonly known as **Red Rubies)**

◊ **Devon and Cornwall Longwool** *noun* breed of sheep with long curly, high-quality fleece; the lambs have a fine soft white wool

◊ **Devon Closewool** *noun* breed of medium-sized sheep, the product of crosses between the Devon Longwool and the Exmoor Horn

dew [djuː] *noun* water which condenses on surfaces in the open air as humid air cools at night; **dew point** = temperature at which dew forms on grass *or* leaves, etc.

◊ **dew claw** ['djuː 'klɔː] *noun* rudimentary fifth digit found on the heels of dogs, pigs and cattle

◊ **dewpond** ['djuː 'pɒnd] *noun* small pond which forms on high chalky soil (it is formed from rainwater not dew); small pond made to maintain a supply of drinking water for livestock

COMMENT: dewponds are found in areas of chalk or limestone country. To make a dewpond, a hollow is scooped out and lined with clay. The pond is kept full by rainwater

dewatering [diː'wɔːtərɪŋ] *noun* extraction of water from a crop by pressing (this reduces the cost of artificial drying)

dewlap ['djuːlæp] *noun* fold of loose skin hanging from the throat of cattle

Dexter ['dekstə] *noun* rare breed of cattle, originating from the west of Ireland; the animals are small in size, coloured black or red

dextrose ['dekstrəuz] *noun* simple sugar found in fruit; also extracted from corn starch

dhal *or* **dal** [daːl] *noun* Indian term for pulses, such as lentils and pigeon peas; also used for curries and soups prepared from these pulses

Diamonds disease ['daɪəməndz dɪ'ziːz] *see* ERYSIPELAS

diarrhoea [daɪə'riːə] *noun* condition where an animal frequently passes liquid faeces (NOTE: also called **scours)**

dibber ['dɪbə] *noun* hand tool for making holes in soil to plant small plants

dichlorodiphenyltrichloroethane (DDT) [daɪklɒrəudaɪfenɪltraɪklɒrəu'iːθeɪn] *noun* highly toxic insecticide, no longer recommended for use

COMMENT: DDT is an organochlorine insecticide, which remains for a long time as a deposit in animal organisms. It accumulates as a breakdown product in fatty material and gradually builds up in the food chain as smaller animals are eaten by larger ones, with the result that raptors (such as eagles and hawks) can be killed by the DDT which builds up in their systems from the bodies of smaller animals which they have eaten

dicotyledon [daɪkɒtɪ'liːdən] *noun* plant of which the cotyledon has two leaves; *dicotyledons form the largest group of plants compare* COTYLEDON, MONOCOTYLEDON

die back ['daɪ 'bæk] *verb (of a branch or shoot of a plant)* to die; **branches of apricot are known to die back from time to time**

◊ **dieback** ['daɪbæk] *noun* **(a)** fungus disease of certain trees which kills shoots or branches **(b)** gradual dying of trees caused by acid rain (starting at the ends of branches); *half the trees in the forest are showing signs of dieback*

COMMENT: there are many theories to try to explain the causes of dieback: sulphur dioxide, nitrogen oxides, ozone, are all possible causes; also acidification of the soil or acid rain on leaves

dieldrin [daɪ'eldrɪn] *noun* organochlorine insecticide which kills on contact. It is very persistent and can kill fish, birds and small mammals when it enters the food chain

diet ['daɪət] *noun* amount and type of food eaten; **balanced diet** = diet which contains the right quantities of basic nutrients

◊ **dietary** ['daɪətri] *adjective* referring to diet; **dietary fibre** = fibrous material in feedingstuff, which cannot be digested

dig [dɪg] *verb* to turn over ground with a fork or spade

◊ **digger** ['dɪgə] *noun* type of plough body with a short, sharply curved mouldboard; used for deep ploughing, especially to prepare for root crops or for land reclamation

◊ **digging stick** ['dɪgɪŋ 'stɪk] *noun* one of the earliest agricultural implements, still used in areas where shifting cultivation is practised; the stick has a sharpened end, sometimes with a metal tip, and is used to dig holes to plant crops

digest [daɪ'dʒest] *verb* **(a)** to break down food in the alimentary tract and convert it into elements which are absorbed by the body **(b)** to process waste, especially organic waste such as manure, to produce biogas; *55% of UK sewage sludge is digested; wastes from food processing plants can be anaerobically digested*

◊ **digester** [daɪ'dʒestə] *noun* machine which takes refuse and produces gas such as methane from it; *anaerobic digesters can be used to convert cattle manure into gas*

◊ **digestible** [daɪ'dʒestəbl] *adjective* which can be digested; *glucose is an easily digestible form of sugar*

◊ **digestibility** [dɪdʒestɪ'bɪlɪti] *noun* the proportion of food which is digested and is therefore of value to the animal which eats it; **digestibility coefficient** = the proportion of food digested and not excreted, shown as a percentage of the total food eaten; digestibility coefficients can be calculated for individual nutrients in foods; **digestibility trial** = test to measure the digestibility of a known food by recording the weight of food eaten, and then excreted; **digestibility value (D value)** = amount of digestible organic matter in the dry matter of plants

◊ **digestion** [daɪ'dʒestʃn] *noun* (i) process by which food is broken down in the alimentary tract into elements which can be absorbed by the body; (ii) process which produces biogas from refuse; **salivary digestion** = the first part of the digestive process, which is activated by the saliva; **stomach digestion** = the part of the digestive process which takes place in the animal's stomach

◊ **digestive** [daɪ'dʒestɪv] *adjective* referring to digestion; **digestive enzymes** = enzymes which speed up the process of digestion; **digestive juices** = juices in an animal's digestive tract, which convert food into a form which is absorbed into the body; **digestive system** = all the organs in the body (such as the liver and pancreas) which are associated with the digestion of food

QUOTE the digester was developed by a group of scientists concerned with technological efficiency
Appropriate Technology

dill [dɪl] *noun* common aromatic herb (*Anethum graveolens*) used in cooking and in medicine

dinitro-ortho-cresol (DNOC) [daɪnaɪtrəʊɔ:θəʊ'kri:sɒl] *noun* chemical used in a compound with petroleum oil to destroy the eggs of fruit tree pests; DNOC is a poisonous yellow dye, and can also be used as a weedkiller

dioecious [daɪəʊ'i:ʃəs] *adjective* with male and female flowers on different plants (such as hops or cucumbers)

dioxide [daɪ'ɒksaɪd] *see* CARBON

dip [dɪp] **1** *noun* chemical which is dissolved in water, used for dipping animals, mainly sheep, to remove lice and ticks; dips usually contain organophosphorous chemicals **2** *verb* to plunge an animal in a dip (for about thirty seconds)

◊ **dipper** ['dɪpə] *noun* deep trench into which sheep are guided to be dipped

◊ **dipping** ['dɪpɪŋ] *noun* the process of plunging an animal in a chemical solution to remove ticks, etc.; **dipping bath** = dipper, a trench in which animals are dipped; **compulsory dipping period** = period of time (usually some weeks) during which all sheep in the country must be dipped

COMMENT: sheep are dipped to eradicate parasites such as lice and ticks, and to prevent sheep scab. Dipping varies from region to region according to custom, breed and climate. Dipping may be ordered by the MAFF to control outbreaks of disease, and in certain cases it has to be witnessed by a local authority inspector

diphtheria [dɪf'θɪːrɪə] *noun* serious infectious disease where a membrane forms in the throat passages of an animal (as in calf diphtheria)

diploid ['dɪplɔɪd] *noun* organism with twice the usual number of chromosomes; *diploid grasses are less palatable than tetraploids*

dipterous ['dɪptərəs] *adjective* (insect) (such as a fly) with two wings

direct drilling ['daɪrekt 'drɪlɪŋ] *noun* form of minimal cultivation, where the seed is sown directly into the field without previous cultivation; several types of drill are used, with heavy discs for cutting narrow drills, or strong cultivator tines

QUOTE over the past years a number of successful demonstrations of how lab lab can be direct drilled into native pastures to improve the quality and quantity of autumn stock feed have been planted on the eastern Darling Downs
Queensland Agricultural Journal

◊ **direct milk sales** *noun* selling of milk from the producer direct to the consumer. The product is known as 'green top milk'

◊ **direct proportional application (DPA)** *noun* system of making sure that the output from a sprayer is proportional to the speed at which it moves forward; **direct reseeding** = sowing grass seed without a cover crop; **direct sowing** = sowing grass seed on a prepared seed bed; grass and clover seeds are small, and therefore must be sown shallow and a fine firm seedbed is necessary

directive [dɪ'rektɪv] *noun* order from the Council of Ministers *or* Commission of the European Union, referring to a particular problem

dirt tares ['dɜːt 'teəz] *noun* tares which grow among sugar beet plants

disadvantaged area [dɪsəd'vɑːntɪdʒd 'eəriə] *noun* name for land in mountainous and hilly areas, which is capable of improvement and use as breeding and rearing land for sheep and cattle; divided into Disadvantaged or Severely Disadvantaged Areas; the EU recognizes such areas and gives financial help to farmers in them

disbud [dɪs'bʌd] *verb* **(a)** to remove the horn buds from calves, soon after birth **(b)** to remove small flower buds from a plant, to allow the main flower to develop more strongly (as when growing chrysanthemums or paeonies)

disc [dɪsk] *noun* one of the heavy round metal plates, used in harrows and ploughs to cultivate the soil; **disc coulter** = part of a plough which cuts the side of the furrow about to be turned; **disc harrow** = type of harrow, with two or more sets of saucer-shaped discs fixed to a frame; the disc angle can be changed and the working depth varied. Disc harrows are used in seedbed work, especially after ploughing in grass; they are also used for cutting up crop remains, such as the stalks of kale; **disc plough** = type of plough with large rotating discs in place of the mouldboard; used for deep cultivation, but not common in Great Britain

QUOTE a type of disc plough has become popular in some dry counties of Ireland. A large revolving disc replaces the body on this plough. The disc is very good in very hard dry soils and land with a lot of straw trash lying around. They break up the soil well, but do not bury the rubbish completely
Practical Farmer

discolour *or US* **discolor** [dɪs'kʌlə] *verb* to change the colour of something; *fruit can be discoloured by the use of sprays*

◊ **discoloration** *or* **discolouration** [dɪskʌlə'reɪʃn] *noun* change of colour, especially a change of colour of fruit

disease [dɪ'ziːz] *noun* illness (of people *or* animals *or* plants, etc.) where the body functions abnormally; *the herd caught the disease from infected feedingstuff; the rice crop has been affected by a virus disease see also* NOTIFIABLE

◊ **diseased** [dɪ'ziːzd] *adjective* (plant *or* person *or* part of the body) affected by an illness *or* not whole or normal; *to treat dieback, diseased branches should be cut back to healthy wood*

◊ **Diseases of Animals Act (1950)** Act of Parliament covering the diseases listed under notifiable diseases

Dishley Leicester ['dɪʃli 'lestə] *noun* breed of improved Leicester sheep; the breed was used by Bakewell in the 18th century

disinfect [dɪsɪn'fekt] *verb* to make a place free from germs *or* bacteria

◊ **disinfectant** [dɪsɪn'fektənt] *noun* substance used to kill germs *or* bacteria

◊ **disinfection** [dɪsɪn'fekʃn] *noun* removing infection caused by germs *or* bacteria (NOTE: the words **disinfect** and **disinfectant** are used for substances which destroy germs on instruments, objects or the skin; substances used to kill germs inside infected animals or people are **antibiotics, drugs, etc.)**

COMMENT: disinfection is a necessary process affecting buildings such as stables, and implements, after infection has been present. It may involve removing litter and dung, and cleaning floors and partitions; implements and tools should also be treated. Methods of disinfection include the use of approved chemical solutions, steam cleaning and fumigation with powerful antiseptics

disorder [dɪs'ɔːdə] *noun* (i) illness *or* sickness; (ii) state where part of the body is not functioning correctly; **congenital disorder** = disorder which is present at birth

COMMENT: an animal may be abnormal at birth because of a genetic defect (a congenital disorder); such disorders may be misshapen heads of calves; other congenital disorders such as swayback in lambs, may be caused by deficiencies in the mother (in the case of swayback, maternal copper deficiency)

dispatcher [dɪs'pætʃə] *noun* implement used to kill chickens

disperse [dɪs'pɜːs] *verb (of organisms)* to spread over a wide area

◊ **dispersal** [dɪs'pɜːsəl] *noun* spreading of individual plants *or* animals into or from an area; *aphids breed in large numbers and spread by dispersal in wind currents;* **water dispersal** *or* **wind dispersal** = spreading of plants by seed carried away by water *or* blown by the wind

◊ **dispersing agent** [dɪs'pɜːsɪŋ 'eɪdʒənt] *noun* chemical added to a fungicide/bactericide formulation to aid the efficient distribution of particles of the active agent

◊ **dispersion** [dɪs'pɜːʃn] *noun* pattern in which animals *or* plants are found over a wide area

distil [dɪs'tɪl] *verb* to produce an alcohol or essential oil by heating and condensing the vapour

◊ **distillation** [dɪstɪ'leɪʃn] *noun* process of producing a pure liquid by heating and then condensing the vapour, used in the production of alcohol and essential oils

◊ **distiller** [dɪ'stɪlə] *noun* person who

distils alcohol from grain; **distillers' grains** = by-product of whisky production, which consists of the remains of malted barley, used as a valuable cattle food which may be fed wet or dry

distort [dɪs'tɔːt] *verb* to change the shape of something, so that it does not look normal, as when brassica roots are distorted by club root disease

ditch [dɪtʃ] **1** *noun* channel dug to take away rainwater **2** *verb* to dig channels for land drainage

◊ **ditcher** ['dɪtʃə] *noun* mechanical excavator used in ditching

COMMENT: ditches may be enough to drain an area by themselves, but usually they serve as outlets for underground drains. Ditches can deal with large quantities of water in very wet periods. They should be kept cleaned to their original depth, a process carried out usually once a year. Many types of machine are now available for making new ditches or cleaning neglected ones

diversify [daɪ'vɜːsɪfaɪ] *verb* to do a number of different things; *farmers are being encouraged to diversify land use by using it for woodlands or for recreational facilities*

◊ **diversification** [daɪvɜːsɪfɪ'keɪʃn] *noun* changing to a different way of working

COMMENT: the main alternative enterprises undertaken by farmers are: farm holidays and bed-and-breakfast; farm shops, selling produce from the farm; camping and caravan sites; country sports, such as horse riding, pony-trekking and fishing.

QUOTE The global anti-golf movement launched a world no-golf day in 1994 and has called for an immediate moratorium on golf course development. GAGM is calling for a halt to the golf green revolution which is turning land into a monoculture of exotic grasses, dependent on huge quantities of water and chemicals In 1991, developers submitted applications for 1,800 new golf courses in Britain to add to the existing 2,000

Ecologist

QUOTE the scheme to help farmers diversify out of agriculture and so help farm income and rural development

Environment Now

DM = DRY MATTER

DNOC = DINITRO-ORTHO-CRESOL chemical used in a compound with petroleum oil to destroy the eggs of fruit tree pests; DNOC is a poisonous yellow dye, and can also be used as a weedkiller

DNQ = DENATURABLE QUALITY WHEAT

docile ['dəʊsaɪl] *adjective (of animals)* quiet and easy to handle

dock [dɒk] **1** *noun* broadleaved or curled weed *(Rumex)* with a long tap root, making it difficult to remove **2** *verb* to cut off the tail of an animal; lowland breeds of sheep are often docked to prevent dirt and faeces accumulating on the tail; **docking disorder** = disorder of sugar beet, caused by eelworms, found on sandy soils in East Anglia, causing irregularly stunted plants with split root growth

◊ **dockage** ['dɒkɪdʒ] *noun* waste material which is removed from grain as it is being processed before milling

doe [dəʊ] *noun* female of deer, goat, rabbit or hare

DoE *GB* = DEPARTMENT OF THE ENVIRONMENT

dogdaisy ['dɒgdeɪsi] *see* MAYWEED

dolly ['dɒli] *noun* **corn dolly** = ornament woven from straw, used formerly as decoration on farms at harvest time

domestic [də'mestɪk] *adjective* referring to the home; **domestic animal** = animal (such as a dog *or* cat) which lives with human beings; **domestic livestock** = farm animals which live near the farmhouse

◊ **domesticate** [də'mestɪkeɪt] *verb* (i) to breed wild animals so that they become tame and can fill human needs; (ii) to breed wild plants, selecting the best strains so that they become useful for food or decoration

◊ **domesticated** [də'mestɪkeɪtɪd] *adjective* (i) (wild animal, such as a dog or cat) which has been trained to live near a house and not be frightened of humans; (ii) (species of animal or plant) which was formerly wild, but has been bred to fill human needs

◊ **domestication** [dəmestɪ'keɪʃn] *noun*

breeding wild animals or plants for use by humans; *the domestication of wheat, barley and other grains took place in the Middle East about 10,000 years ago*

dominance ['dɒmɪnəns] *noun* state where one group in society rules other groups

◊ **dominant** ['dɒmɪnənt] **1** *noun* the plant *or* species which has most influence on the composition and distribution of other species of plants; the plant which is most prevalent in a vegetation community **2** *adjective* (genetic trait) which is more powerful than other recessive genes

COMMENT: since each physical trait is governed by two genes, if one is recessive and the other dominant, the resulting trait will be that of the dominant gene, and only if two recessive genes are present will the recessive trait be shown

Dorking ['dɔːkɪŋ] *noun* breed of fowl, with dark and silver-grey plumage; also a silver-grey breed of bantam

dormancy ['dɔːmənsi] *noun* state of an animal *or* plant, where metabolism is slowed during the winter months; **seed dormancy** = period when a seed is not active

◊ **dormant** ['dɔːmənt] *adjective* not actively growing; **dormant plant** = plant which is not actively growing (as during the winter)

Dorset Down ['dɔːsət 'daʊn] *noun* medium-sized down breed of sheep; with a brown face, and wool growing over the forehead; providing a good-quality fine stringy fleece

◊ **Dorset Horn** *noun* breed of sheep in the south-west of England; both rams and ewes have long curly horns; it produces a fine white clear wool, and is unique among British breeds in that it can lamb at any time of the year

◊ **Dorset wedge silage** *noun* method of storing silage in wedge-shaped layers, usually covered with polythene sheeting. The first loads are tipped against the end wall and further loads are built up with a buckrake to form a wedge

dose [dəʊs] **1** *noun* amount of medicine given to an animal to cure it of a disorder **2** *verb* to give an animal medicine; **dosing gun** = device used to give an animal medicine in the form of pellets; the pellet is forced into the back of animal's throat

double ['dʌbl] **1** *noun* **(a) doubles** = twins of animals, especially lambs **2** *adjective* **(a) double lows** = varieties of oilseed rape with low erucic acid and glucosinolate contents **(b) double flower** = flower with two series of petals (as opposed to a single flower)

◊ **double chop harvester** ['dʌbl tʃɒp 'hɑːvəstə] *noun* type of forage harvester, which chops the crop into short lengths rather than just lacerating it; the chopping unit is a vertical rotating disc, usually with three knives and three fan blades

◊ **double cropping** [dʌbl 'krɒpɪŋ] *noun* multi-cropping, taking more than one crop off a piece of land in one year

◊ **double digging** [dʌbl 'dɪgɪŋ] *noun* cultivation technique, where a spit is dug out, the soil placed on one side, and a second spit dug; this loosens the soil at a deeper level than normal digging

◊ **double suckling** [dʌbl 'sʌklɪŋ] *noun* method of raising beef calves, where a second calf is placed with the cow's own calf and allowed to suckle

dough [dəʊ] *noun* mixture of flour and water prepared to make bread; it can have salt and yeast added. Bread (such as chapatis) made from dough without yeast is 'unleavened'

Douglas fir ['dʌgləs 'fɜː] *noun* important North American softwood tree (*Pseudotsuga menziesii*) widely planted throughout the world, and yielding vast amounts of strong timber

Doum palm ['duːm pɑːm] *noun* palm (*Hyphaene thebaica*) with fan-shaped leaves found in tropical Africa. The seeds (known as vegetable ivory) are used for beads. The fruit are used for syrup and the leaves for matting

dove [dʌv] *noun* white domesticated pigeon; **dove cote** = special building for keeping doves

down [daʊn] *noun* **(a)** soft first plumage of young birds; an adult bird's under plumage, used to make pillows **(b)** undercoat of very soft hair on a goat

◊ **downs** [daʊnz] *noun* rolling grass-covered chalky hills characterized by low bushes and few trees, used for cereal growing and sheep grazing; **Down breeds** = breeds of short-wooled sheep, giving wool of a creamy colour; the sheep have dark faces and legs, and are hornless. They are found in hilly areas, and include the Southdown, Hampshire Down, Dorset Down and Suffolk

◊ **down-calver** ['daʊnkɑːvə] *noun* cow or heifer about to calve

◊ **downland** ['daʊnlænd] *noun* area of downs

◊ **downy** ['daʊni] *adjective* very soft (plumage)

◊ **downy mildew** ['daʊni 'mɪldjuː] *noun* disease (*Perenospora brassica*), causing white bloom on the under surface of leaves, most damaging to Brassica seedlings

Doyenné du Comice ['dwæjeneɪ djuː 'kɒmiːs] *noun* variety of dessert pear, originating in France; the fruit are very round and mature slowly

DPA = DIRECT PROPORTIONAL APPLICATION

draft ewe ['drɑːft 'juː] *noun* ewe sold from a breeding flock of sheep while still young enough to produce lambs; the main source of mature purebred ewes is the hill and upland farms

◊ **draft off** ['drɑːft 'ɒf] *verb* to remove certain animals from a herd or flock

QUOTE farmers are busy drafting lambs in anticipation of a long dry period
New Zealand Farmer

drag harrow ['dræg 'hærəʊ] *noun* heavy type of harrow, used in the preparation of seedbeds

drain [dreɪn] **1** *noun* open channel for taking away surplus water from farmland; **mole drain** = underground drain formed by a mole plough as it is pulled across a field; mole drains are normally drawn 3 to 4 metres apart, and are used in fields with a clay subsoil; **pipe drain** = underground drain made of lengths of clay tiles linked together; they may also be made of concrete or plastic pipes. The spacing of the pipes will depend on the nature of the soil; **storm drain** = specially wide channel for taking away large amounts of rainwater which fall during tropical storms **2** *verb* **(a)** to remove liquid; *silage liquor drains into a slurry tank* **(b)** to remove water from farmland; on most types of farmland, except free-draining soils, some sort of artificial drainage is necessary to carry away surplus water and so keep the water table at a reasonable level; **draining pen** = pen for sheep to go in after dipping, where surplus liquid can drain off the wet fleece and go back into the sheep dip

◊ **drainage** ['dreinidʒ] *noun* removing of water by laying drains in or under fields; **drainage basin** = catchment area *or* area of land which collects and drains away the rainwater which falls on it

COMMENT: the main methods of drainage are open channels (ditches) and underground pipe drains and mole drains. Signs of bad drainage may be: machinery getting bogged down in mud; poaching by stock in grazing pastures; water lying in pools after heavy rain; weeds such as rushes, sedges and horsetail common in grassland; young plants being pale green or yellow; subsoil is various shades of blue or grey

QUOTE dairy farms are responsible for most water pollution incidents caused by inadequate facilities for dirty water disposal; effective systems can be designed to handle yard drainage
Farmers Weekly

drake [dreik] *noun* male duck

Drakensberger ['dreikənzbɜːgə] *noun* breed of cattle found in South Africa, derived from the native black cattle crossed with Groningen bulls from the Netherlands. They are black in colour with short white horns, and are a triple-purpose breed (i.e., raised for meat, milk and as draught animals)

draught [drɑːft] *noun* the effort needed to pull an implement through the soil; **draught control system** = system of preventing damage to an implement such as a harrow, as it is being pulled through the soil (when the draught reaches a set level, the implement is automatically raised out of the soil); **draught animals** = animals which are used to pull vehicles or carry heavy loads (NOTE: also sometimes called **draughts)**

COMMENT: considerable use is made of draught animals in many areas of the world. Oxen, buffaloes, yaks, camels, elephants, donkeys, horses are all used as draught animals. The advantages of using animals are many: they produce young so do not always have to be bought; they are cheaper to buy than machines; they do not use

expensive fuel, even though they eat large quantities of food; they may be slower than machines but they can work in difficult terrains. Their most important advantage is that they are appropriate to the local conditions

drawbar ['drɔːbɑː] *noun* metal bar at the back of a tractor, used to pull trailed implements; some tractors have a drawbar which can be attached to the hydraulic linkage; **swinging drawbar** = moveable drawbar which is used to pull heavy trailed equipment; **drawbar power** = power available to pull an implement, as opposed to the brake horsepower of a tractor. Under field conditions, not all brake horsepower will be available to pull implements, because some of it is needed to make the tractor itself move forwards and overcome the resistance of the bearings and the soil on the wheels

draw hoe ['drɔː 'həu] *noun* hoe whose blade is at right angles to the handle and is pulled backwards towards the worker

dray [drei] *noun* flat cart without sides

dredge corn ['dredʒ 'kɔːn] *noun* mixture of cereals grown together and used for livestock feeding; the commonest type is a mixture of barley and oats; sometimes cereals and pulses are mixed

drench [drentʃ] **1** *noun* method of applying a liquid medicine, by passing it into the stomach through a tube **2** *verb* to soak with a liquid, as when spraying with a coarse nozzle

dress [dres] *verb* to clean or prepare

◊ **dressing** ['dresin] *noun* treating seeds before sowing, to control disease; **top dressing** = (i) application of fertilizer to the surface of the soil; (ii) the fertilizer applied onto the surface of the soil

dressage [dre'sɑːʒ] *noun* A competitive equestrian sport. In Britain national tests are graded in order of difficulty (NOTE: the word comes from the French verb **dresser** meaning 'to train')

dribble bar ['dribl 'bɑː] *noun* attachment which applies a liquid top dressing to a crop through trailing pipes from a boom

dried [draid] *adjective* (foodstuffs) preserved by dehydration (this removes water and so slows down deterioration)

◊ **dried blood** [draɪd 'blʌd] *noun* organic fertilizer with a nitrogen content of 10% - 13%; a soluble quick-acting fertilizer, used mainly by horticulturists

◊ **dried fruit** [draɪd 'fruːt] *noun* fruit that has been dehydrated to preserve it for later use (currants and raisins are dried grapes)

◊ **dried grass** [draɪd 'grɑːs] *noun* grass which has been artificially dried; used as an animal feed of high nutritional value (dried grass has a high carotene content, and a crude protein content of 10% - 20%)

◊ **dried milk** [draɪd 'mɪlk] *noun* milk powder produced by removing water from liquid milk; the techniques involved include roller-drying and spray-drying

◊ **drier** ['draɪə] *noun* machine used to dry a crop, usually grain; **grain drier** = machine which dries moist grain before storage; the grain is dried under a blast of hot or warm air

drift [drɪft] *verb (of spray)* to float in the air onto areas which are not to be sprayed; **spray drift** = blowing away of spray in windy conditions; vibrojets are used where spray drift is a problem; otherwise low pressure or rapidly vibrating sprayers may help to cut down drift

drill [drɪl] **1** *noun* **(a)** implement used to sow seed; a drill consists of a hopper carried on wheels, with a feed mechanism which feeds the seed into seed tubes; **combine drill** = drill which sows grain and fertilizer at the same time; some drills have separate tubes for the seed and fertilizer, others have one tube for both **(b)** little furrow for sowing seed **2** *verb* to sow seed in drills

COMMENT: most seed drills are designed to plant seed in prepared seed beds; drills for cereals and grasses sow the seed at random while precision drills, used mainly for sugar beet and vegetable crops, place seed at preset intervals in the rows; these are also called 'seeder units'. Grain drill feed mechanisms may be internal force feed, external force feed or studded roller. Some drills have a hopper divided into two parts: one contains the seed, the other fertilizer; this is the combine drill which drills grain and fertilizer at the same time

drip irrigation ['drɪp ɪrɪ'geɪʃn] *noun* irrigation system where water is brought to the base of each plant and drips slowly into the soil, also called 'trickle system'

dromedary ['drɒmədəri] *noun* one of the two types of camel; it has one hump, and is found in the Middle East, North Africa and India

drone [drəʊn] *noun* male bee

drop [drɒp] **1** *noun* **(a)** small quantity of liquid; *a drop of acid burned through the material; drops of rain soon began to fall* **(b)** fall (of fruit); **June drop** = fall of small fruit in early summer, which allows the remaining fruit to grow larger **2** *verb (of ewe)* to give birth to a lamb

◊ **droppings** ['drɒpɪŋz] *noun* excreta from animals; *the grass was covered with rabbit and sheep droppings;* **droppings board** = a bench, under the perches in smaller poultry houses, on which bird droppings collect

drought [draʊt] *noun* long period without rain at a time when rain normally falls

◊ **droughtmaster** ['draʊtmɑːstə] *noun* Australian breed of beef cattle derived from the Indian Brahman and the British Shorthorn. The animals are red, with a characteristic small hump

drove [drəʊv] *noun* number of cattle or sheep being driven from one place to another; **drove road** = track along which sheep or cattle are regularly driven

◊ **drover** ['drəʊvə] *noun* person in charge of a flock or herd which is being moved from one place to another

drum [drʌm] *noun* the cylinder of a combine harvester, which rotates and has rasp-like beater bars which thresh the grain

drupe [druːp] *noun* fruit with a single hard seed and a fleshy body (such as a peach)

dry [draɪ] **1** *adjective* **dry adiabatic lapse rate** = rate of temperature change in rising dry air; **dry cow** = cow not giving milk, between lactations; **dry curing** = curing meat in salt, as opposed to brine; **dry farming** = system of extensive agriculture, producing crops in areas of limited rainfall, without using irrigation; soil moisture is retained by mulching, frequent fallowing and removing the weeds that would take up some of the moisture; **dry feeding** = feeding of meal to animals without the addition of water; this may cause problems with pigs and poultry; **dry deposition** = falling of dry particles from polluted air (they form a harmful deposit on surfaces such as buildings or leaves of trees); **dry matter (DM)** = components of a feed left after drying; **dry period** = in cattle, a period of six to eight weeks between lactations, when a cow is rested from giving milk; **dry rot** = fungus disease of potatoes, which makes the tubers shrink; blue-pink or white pustules appear on the skin; **dry roughage** = dry bulky foodstuffs, such as hay or straw; **the dry season** = period in some countries when very little rain falls (as opposed to the rainy season) **2** *verb* to remove moisture from something; *see also* DRIED

◊ **drying** ['draɪɪŋ] *noun* method of preserving food by removing moisture (either by leaving it in the sun, as for dried fruit, or by passing it through an industrial process); **drying off** = gradual reduction in the quantity of milk taken from a cow, so as to make it stop lactating

◊ **dryness** ['draɪnəs] *noun* state of being dry

◊ **drystone wall** ['draɪstəun 'wɔːl] *noun* wall made of stones carefully placed one on top of the other without using any mortar; these are features of much of Britain's hill country. Most walls are wider at the base than at the top

> QUOTE hayage is harvested at a higher dry matter and then compressed to give a finished 60% dry matter product
> *Farmers Weekly*

Drysdale ['draɪzdeɪl] *noun* breed of New Zealand sheep, a crossbreed from Romney and Cheviot

dual-purpose breed [dju:əl'pɜːpəs 'briːd] *noun* breed of animal valuable for more than one product; **dual-purpose cattle** (such as the South Devon) are bred for milk and meat; **dual-purpose poultry** are bred for egg-laying and as table birds

dubbin ['dʌbɪn] *noun* prepared grease used for waterproofing and softening leather

duck [dʌk] *noun* bird reared for both egg and meat production; the male is a 'drake'

◊ **duckling** ['dʌklɪŋ] *noun* young duck

dug [dʌg] *noun* teat or udder of an animal, especially of a cow

dump [dʌmp] **1** *noun* place where waste is thrown away; **dump box** = large hopper on wheels with a floor conveyor which receives

silage from trailers and from which the crop is discharged into the silo **2** *verb* **(a)** to throw away waste **(b)** to get rid of large quantities of excess farm products cheaply in an overseas market

◊ **dumping** ['dʌmpɪŋ] *noun* selling of agricultural products at a price below the true cost, to get rid of excess produce cheaply, usually in an overseas market

dunes [djuːnz] *noun* area of sand blown by the wind into ridges which have very little soil or vegetation; dunes can be stabilized by planting with marram grass

dung [dʌŋ] **1** *noun* solid waste excreta from animals (especially horses and cattle), often used as fertilizer **2** *verb* **(a)** to spread dung on the land as a fertilizer **(b)** to clear away animals' dung; **dunging passage** = passage at the back of a cow shed, into which dung can be washed with water

> COMMENT: in some areas of the world, in particular in India, dried dung is used as a cooking fuel; this has the effect of preventing the dung being returned to the soil and so leads to depletion of soil nutrients

dungleweed ['dʌŋgəlwiːd] *noun* another name for orache

duramen [djuː'rɑːmən] *noun* heartwood, the hard wood in the centre of a tree's trunk

Durham ['dʌrəm] *noun* breed of dairy shorthorn cattle, developed in the Tees valley of County Durham

durian ['duːrɪæn] *noun* tropical fruit from a tree *(Durio zibethinus)* native to Malaysia. The fruit has an extremely unpleasant smell, but is highly regarded as a dessert fruit in south-east Asia

Duroc [djuː'rɒk] *noun* breed of pig, originating in the eastern USA, imported into the UK for cross-breeding; the pigs are red in colour

durum ['djuːrəm] *noun* hard type of wheat *(Triticum durum)* grown in southern Europe and used in making semolina for processing into pasta; also grown in the USA, it is one of the wheat classes used in classifying US wheat exports

dust [dʌst] *noun* fine powder made of particles of dry dirt *or* sand; **dust bowl** =

large area of land where the strong winds blow away the dry topsoil (used to describe areas of Kansas, Oklahoma and Texas which lost topsoil through wind erosion in the 1930s); **dust devil** = little whirlwind in the desert

◊ **dusting** ['dʌstɪŋ] *noun* using dry powdered fungicide or insecticide on crops

◊ **dust storm** ['dʌst stɔːm] *noun* storm of wind which blows dust and sand with it, common in North Africa

QUOTE the spectre of a new dust bowl and worldwide food shortages was evoked by US ecologists as dust clouds hundreds of miles wide swept across the Midwest, whipped up by 60 mph winds from the drought-parched soil. The National Agricultural Statistical Service runs regular surveys of crop conditions, and this year across the Midwest most of the winter wheat is rated poor to fair
Guardian

Dutch barn ['dʌtʃ 'bɑːn] *noun* type of farmyard building used for storage of hay, loose or baled, corn crops and agricultural implements; the older types of Dutch barn were built of iron with no enclosing side walls; modern designs incorporate precast concrete, asbestos-cement sheeting with curved roofs; the sides may be partly or completely covered

Dutch elm disease [dʌtʃ 'elm dɪ'ziːz] *noun* fungus disease caused by *Ceratocystis ulmi*. It kills elm trees and is spread by a bark beetle

Dutch harrow ['dʌtʃ 'hærəʊ] *noun* implement with metal or wooden frame, with heavy tine bars almost at right angles to the direction of travel; the tines loosen the soil and the heavy bars level the surface (NOTE: sometimes also called a **float**)

◊ **Dutch hoe** ['dʌtʃ 'həʊ] *noun* implement with a long handle and a more or less straight D-shaped blade, used with a push-pull action

D value ['diː 'væljuː] *noun* = DIGESTIBILITY VALUE amount of digestible organic matter in the dry matter of plants

QUOTE the feeding value of late silage rarely matches the analyses which can average 68 D value and 22% of dry matter
Farmers Weekly

DV = DISTRICT VALUER

DVO = DIVISIONAL VETERINARY OFFICER

dwarf beans ['dwɔːf 'biːnz] *noun* term used for French or kidney beans, which make a bushy plant, as opposed to runner beans which climb

dwarfing rootstock ['dwɔːfɪŋ 'rʊtstɒk] *noun* rootstock used for grafting fruit trees; it produces a tree which does not grow very tall, and so makes fruit-picking easier

dyke [daɪk] **1** *noun* **(a)** long wall of earth built to keep water out **(b)** ditch for drainage **2** *verb* to build walls of earth to help prevent water from flooding land

COMMENT: dyke pond farming is a system of organic agriculture combining crop growing on the dykes which surround ponds in which fish are bred; it is common in China

dysentery ['dɪsəntri] *noun* infection and inflammation of the colon, caused either by bacilli or by parasites

dystocia [dɪs'təʊsiə] *noun* difficulty in the process of giving birth

Ee

E Excellent (in the EUROP carcass conformation classification system)

E number ['i: nʌmbə] classification of additives to food, according to the European Union

COMMENT: additives are classified as follows: colouring substances: E100–E180; preservatives: E200–E297; antioxidants: E300–E321; emulsifiers and stabilizers: E322–E495; acids and bases: E500–E529; anti-caking additives: E530–E578; flavour enhancers and sweeteners: E620–E637

Vitamin E ['vɪtəmɪn 'iː] noun vitamin found in vegetables, vegetable oils, eggs and wholemeal bread. Essential to fertility of animals

ear [ɪə] noun head or spike of a cereal, containing the flowers or seeds; **ear emergence** = the main date for heading; in the case of a sward, this is the date at which 50% of the inflorescences have appeared

◊ **earmarking** ['ɪəmɑːkɪŋ] noun identifying an animal by attaching a tag to its ear; also sometimes done by cutting a notch in the edge of the ear

earlies ['ɜːlɪz] noun potatoes grown for harvesting early in the season, as opposed to 'maincrop potatoes'; see also POTATO

early bite ['ɜːli 'baɪt] noun grazing in the spring, provided by new growths of grass which sprout when the weather gets warmer

◊ **early weaning** ['ɜːli 'wiːnɪŋ] noun removing young from the dam earlier than is usual

earth [ɜːθ] noun soil, a soft substance formed of mineral particles, decayed organic matter, chemicals and water in which plants can grow

◊ **earth up** ['ɜːθ 'ʌp] verb to move soil to make a ridge, in which a crop such as potatoes or celery can grow

COMMENT: plants are earthed up to protect the tender stems from frost, or to make them white. Potatoes are earthed up to prevent the tubers from turning green and tasting bitter

◊ **earthworm** ['ɜːθwɜːm] noun kind of invertebrate animal (Lumbricidae), with a long thin body, living in large numbers in the soil

COMMENT: earthworms provide a useful service by aerating the soil as they tunnel. They also eat organic matter and help increase the soil's fertility. They help stabilize the soil structure by compressing material and mixing it with organic matter and calcium. It is believed that they also secrete a hormone which encourages rooting by plants

easement ['iːzmənt] noun the right of access or use over land

East Friesland ['iːst 'friːzlənd] noun breed of sheep introduced into the UK from Holland; a large long slim-bodied breed much valued for its high milk yield

easy feed ['iːzi 'fiːd] noun means of feeding livestock which allows easy access to feed by means of hoppers or feeding passages

ebony ['ebəni] noun black tropical hardwood, now becoming scarce

ecology [ɪ'kɒlədʒi] noun study of the relationships among organisms and the relationship between them and their physical environment; **human ecology** = study of man and communities, the place which they occupy in the natural world, and the ways in which they adapt to or change the environment

◊ **ecological** [iːkə'lɒdʒɪkl] adjective referring to ecology; **ecological balance** = situation where relative numbers of organisms remain more or less static (by polluting the environment, destroying habitats, and increasing their own numbers, people can cause major changes in populations which may be irreversible); **ecological disaster** = disaster which seriously disturbs the balance of the environment; **ecological factors** = factors which influence the distribution of a plant species in a habitat; **ecological succession** = series of communities of organisms which follow on one after the other, until a climax community is established

◊ **ecologist** [i:ˈkɒlədʒɪst] *noun* **(a)** scientist who studies ecology **(b)** person who believes that a balance should be maintained between living things and the environment in which they live, and so improving the life of all organisms

economist [ɪˈkɒnəmɪst] *noun* person who specializes in the study of economics; **agricultural economist** = person who studies the economics of the agricultural industry

ecoparasite [i:kəʊˈpærəsaɪt] *noun* parasite which is adapted to a specific host; *compare* ECTOPARASITE, ENDOPARASITE

ecosystem [ˈi:kəʊsɪstəm] *noun* system which includes all the organisms of an area and the environment in which they live; European wetlands are examples of ecosystems that have been shaped by humans. For thousands of years they have provided pasture for livestock, reeds for thatching and fuel, silage and bedding for livestock

COMMENT: an ecosystem can be any size, from a flower bud to the whole biosphere. The term was first used in the 1930s to describe the interdependence of organisms among themselves and with the living and non-living environment

ecotone [ˈi:kəʊtəʊn] *noun* area between two different types of vegetation (such as the border between forest and moorland), which may share the characteristics of both

ectoparasite [ektəʊˈpærəsaɪt] *noun* parasite which lives on the skin or outer surface of the host, but feeds on the inside by making a hole in the skin; these are mainly arthropod species; **ectoparasite disease** = disease caused by lice and other insects, usually with intense irritation; *compare* ECOPARASITE, ENDOPARASITE

ECU [ˈekju:] the European Currency Unit, the unit of account used as a unit of trade within the EU

edaphic [ɪˈdæfɪk] *adjective* referring to environmental conditions determined by soil characteristics; **edaphic climax** = climax caused by the soil in the area; **edaphic factors** = different soil conditions which affect the organisms living in a certain area

edging [ˈedʒɪŋ] *noun* preparing the edge of a lawn, by trimming long grass using a special tool called an 'edger'

edible [ˈedəbl] *adjective* which can be eaten; **edible fungi** = fungi which can be eaten and are not poisonous

eelworm [ˈi:lwɜ:m] *noun* minute worm-like animal (*Nematode*) which attacks a great variety of food crops; the potato cyst eelworm is a serious pest; *see also* STEM EELWORM

EFA = ESSENTIAL FATTY ACID

effective field capacity [ɪˈfektɪv ˈfi:ld kəpæsɪti] *noun* the actual average rate of work achieved by a machine, usually expressed in acres or hectares per hour

effluent [ˈefluənt] *noun* liquid, solid or gas waste material such as slurry or silage effluent from a farm; **effluent standard** = amount of sewage which is allowed to be discharged into a river *or* the sea

egg [eg] *noun* round object laid by female birds, with a hard calcareous shell forming a case containing albumen and yolk; the young bird grows inside the egg until it hatches; **egg binding** = unsuccessful attempt by a hen to lay an egg; **egg classes** = the grading of eggs under EU regulations, into Class A (fresh eggs), Class B (preserved eggs) and Class C (eggs for use in food processing); **egg eating** = action of intensively housed poultry, where birds eat their own eggs; it may be due to eggs being broken because of thin shells

◊ **eggbound** [ˈegbaʊnd] *adjective* condition of poultry where the egg is formed but the hen is unable to lay it

COMMENT: the average hen's egg weighs about 60g, of which about 20g is yolk, 35g white and the rest shell and membranes. Eggs contain protein, fat, iron and vitamins A, B, D and E. Total egg consumption in the

UK in 1993 was 9,852 million, and the annual egg consumption per capita was 171. 87% of egg production came from intensive systems, 11% was freerange and 4% came from other sources. The value of retail sales of eggs in 1993 was £577 million and £124 million were sold for manufacturing and processing

eggplant ['egplɑːnt] *noun* plant with purple fruit *(Solanum melongena),* used as a vegetable. A native of tropical Asia, it is also called 'aubergine'; its Indian name is 'brinjal'

Egyptian [ɪ'dʒɪpʃn] *noun* breeds of cattle found in Egypt, including the Damietta, Baladi and Saidi types. They are mainly used as draught animals, and are usually red in colour

EHF = EXPERIMENTAL HUSBANDRY FARM

EHS = EXPERIMENTAL HORTICULTURE STATION

eject [ɪ'dʒekt] *verb* to throw out

◊ **ejector** [ɪ'dʒektə] *noun* mechanism at the back of a farmyard manure spreader, which throws out the manure over a wide area

ejido [e'dʒiːdəu] *noun* communal farm or village in Mexico, formed after land redistribution

elder ['eldə] *noun* small tree *(Sambucus nigra)* with little black berries used to make wine

electric dog [ɪ'lektrɪk 'dɒg] *noun* electric wire at the side of the fence at the entrance to a milking parlour, which encourages the cows to go into the parlour

◊ **electric fence** [ɪ'lektrɪk 'fens] *noun* thin wires supported by posts, the wires being able to carry an electric current; such a fence is easily moved around the farm, and makes strip grazing on limited areas possible

element ['elɪmənt] *noun* chemical substance which cannot be broken down to a simpler substance; **essential element** = chemical element (such as carbon, hydrogen, oxygen) which is necessary to an organism's growth and function; **trace element** = substance which is essential to

organic growth, but only in very small quantities

elevator ['elɪveɪtə] *noun* **(a)** machine for carrying grain or silage to the top of a storage unit **(b)** very large storage unit for grain, in the US and Canadian prairies; **public elevator** = elevator which is used for storage by several farmers, and does not belong to one farmer alone **(c) elevator digger** = machine for harvesting crops such as potatoes, which can be adapted to harvest carrots, onions or flower bulbs. Most elevator diggers are semi-mounted; the soil and the potatoes are lifted onto a rod-link conveyor and the soil falls back to the ground as the conveyor passes over a series of agitators

elm [elm] *noun* large hardwood trees *(Ulmus* spp.) which grow in temperate areas; all species reproduce by root suckers, except the wych elm which is propagated by seed; *see also* DUTCH ELM DISEASE

ELMS = ENVIRONMENT LAND MANAGEMENT SERVICES

ELO = EUROPEAN LANDOWNING ORGANIZATIONS

elt [elt] *noun* young sow (NOTE: not a common word)

emaciation [ɪmeɪʃi'eɪʃn] *noun* becoming extremely thin; *scab causes emaciation in sheep*

Embden ['emdən] *noun* a heavy white breed of goose, with blue eyes

emblements ['embləmənts] *noun* crops which are cultivated, as distinct from tree crops

embryo ['embriəu] *noun* living organism which develops from a fertilized egg or seed (such as an animal in the first weeks of gestation or a seedling plant with cotyledons and a root); **embryo transfer** *or* **embryo transplant (ET)** = form of artificial insemination, in which the dam is impregnated with an embryo, rather than with semen

◊ **embryonic** [embri'ɒnɪk] *adjective* referring to an embryo; (something) in the first stages of its development

emerge [ɪ'mɜːdʒ] *verb (of seedling)* to come out of the soil

◊ **emergence** [ɪ'mɜ:dʒns] *noun* stage in the growth of a plant, when the new shoot or stalk appears through the surface of the soil

Emmer ['emə] *noun* species of wheat *(Triticum dicoccoides)* which is a natural hybrid of wild wheat and a goat grass; a further crossing between Emmer and a goat grass produced wheat *(Triticum aestivum)*

emulsify [ɪ'mʌlsɪfaɪ] *verb* to mix two liquids so thoroughly that they will not separate

◊ **emulsifier** *or* **emulsifying agent** [ɪ'mʌlsɪfaɪə *or* ɪ'mʌlsɪfaɪɪŋ 'eɪdʒənt] *noun* substance added to mixtures of food (such as water and oil) to hold them together (used in sauces, etc.), and also added to meat to increase the water content, so that the meat is heavier (in the EU, emulsifiers and stabilizers have E numbers E322 to E495): *see also* STABILIZER

encephalopathy [enʋefə'lɒpəθi] *see* BOVINE SPONGIFORM ENCEPHALOPATHY

enclosure [ɪŋ'kləʊʒə] *noun* the action of enclosing open land: used in England to refer especially to the enclosure of common land in the 16th and 18th centuries, when rights to common land were removed and ditches, fences, hedgerows and walls were used to mark the boundaries of land which was owned freehold by major landowners; *see also* FIELD

encroach on [ɪn'krəʊtʃ 'ɒn] *verb* to come close to and gradually cover (something); *the dunes are encroaching on the pasture land; trees are spreading down the mountain and encroaching on the lower more fertile land in the valleys*

◊ **encroachment** [ɪn'krəʊtʃmənt] *noun* the action of encroaching

endangered species [ɪn'deɪndʒəd 'spi:ʃi:z] *noun* any species at risk of extinction

COMMENT: The categories of the degree of risk of extinction as drawn up by the IUCN are: 1. Endangered: in danger of extinction and unlikely to survive if the causal factors continue to operate. 2. Vulnerable: will move into the endangered category in the near future if the causal factors continue to operate. 3. Rare: not endangered or vulnerable, but at risk of becoming so

endemic [ɪn'demɪk] *adjective* **(a)** (plant *or* animal) that exists in a certain area; *the isolation of the islands has led to the evolution of endemic forms; the northern part of the island is inhabited by many endemic mammals and birds* **(b)** (pest *or* disease) which is very common in a certain area; *this disease is endemic to Mediterranean countries; see also* EPIDEMIC, PANDEMIC

endive ['endɪv] *noun* salad plant *(Cichorium endiva)*

endo- ['endəʊ] *prefix meaning* inside

◊ **endoparasite** [endəʊ'pærəsaɪt] *noun* parasite which lives inside the host; *compare* ECOPARASITE, ECTOPARASITE

◊ **endosperm** ['endəʊspɜ:m] *noun* albumen enclosed with the embryo in seeds; the interior of wheat germ is formed of floury endosperm which is a valuable food source

◊ **endotoxin** [endəʊ'tɒksɪn] *noun* poison from bacteria which pass into the body when contaminated food is eaten

energy ['enədʒi] *noun* (i) force *or* strength to carry out activities; (ii) specifically, electricity; **energy conservation** = avoiding wasting energy; **energy consumption** = amount of energy consumed by a person *or* an apparatus, shown as a unit; **energy crops** = crops, such as fast-growing trees, which are grown to be used to provide energy; **energy value** *or* **calorific value** = heat value of food *or* the number of Calories which a certain amount of a certain food contains (NOTE: energy is measured in **joules** or **calories**)

English Leicester ['ɪŋglɪʃ 'lestə] *noun* breed of sheep derived from Robert Bakewell's flock, used for breeding of many other longwool breeds; it produces a heavy fleece and is now a rare breed

enhancer [ɪn'hɑːnsə] *noun* artificial substance which increases the flavour of food, or even the flavour of artificial flavouring that has been added to food (in the EU, flavour enhancers added to food have the E numbers E620 to 637)

ensile [ɪn'saɪl] *verb* to make silage; *we ensile lush young material*

◊ **ensilage** *or* **ensiling** [ɪn'saɪlɪdʒ *or* ɪn'saɪlɪŋ] *noun* process of making silage for livestock by cutting grass and other green plants and storing in silos

enter- *or* **entero-** ['entərəʊ] *prefix* referring to the intestine

◊ **enteric** [ɪn'terɪk] *adjective* referring to the intestine; **calf enteric disease** = disease of calves causing severe diarrhoea

◊ **enteritis** [entə'raɪtɪs] *noun* inflammation of the mucous membrane of the intestine; **infective enteritis** = enteritis caused by bacteria

◊ **Enterobacteria** [entərəʊbæk'tɪərɪə] *noun* important family of bacteria, including Salmonella and Escherichia

◊ **Enterobius** [entə'rəʊbɪəs] *noun* threadworm, small thin nematode which infests the intestine

◊ **enterotoxin** [entərəʊ'tɒksɪn] *noun* bacterial exotoxin which particularly affects the intestine

◊ **enterovirus** ['entərəʊvaɪrəs] *noun* virus which prefers to live in the intestine

COMMENT: the enteroviruses are an important group of viruses, and one causes Teschen disease in pigs

entire [ɪn'taɪə] *adjective* (animal) which has not been castrated

entomology [entə'mɒlədʒi] *noun* study of insects

◊ **entomological** [entəmə'lɒdʒɪkl] *adjective* referring to insects

◊ **entomologist** [entə'mɒlədʒɪst] *noun* scientist who specializes in the study of insects; **medical entomologist** = scientist who studies insects which may carry diseases

environment [ɪn'vaɪrənmənt] *noun* the surroundings of any organism, including the physical world and other organisms; **environment protection** = act of protecting the environment by regulating the discharge of waste, the emission of pollutants, and other human activities

COMMENT: the environment is anything outside an organism in which the organism lives. It can be a geographical region, a

certain climatic condition, the pollutants or the noise which surrounds an organism. Man's environment will include the country or region or town or house or room in which he lives; a parasite's environment will include the intestine of the host; a plant's environment will include a type of soil at a certain altitude

environmental [ɪnvaɪrən'mentl] *adjective* referring to the environment; **environmental protection** = ENVIRONMENT PROTECTION **Environmental Protection Agency (EPA)** = administrative body in the USA, which deals with pollution; **environmental quality standards** = amount of an effluent *or* pollutant which is accepted in a certain environment, such as the amount of trace elements in drinking water or the amount of additives in food; **environmental set-aside** = set-aside schemes which may improve the environment; for example, paying farmers to sow grass seed mixtures for a green cover and cutting a 15m-wide strip along the side of footpaths at the edge of cereal fields

◊ **environmentalist** [ɪnvaɪrən'mentəlɪst] *noun* person who is concerned with protecting the environment to keep it healthy

◊ **Environmentally Sensitive Area (ESA)** *noun* area designated by the Minister of Agriculture on the basis of recommendations made by the Countryside Commission and the Nature Conservancy Council

COMMENT: ESAs were introduced under the 1986 Agriculture Act; they are selected for their landscape and wildlife value, and where traditional farming methods would help to maintain this value. Payments may be made to farmers within these areas who agree to farm according to the practices set out in guidelines by the MAFF. ESAs are a means of slowing down and preventing agricultural landscape change, but also act as a mechanism for more positive change through the conversion of arable land to grass

enzootic disease [enzəʊ'ɒtɪk dɪ'ziːz] *noun* an outbreak of disease among certain species of animals in a certain area; **enzootic abortion** = virus infection of sheep causing abortion about two weeks before lambing; **enzootic bovine leucosis** = a blood cancer disease of cattle; *compare* EPIZOOTIC

enzyme ['enzaɪm] *noun* protein substance

produced by living cells which catalyzes a biochemical reaction in living organisms (NOTE: the names of enzymes mostly end with the suffix **-ase)**

COMMENT: many different enzymes exist in organisms, working in the digestive system, in the metabolic processes and helping the synthesis of certain compounds. Some pesticides and herbicides work by interfering with enzyme systems, or by destroying them altogether

EPA = ENVIRONMENTAL PROTECTION AGENCY

ephemeral [ɪ'fiːmərəl] *noun* plant with such a short life cycle that the cycle may be completed many times in one growing season

epicarp ['epɪkɑːp] *noun* the exocarp, the outer layer of skin on a fruit

epidemic [epɪ'demɪk] *adjective & noun* (infectious disease) which spreads quickly through a large number of animals; *see also* ENDEMIC, PANDEMIC

epidermis [epɪ'dɜːmɪs] *noun* outer layer of skin of plants or animals, including dead skin on the surface

epilimnion [epɪ'lɪmnɪən] *noun* top level of water in a lake

epiphyte ['epɪfaɪt] *noun* plant which grows on another plant but does not derive nourishment from it

epizoon [epɪ'zəʊɒn] *noun* animal which lives on another animal

epizootic disease [epɪzəʊ'ɒtɪk dɪ'ziːz] *noun* disease which spreads to large numbers of animals over a large area; *compare* ENZOOTIC

eradicate [ɪ'rædɪkeɪt] *verb* to remove completely; *the use of herbicides to eradicate weeds*

◊ **eradication** [ɪrædɪ'keɪʃn] *noun* removing completely; **eradication area** = area from which a particular animal disease is eradicated, usually involving the slaughter of infected animals

erect [ɪ'rekt] *adjective* upright, not lying down; **erect habit** = habit of a plant which grows upright, and does not lie on the ground

ergot ['ɜːgət] *noun* fungus which grows on cereals, especially rye

◊ **ergotamine** [ɜː'gɒtəmiːn] *noun* the poison which causes ergotism

◊ **ergotism** *or* **ergot poisoning** ['ɜːgətɪzm *or* 'ɜːgət 'pɔɪznɪŋ] *noun* poisoning by eating cereals *or* bread which have been contaminated by ergot

erode [ɪ'rəʊd] *verb* to wear away

◊ **erosion** [ɪ'rəʊʒn] *noun* wearing away of earth *or* rock by the effect of rain, wind, sea and the action of toxic substances; *grass cover provides some protection against soil erosion*

COMMENT: accelerated erosion is that caused by human activity which is in addition to the natural rate of erosion. Cleared land in drought-stricken areas can produce dry soil which can blow away; felling trees removes the roots which bind the soil particles together and so exposes the soil to erosion by rainwater; ploughing up and down slopes (as opposed to contour ploughing) can lead to the formation of rills and serious soil erosion

erucic acid [ɪ'ruːsɪk 'æsɪd] *noun* a fatty acid found in rape oil, which is linked to heart disease. Varieties of oilseed rape with low erucic acid content are considered the best

Erysipelas [erɪ'sɪpələs] *noun* an infectious disease mainly affecting pigs and also turkeys; in pigs, the symptoms are reddish inflammations on the skin and a high fever. It may cause infertility or abortion and manifests itself in three forms: acute, sub-acute and chronic (NOTE: also called **Diamonds disease)**

erythromycin [ɪrɪθrəʊ'maɪsɪn] *noun* antibiotic used to combat bacterial infections

Escherichia [eʃə'rɪkɪə] *noun* one of the Enterobacteria commonly found in faeces; **Escherichia coli** = bacillus associated with mastitis in cows, sheep and pigs

ESA = ENVIRONMENTALLY SENSITIVE AREA

ESF = ELECTRONIC SOW FEEDERS

espalier [ɪ'spælieɪ] *noun* (i) method of training a fruit tree; (ii) a tree (especially apple or pear) trained in this way; *compare* CORDON

COMMENT: from a vertical trunk pairs of branches are trained horizontally about 50cm apart, resulting in a flat tree, which is often trained against a wall

esparto [ɪ'spɑːtəʊ] *noun* species of grass which yields fibres used mainly in making paper; originally coming from North Africa and Southern Spain

essence ['esəns] *noun* concentrated oil from a plant, used in cosmetics, and sometimes in pharmacy as analgesics or antiseptics

◊ **essential** [ɪ'senʃəl] *adjective* extremely important *or* necessary; **essential amino acid** = amino acid which is necessary for growth but which cannot be synthesized and has to be obtained from the food supply; **essential elements** = chemical elements (such as carbon, oxygen, hydrogen, nitrogen and many others) which are necessary to an organism's growth or function; **essential fatty acid (EFA)** = unsaturated fatty acid which is necessary for growth and health; **essential oils** *or* **volatile oils** = concentrated oils from a scented plant used in cosmetics or as antiseptics

COMMENT: the eight essential amino acids are: isoleucine, leucine, lysine, methionine, phenylalanine, threonine, tryptophan and valine. The essential fatty acids are linoleic acid, linolenic acid and arachidonic acid

Essex Saddleback ['eseks 'sædlbæk] *noun* breed of pig which has been bred with the Wessex Saddleback to form the British Saddleback

establish [ɪ'stæblɪʃ] *verb* to come to live permanently; *the starling has become established in all parts of the USA; even established trees have been attacked by the disease*

◊ **establishment** [ɪ'stæblɪʃmənt] *noun* **(a)** germination and emergence of seedlings; *there was a good crop establishment* **(b)** period when a newly seeded sward is becoming established

estancia [ɪ'stænsiə] *noun* large farm in South America, especially in Argentina, used for ranching on an intensive scale

estate [ɪ'steɪt] *noun* **(a)** landed property,

of varying size, usually with parts let to tenants; **estate village** = planned village built within an estate **(b)** *(in tropical countries)* plantation

ET = EMBRYO TRANSFER, EMBRYO TRANSPLANT

ETA = ESTIMATED TRANSMITTING ABILITY

ethylene ['eθiliːn] *noun* gas (CH_2CH_2) used as a ripening agent on fruit

etiolation [iːtiə'leɪʃn] *noun* process by which a green plant grown without sufficient light becomes yellow and grows long shoots

eucalyptus [juːkə'lɪptəs] *noun* Australian hardwood tree *(Eucalyptus* spp.), with strong-smelling resin; the trees are quick-growing and often used for afforestation. They are susceptible to fire

Euro-fraud [juːrəʊ'frɔːd] *noun* farming fraud in the European Union

QUOTE nearly £700 million has been stolen from EU farm coffers in the past 20 years. Italy tops the fraud league with more than £400 million. Only France and Ireland systematically inspect production
Farmers Weekly

EUROP lettering for the conformation classes in the Carcass Classification System

European Currency Unit (ECU) the unit of account used as a unit for trade within the EU

European Union (EU) [juərə'piːən 'juːniən] *noun* an organization established in 1957 (as the Common Market or European Community) by the Treaty of Rome. The original six members were Belgium, Netherlands, France, Italy, Luxembourg and West Germany. Denmark, Ireland and the United Kingdom joined in 1973, Greece in 1981; Spain and Portugal joined in 1986. Sweden, Finland and Austria in 1994. The organization is responsible for agricultural policy among its member states. See Common Agricultural Policy (CAP)

euthanasia ['juːθə'neɪziə] *noun* killing of sick animals

eutrophy ['juːtrəfi] *verb* to fill up with nutrients

◊ **eutrophic** [juˈtrɒfɪk] *adjective* (water) which is full of dissolved nutrients, being best for plant and animal growth; **eutrophic lake** = lake which has a high decay rate in the epilimnion (the top layer of water), and so contains little oxygen at the lowest levels (it has few fish but is rich in algae); *compare* OLIGOTROPHIC

◊ **eutrophication** [jutrɒfɪˈkeɪʃn] *noun* process by which a river or lake becomes full of phosphates and other nutrients which encourage the growth of algae, which can kill off other organisms

evaporate [ɪˈvæpəreɪt] *verb* (i) to change liquid into vapour by heating; (ii) *(of liquid)* to change into vapour; **evaporated milk** = milk which has been made thick and rich by evaporating some of its water content

◊ **evaporation** [ɪvæpəˈreɪʃn] *noun* changing liquid into vapour; *the evaporation of water from the surface of a lake*

◊ **evapotranspire** [ɪvæpəʊˈtrænspaɪə] *verb* to lose water into the atmosphere by evaporation and transpiration

◊ **evapotranspiration** [ɪvæpəʊtrænspɪˈreɪʃn] *noun* loss of water in an area through evaporation from the soil and transpiration by plants

> QUOTE deforestation is already disrupting the hydrological cycles that control rainfall. At least half of the rainwater that falls on most tropical forests is returned back to the atmosphere through evapotranspiration, hence the perpetual cloud that hangs over the world's great rainforests. The evaporated moisture is then carried by the wind to fall as rain in areas often many thousands of miles away
> *Ecologist*

evergreen ['evəgriːn] **1** *adjective* (plant) which does not lose its leaves in winter **2** *noun* tree *or* shrub which does not lose its leaves in winter; most conifers are evergreens; *compare* DECIDUOUS

eviscerate [ɪˈvɪsəreɪt] *verb* to remove the intestines and offal from a carcass

ewe [juː] *noun* adult female sheep; **ewe lamb** = female lamb less than six months old

excavator ['ekskəveɪtə] *noun* large machine for digging holes, as for laying drainage pipes

exceed [ɪkˈsiːd] *verb* to be more than (a certain limit); *the concentration of radioactive material in the waste exceeded the government limits;* **it is dangerous to exceed the stated application rate** = do not apply more than the recommended amount

excess [ɪkˈses] *noun & adjective* too much of a substance; *excess phosphates run off into the river system*

◊ **excessive** [ɪkˈsesɪv] *adjective* more than normal; *excessive application of fertilizers can cause nitrate runoff*

excrement ['ekskrɪmənt] *noun* faeces

excrete [ɪkˈskriːt] *verb* to pass waste matter out of the body, especially to discharge faeces; *the urinary system separates waste liquids from the blood and excretes them as urine*

◊ **excreta** [ɪkˈskriːtə] *plural noun* waste material from the body (such as faeces)

◊ **excretion** [ɪkˈskriːʃn] *noun* passing waste matter (faeces *or* urine *or* carbon dioxide) out of the system; **excretion rate** = rate at which a substance, such as nitrogen, is excreted by an animal; *compare* SECRETE, SECRETION

ex-farm ['eks 'faːm] *adverb* price for a product which does not include transport from the farm to the buyer's warehouse

exhaust [ɪgˈzɔːst] *verb* to use up completely; **exhausted fallow** = fallow land which is no longer fertile

Exmoor Horn ['eksmɔː 'hɔːn] *noun* stock fat sheep, with a broad head, curled horns and dense fleece. Mainly found on Exmoor, the breed has been crossed with the Devon Longwool to create the Devon Closewool

exocarp ['eksəʊkaːp] *noun* the skin of a fruit

exotic [ɪgˈzɒtɪk] *adjective* plant or animal introduced into an area from outside

exotoxin [eksəʊˈtɒksɪn] *noun* poison produced by bacteria, which affects parts of the body away from the place of infection (such as the toxins which cause botulism)

expander [ɪk'spændə] *noun* device behind a mole which widens a drain

experimental farm [ɪkspɛrɪ'mentəl 'fɑːm] *noun* farm which is used to experiment with new farming techniques, rather than being run as a commercial enterprise; such farms can be classified as Experimental Husbandry Farms (EHF) or as Experimental Horticulture Stations (EHS)

exploit [ɪk'splɔɪt] *verb* to take advantage of something; *large companies ruined the environment by exploiting the natural wealth of the forest; the birds have exploited the sudden increase in the numbers of insects*

◊ **exploitation** [ɪksplɔɪ'teɪʃn] *noun* taking advantage of something to make money; *further exploitation of the coal deposits in not economic*

export 1 *noun* ['ekspɔːt] crop or produce which is sold to a foreign country; **export quotas** = limits set to the amount of a type of produce which can be exported; **export refunds** = refunds made by the EU to farmers to compensate for a lower export price for produce **2** *verb* [ɪk'spɔːt] to send and sell crops or produce to foreign countries

extender [ɪk'stendə] *noun* food additive which makes the food bigger or heavier (without adding to its food value)

extensive [ɪk'stensɪv] *adjective* **extensive farming** = way of farming which is characterized by a low level of inputs per unit of land; yields per unit area are very low. Extensive farms are usually large in size; **extensive systems** = systems which use a large amount of land per unit of stock or output

◊ **extensification** [ɪkstensɪfɪ'keɪʃn] *noun* using less intensive ways of farming; payment to farmers to encourage them to farm less intensively; **extensification schemes** = pilot schemes for beef cattle and sheep which were begun in 1990 to offer compensation to farmers who reduced their beef output or the number of sheep by at least 20% and maintained this reduction over a 5 year period. The schemes should lead to a less intensive use of land and reduction in use of pesticides and fertilisers

COMMENT: less intensive use of farming involves using fewer chemical fertilizers, leaving uncultivated areas at the edges of fields, reducing sizes of herds of cattle, etc. This allows lower yields from the same area of farmland, which is necessary if production levels are too high (as they are in the EU)

extract [ɪk'strækt] *verb* to produce a substance from another; *coconut oil is extracted from copra*

◊ **extraction** [ɪk'strækʃn] *noun* action of producing a substance out of another; *the extraction of sugar from cane;* **extraction rate** = the percentage of flour produced as a result of milling grain

eye [aɪ] *noun* **(a)** a growth bud which has not developed, such as the bud of a potato tuber which develops shoots when the tuber is planted **(b)** the instinctive action of a sheepdog when working sheep

◊ **eyespot** ['aɪspɒt] *noun* disease of cereals *(Cercosporella herpotrichoides)*, which causes lesions to form on the stem surface and grey mould inside the stem; it can be treated with benomyl; *compare* SHARP EYESPOT

Ff

F 1 *abbreviation for* Fahrenheit **2** *chemical symbol for* fluorine

◊ **F₁** ['ef 'wʌn] *noun (in breeding experiments)* the first generation of offspring; **F₁ hybrid** = plant produced by breeding two parent plants, which is stronger than the parents, but which will not itself breed true

COMMENT: F₁ hybrids can be crossbred to produce F₂ hybrids and the process can be continued for many generations

FABLS = FARM ASSURED BEEF AND LAMB SCHEME

FAC = FEDERAL AGRICULTURAL COOPERATIVES

factory farming ['fæktəri 'fɑːmɪŋ] *noun* highly intensive rearing of animals characterized by keeping large numbers of animals indoors in relatively small areas and feeding them concentrated foodstuffs, with frequent use of drugs to control diseases which are a constant threat under these conditions

faeces ['fiːsiːz] *plural noun* excreta or dung, solid waste matter passed from the bowels of animals after food has been eaten and digested

◊ **faecal** ['fiːkəl] *adjective* referring to faeces; **faecal matter** = solid waste matter from the bowels (NOTE: also spelt **feces, fecal** especially in the USA)

Fagopyrum [fægəʊ'paɪrəm] *Latin name for* buckwheat

Fagus ['feɪgəs] *Latin name for* beech

Fahrenheit ['færənhaɪt] *noun* scale of temperatures where the freezing and boiling points of water are 32° and 212°; *compare* CELSIUS, CENTIGRADE (NOTE: used in the USA, but less common in the UK. Normally written with an **F** after the degree sign: **32°F** (say: 'thirty-two degrees Fahrenheit'))

COMMENT: to convert Fahrenheit temperatures to Celsius, subtract 32, multiply by 5 and divide by 9: so 68°F equals 20°C. As a quick rough estimate, subtract 30 and divide by two

fair [feə] **1** *noun* regular meeting for the sale of goods or animals, often with sideshows and other entertainments; fairs can be specialized, such as horse fairs or cheese fairs, or can cover all types of farm animals **2** *adjective* which is reasonable; **fair average quality (FAQ)** = average quality of agricultural produce based on samples taken from bulk

fairy ring [feəri 'rɪŋ] *noun* circle of darker coloured grass in a pasture, which is caused by fungi

falling time ['fɔːlɪŋ 'taɪm] *noun* time taken for wheat grain to fall to the bottom of a container of water, measured by the Hagberg test

fallow ['fæləʊ] **1** *adjective* (land) which is not used for growing crops for a period so that the nutrients can build up again in the soil; **to let land lie fallow** = to allow land to stand without being cultivated **2** *noun* period when land is not used for cultivation; *shifting cultivation is characterized by short cropping periods and long fallows;* **fallow crop** = crop grown in widely spaced rows, so that it is possible to hoe and cultivate between the rows; **fallow cultivation** = type of shifting cultivation: the period under crops is increased and the length of the fallow is reduced; **fallow length** = period of time between cultivation periods (as population density increases and land becomes scarce, food supply can be increased by making the period under crops longer and the length of fallow shorter); **bush fallow** = subsistence type of agriculture in which land is cultivated for a few years until its natural fertility is exhausted, then allowed to rest for a considerable period during which the natural vegetation regenerates itself, after which the land is cleared and cultivated again

◊ **fallowing** ['fæləʊɪŋ] *noun* allowing land to lie fallow for a period

false seedbed ['fɔːls 'siːdbed] *noun* seedbed prepared in spring to allow weed seeds to germinate; these are then killed by cultivation before sowing root crops

false staggers ['fɔːls 'stægəz] *noun* disease of sheep caused by maggots which cause inflammation of nostrils and head; the sheep appear dazed

family ['fæməli] *noun* group of genera which have certain characteristics in common (several families form an order) (NOTE: names of families of animals end in **-idae** and of families of plants in **-ae)**

◊ **family farm** ['fæmɪli 'fɑːm] *see* FARM

famine ['fæmɪn] *noun* period of severe shortage of food; *when the monsoon failed for the second year, the threat of famine became more likely;* **famine relief** = sending supplies of basic food to help people who are starving

fancy breed ['fænsi 'briːd] *noun* breed reared for decoration or show, rather than for produce

FAO = FOOD AND AGRICULTURE ORGANIZATION an agency of the United Nations established with the purpose of encouraging higher standards of nutrition and eradicating malnutrition and hunger

FAQ = FAIR AVERAGE QUALITY

farinha *or* **farina** [fə'riːnə] *noun* coarse meal made from cassava, an important food in parts of South America; *compare* GARI

farm [fɑːm] **1** *noun* area of land which may vary in size, and the buildings on it, owned or rented by one management and used to grow crops and/or raise livestock; **family farm** = a farm unit which ideally supports one family; **farm buildings** = buildings on a farm: such buildings are needed to shelter animals, get stock off the ground to improve grass production, provide an artificial environment for stock or crops, improve working conditions for staff, house machinery, and protect stored materials from weather and vermin; **farm consolidation** = joining small plots of land together to form larger farms or bringing scattered units together to form large fields (consolidation is promoted by most governments to correct the inefficiencies associated with fragmentation in small scattered land holdings); **farm fragmentation** = situation where the fields of a farm are scattered over an area, so that the holding is not made up of a single unit of land; it can be the result of inheritance practices where the land of a person who has died is split between all the children, of land reclamation schemes where the land is reclaimed piece by piece, or because old open-field systems have been kept; **farm fresh eggs** = term used in the EU to describe Class A eggs; **farm gate prices** = prices which a farmer receives for his produce; **farm rent** = rent paid by a tenant farmer to a landlord on a regular basis for the use of the farm holding **2** *verb* to run a farm; to keep animals for sale *or* to grow crops for sale

◊ **Farm and Conservation Grant Scheme** British government scheme (1989) which makes grants for improvements such as planting of hedges, installation of waste and effluent handling systems, building of walls

◊ **Farm Animal Welfare Council (FAWC)** agency established by the British government in 1979 to keep under review the welfare of farm animals on agricultural land, at markets, in transit and at the place of slaughter

◊ **Farm Capital Grant Scheme** British government scheme begun in 1970, providing grants towards capital expenditure on farms for improvements such as field drainage and farm buildings

◊ **Farm Diversification Grant Scheme** British government scheme which offers assistance to farmers for diversifying into non-agricultural profit-making activities on the farm, for example, processing of farm produce, holiday accommodation, farm machinery repair, livery and craft manufacture

◊ **Farm Support Scheme** article 39 of the Treaty of Rome provides the framework of the Common Agricultural Policy. Each member state contributes to the European Agricultural Guarantee and Guidance Fund. Payments are made for structural changes under the guidance fund and much larger payments under the guarantee section

◊ **Farm Watch Scheme** scheme which

organises networks of farmers to be in touch with one another and the local police to report suspicious activity. Theft from farms has increased markedly and in 1993 was estimated at £13.6 million

◊ **Farm Woodland Scheme** scheme started in 1988 which pays farmers to replace land in agricultural production by tree planting. Planting is permitted up to a limit of 40 hectares per holding

farmer ['fɑ:mə] *noun* person who makes a living from agriculture by managing and cultivating a farm, whether as a tenant or owner; **dairy farmer** = farmer who raises cattle for milk; **farmer's lung** = form of asthma caused by an allergy to rotting hay; **farmers' list** = list of veterinary medicines which can be obtained from agricultural merchants with a prescription from a veterinary surgeon

◊ **farmhand** ['fɑ:mhænd] *noun* person who works on a farm

◊ **farmhouse** ['fɑ:mhaʊs] *noun* house where a farmer and his family live

◊ **farming** ['fɑ:mɪŋ] *noun* working a farm by growing crops *or* keeping animals for sale, and producing cereals, vegetables, meat, dairy products, etc.; **farming systems** = the classification of different farming types: such as shifting cultivation systems, ley systems, systems with permanent upland cultivation, fallow systems, grazing systems and systems with perennial crops; **contour farming** = method of cultivating sloping land, where the land is ploughed along a level across the slope, to prevent soil erosion; *see also* ARABLE, DAIRY, FACTORY FARMING, FISH FARMING

◊ **farmland** ['fɑ:mlænd] *noun* cultivated land, land which is used for raising crops or animals for food

◊ **farmscape** ['fɑ:mskeɪp] *noun* landscape dominated by agriculture; farmland is the main element in farmscape, though non-agricultural uses may be included; agricultural villages are part of the farmscape

◊ **farmstead** ['fɑ:msted] *noun* farmhouse and the farm buildings around it

◊ **farmworker** ['fɑ:mwɜ:kə] *noun* person who works on a farm

◊ **farmyard** ['fɑ:mjɑ:d] *noun* area around the farm buildings; **farmyard manure (fym)** = manure formed of cattle excreta mixed with straw, used as a fertilizer; **farmyard**

manure spreader = machine for spreading manure, basically a trailer with a moving floor conveyor and a combined shredding and spreading mechanism which distributes the material: there are two types, the rear ejector and the side delivery spreader

Farrand test ['færənd 'test] *noun* a method for determining the alpha amylase content of milling wheat. The amount of alpha amylase enzyme present in wheat is important for making bread. Excessive alpha amylase in flour results in poorer loaves

farrier ['færiə] *noun* a person who makes and fits shoes for horses; *see also* BLACKSMITH

farrowing ['færəʊɪŋ] *noun* parturition in a sow (the act of giving birth to piglets); **farrowing crate** = steel frame which holds the sow during farrowing and helps to prevent the overlying of the piglets; **farrowing rails** = rails which prevent the sow from overlying the piglets

◊ **farrowing fever** ['færəʊɪŋ 'fi:və] *noun* disease of pigs caused by inflammation of the womb. Pigs suffer high temperatures and loss of appetite

fasciation [feɪʃi'eɪʃn] *noun* **(a)** abnormal plant growth in which several stems become fused together **(b)** production of several shoots from the crown of a plant, such as a pineapple; pineapples are propagated by crown slips which shoot from the base of the crown. If several are produced by fasciation, the plant's food supply is reduced and the plant should be destroyed after harvesting

fat [fæt] **1** *adjective* (animal) which has been reared for meat production and which has reached the correct standard for sale in a market; **fat lamb** = lamb in condition satisfactory for slaughter **2** *noun* **(a)** white *or* oily substance in the body which stores energy and protects the body against cold; **fat class** = assessing the amount of external fat present on a beef or sheep carcass; **brown fat** = animal fat which can easily be converted to energy and is believed to offset the effects of ordinary white fat; **saturated fat** = fat which has the largest amount of hydrogen possible; **unsaturated fat** = fat which does not have a large amount of hydrogen, and so can be broken down more easily **(b)** type of food which supplies protein and Vitamins A and D, especially that part of meat which is white and solid substances (like lard *or* butter) produced from animals and used for cooking, or liquid substances like oil

◊ **fat hen** ['fæt 'hen] *noun* weed (*Chenopodium album*) which affects spring cereals, peas and row crops; found especially in rich soils and muck heaps; also called 'dung weed' and 'muck weed'

◊ **fatness** ['fætnəs] *noun* amount of fat on an animal

◊ **fatstock** ['fætstɒk] *noun* animals which are fattened for meat production; **Fatstock Guarantee Scheme** = UK government scheme which provides a guaranteed price for sheep and their carcasses

◊ **fatten** ['fætn] *verb* to give animals more food so as to prepare them for slaughter; *he buys lambs for fattening and then sells them for meat*

◊ **fatty** ['fæti] *adjective* containing fat; **fatty acid** = organic acid (such as stearic acid) which occurs in fat; they are important substances for the maintenance of the body; **essential fatty acid (EFA)** = unsaturated fatty acid which is essential for growth but which cannot be synthesized by the body and has to be obtained from the food supply; **fatty liver (syndrome)** = condition in older cows, where the animal absorbs calcium too slowly and the liver is affected; goats are also affected

> QUOTE modern strains of broiler are not only heavier but also contain a greater proportion of fat at killing age compared with broilers of 20 years ago. Actual levels of fatness can vary considerably between flocks, but body fat contents of 12-25% are not uncommon
> *Poultry Science Symposium 18*

fatal ['feɪtəl] *adjective* which kills; *BSE is a fatal disorder of cattle*

faults [fɔːlts] *noun* proportion of rotten or diseased items in a quantity of produce sold

> QUOTE discussions with the potato industry to ensure that the futures market tenders met the 5% faults allowed in the PMB ware standard
> *Farmers Weekly*

fauna ['fɔːnə] *noun* wild animals and birds which live naturally in a certain area; *see also* FLORA (NOTE: plural is **fauna** or **faunas)**

FAWC = FARM ANIMAL WELFARE COUNCIL

Faverolle ['fævərɒl] *noun* breed of poultry for table consumption

fawn [fɔːn] **1** *noun* young deer **2** *adjective* pale brown colour

fax [fæks] *see* DAIRY FAX

fazenda [fə'zendə] *noun* a large plantation in Brazil; *see also* ESTANCIA

FBS = FARM BUSINESS SURVEY

Fe *chemical symbol for* iron

feather ['feðə] *noun* **(a)** growth on the skin of birds made up of a hard central stem and soft 'hairs' growing along each side of it; feathers are formed at the base of follicles; a bird will moult regularly and lose its old feathers, growing new ones to replace them. Before moulting takes place new feathers start to grow in the follicles. Feathers are removed ('plucked') from carcasses of birds by machine **(b)** growths of hair round the feet of some breeds of shire horse

◊ **feathereating** *or* **featherpecking** *or* **featherpulling** ['feðəiːtɪŋ *or* 'feðəpekɪŋ *or* 'feðəpulɪŋ] *noun* pulling of the feathers of a bird by another bird

> COMMENT: featherpecking occurs in poultry where the birds are penned up and lack exercise and facilities for scratching. It may also be caused by wrong nutrition. Once the birds start pulling feathers, an outbreak of cannibalism may occur

febrile ['febraɪl] *adjective* referring to a fever; **febrile disease** = disease (such as Newcastle disease) which is accompanied by a fever

feces, fecal ['fiːsiːz *or* fiːkəl] *see* FAECES, FAECAL

fecundity [fɪ'kʌndɪti] *noun* measurement of the number of offspring born and reared by a dam

fee [fiː] *noun* money paid to a professional for a service; **stud fees** = money paid to the owner of a stud animal for servicing a female

feed [fiːd] **1** *noun* various types of food

available to animals; *traces of pesticide were found in the cattle feed;* **feed additives** = supplements added to the feed of farm livestock, particularly pigs and poultry, to promote growth; **feed block** = a block of foodstuff left out in the pasture, especially on hill farms, used by sheep to prevent loss of condition; **feed compounds** = a number of different ingredients including major minerals, trace elements and vitamins, mixed and blended to provide properly balanced diets for stock; **feed concentrates** = animal feedingstuffs which have a high food value relative to volume (oats, wheat, maize, oilseed meal, etc. , are used to make feedingstuffs); **feed grains** = grains (such as oats, corn, barley, etc.) used to feed animals; **feed intake** = amount of food eaten by an animal; before a balanced diet can be drawn up it is important to know in advance the expected feed intake of the class of animal for which the diet is designed; **feed preparation** = the milling and crushing of grain, mixing of the ingredients and making into cubes or pellets; **feed ratio** = ratio showing the price of an animal sold on the market against the cost of feeding it; **feed wheat** = wheat which is used as an animal feed, and not for human consumption **2** *verb* to eat food; to give an animal *or* a person food to eat

◊ **feeder** ['fi:də] *noun* container from which livestock are fed; **feeder cattle** = cattle which are being fed to be slaughtered

◊ **feeding** ['fi:dɪŋ] *noun* action of giving animals food to eat; **feeding face** = the area allowed to each animal to feed from under controlled conditions; each cow needs 150mm of feeding face, or less if continuous access is provided (this refers to the self-feeding method of silage); **feeding value** = nutritional value of feedingstuffs

> QUOTE when he switched from self-feeding to forage box feeding of silage mixed with sugar beet pulp and molasses, and cut back on concentrates, he expected individual milk yields to drop
> *Farmers Weekly*

feedingstuffs ['fi:dɪŋstʌfs] *noun* various types of food available for farm animals

feedlot ['fi:dlɒt] *noun* area of land where livestock are kept at a high density, with small pens in which the animals are fattened; all feed is brought into the feedlot from outside sources

COMMENT: a new type of feedlot is an area of land surrounded by an earth embankment, which protects the cattle from cold winds while they are being fed intensively. See also CHILLSHELTER

feedmill ['fi:dmɪl] *noun* mill for preparing animal feed (small transportable mills are being constructed for use on individual farms)

◊ **feed off** ['fi:d 'ɒf] *noun* the practice of allowing stock to feed on a crop while it is still in the ground

◊ **feed ring** ['fi:d 'rɪŋ] *noun* a circular container for forage, from which livestock can feed

feedstuff ['fi:dstʌf] *see* FEEDING STUFFS

Feeke's large scale ['fi:ks lɑ:dʒ 'skeɪl] *noun* a method of determining the growth stage of a crop relying on comparing plant size and leaf arrangement when the plant is young; it is not a very reliable method and has been replaced by Zadoks scale

fell [fel] **1** *noun* high moor and mountain in the North of England **2** *verb* to cut down (a tree); **felling licence** = permission from the Forestry Commission to fell trees. There are some exceptions to this: for example trees in gardens and fruit trees in orchards do not need a licence to be felled; *see also* CLEARFELL

female ['fi:meɪl] *adjective & noun* animal which produces ova and bears young; (flower) which has carpels but not stamens

fen [fen] *noun* large area of flat marshy land, with reeds and mosses growing in alkaline water

◊ **fenland** ['fenlænd] *noun* area of land covered by fens; **fenland rotation** = a system of crop rotation developed on the Fens of East Anglia, using potatoes, sugar beet and wheat in rotation

fenbendazole [fen'bendəzəʊl] *noun* medicinal substance used to worm cattle

fence [fens] **1** *noun* barrier put round a field, either to mark the boundary or to prevent animals entering or leaving; **fence post** = wooden post which supports the wire of a fence **2** *verb* to put a fence round an area of land

COMMENT: various methods are used to fence field boundaries, most commonly woven wire and wooden fence posts. Movable electric fences are an efficient way of limiting areas of a field for grazing purposes

fennel ['fenəl] *noun* aromatic herb *(Foeniculum vulgare)* of Mediterranean origin, used to flavour fish dishes and soups

FEPA = FOOD AND ENVIRONMENTAL PROTECTION ACT

ferment [fə'ment] *verb* to produce alcohol and heat under the effect of yeast

◊ **fermentation** [fɜːmən'teɪʃn] *noun* process whereby carbohydrates are broken down by enzymes from yeast and produce heat and alcohol; in making silage, fermentation is essentially the breaking down of carbohydrates and proteins

fertile ['fɜːtaɪl] *adjective* (i) able to bear fruit; (ii) *(of animal)* able to produce young; (iii) *(of soil)* able to produce good crops

◊ **Fertile Crescent** ['fɜːtaɪl 'krezənt] *noun* an area of South Western Asia (modern Southern Turkey, Iran, Northern Iraq and Syria) with a Mediterranean climate and fertile soil, where man first settled and started to farm. This area was the centre of the domestication of wheat, barley, pulses and other crops which formed the basis of seed agriculture about 10, 000 years ago

◊ **fertility** [fɜː'tɪlɪti] *noun* (i) being fertile; (ii) measure of the ability of a female to conceive and produce young or of the male to fertilize the female; (iii) proportion of fertile eggs which develop into young (NOTE: the opposite is **sterile, sterility)**

fertilization [fɜːtəlaɪ'zeɪʃn] *noun* joining of an ovum and a sperm to form a zygote and so start the development of an embryo

◊ **fertilize** ['fɜːtəlaɪz] *verb* **(a)** to put fertilizer on land **(b)** *(of a sperm)* to join with an ovum **(c)** *(of a male)* to make a female pregnant

◊ **fertilizer** ['fɜːtəlaɪzə] *noun* chemical *or* natural substance spread and mixed with soil to make it richer and stimulate plant growth; **artificial fertilizers** *or* **chemical fertilizers** = fertilizers which are made from chemicals; **nitrogen fertilizers** = fertilizers containing mainly nitrogen; **compound fertilizer** *or* *US* **mixed fertilizer** = fertilizer

that supplies two or more nutrients (fertilizers that supply only one nutrient are called 'straights'); **liquid fertilizer** = simple solution, not kept under pressure, of the normal raw materials of solid fertilizers (as opposed to pressurized solutions, such as aqueous ammonia); *see also* BASIC SLAG

◊ **fertilizer distributor** *noun* the main machines used to spread fertilizers are (i) full width machines which apply high quantities of fertilizer uniformly and may be trailled or mounted; and (ii) broadcast machines which are used to apply material over considerable distances and have spinning discs using hopper containers. Oscillating spout machines and pneumatic machines where the fertilizer falls into an airstream created by a fan operated by the tractor are also in use

COMMENT: organic materials used as fertilizers include manure, slurry, rotted vegetable waste, bonemeal, fishmeal and seaweed. Inorganic fertilizers are also used, such as powdered lime or sulphur. In commercial agriculture, artificially prepared fertilizers (manufactured compounds containing nitrogen, potassium and other chemicals) are most often used, but excessive use of them can cause pollution. This happens when the chemicals are not taken up by plants and the excess is leached out of the soil into rivers and may cause algal bloom

QUOTE excessive amounts of potassium lead to poor fertility in dairy cows grazing intensively fertilized pastures. High levels of nitrate in grass are suspected of causing liver damage in cows
Ecologist

fescue ['feskjuː] *noun* common name for about 100 species of grasses, including many valuable pasture and fodder species

fetlock ['fetlɒk] *noun* the thicker back part of a horse's leg near the hoof (NOTE: the thin part between the fetlock and the hoof is called the **pastern)**

FFA = FREE FATTY ACID

FFB *or* **ffb** = FRESH FRUIT BUNCH bunch of fruitlets from the palm oil tree (each ffb is very large)

QUOTE harvesting and in-field collection of ffb constituted one of the major cost elements in the production of palm oil
The Planter

fibre *or US* **fiber** ['faɪbə] *noun* **(a)** organic structure shaped like a thread (in plants forming the structure of the stems, and in animals forming connective tissue) **(b)** hair (of rabbit, goat, etc.) **(c)** **dietary fibre** = constituent of animal feedingstuffs, including cellulose and lignin; important in livestock diet as an aid to digestion

◊ **fibrous** ['faɪbrəs] *adjective* made of a mass of fibres; **fibrous rooted plants** = plants with roots which are masses of tiny threads, with no major roots like tap roots

field [fiːld] *noun* **(a)** area of cultivated land, usually surrounded by a fence *or* wall *or* hedge, used for growing crops *or* for pasture; **field beans** = *Vicia faba,* used for stock feeding, or for producing broad beans, which are the immature seeds used for human consumption. Field beans are usually grown as a break crop; **field bindweed** = deep-rooted perennial weed (*Convulvulus arvensis*) which causes great problems when combining because of its mass of clinging growths; **field book** = annual record of field utilization and other operations, kept on a farm; **field botanist** = botanist who examines plants in their growing habitat; **field capacity** = the amount of water which is kept in the soil after free drainage has taken place. The amount of moisture will vary according to the texture and structure of the soil; *see also* MOISTURE-HOLDING CAPACITY; **field crop** = crop grown over a wide area, such as most agricultural crops and some market-garden crops; **field drainage** = building drains in or under fields to remove surplus water; **field observations** *or* **observations in the field** = observations taken in the open air, looking at organisms in their natural habitat (as opposed to laboratory observations); **field pansy** = widespread flower (*Viola arvensis*) increasingly found in winter crops, especially cereals; also called 'corn pansy', 'love-in-idleness' and 'cats' faces' **(b)** area of interest or study; *his field of study is pest control*

◊ **fieldman** ['fiːldmæn] *noun* a farmworker whose work is mainly in the fields

COMMENT: Some older types of field can still be seen in the UK:
Celtic fields, small rectangular fields, mainly on chalk uplands, surrounded by banks and walls, dating back to as early as the 6th century B. C.;

enclosure fields, rectangular fields with regular hedgerows, formed after the enclosures of the 16th to 18th centuries; *open fields,* the Saxons had heavier ploughs and used open fields with large furrows in strips, separated by banks of turf or walls of stone. In recent years the removal of many field boundaries in the interest of farm consolidation has led to an increase in the size of the average British field. Hedges have been removed to allow large farm machinery to be used more economically, and the loss of hedgerows has had a marked effect on the wildlife in the countryside

fig [fɪg] *noun* tree (*Ficus* spp) with soft sweet fruit with many small seeds, grown mainly in Mediterranean countries, where it is known as the 'poor man's food'. The fig was one of the earliest fruit trees cultivated by man

fighting bull ['faɪtɪŋ 'bʊl] *noun* breed of black cattle from Spain and Portugal, bred to be aggressive and used in bull fights

filbert ['fɪlbət] *noun* the Kentish cob, a commercially grown hazel-like nut (*Corylus maxima*)

fill-belly ['fɪlbeli] *noun* feed which fills the animal's stomach, without providing any useful nutrients

filly ['fɪli] *noun* young female horse; so called for a year or so, after which she becomes a 'mare'

fine grains ['faɪn 'greɪnz] *noun* high quality grains, such a wheat and rice, as opposed to the coarse grains (oats, maize, etc.)

◊ **fineness count** ['faɪnnəs 'kaʊnt] *noun* scale used to assess the fineness of wool fibres

◊ **fine wool** ['faɪn 'wʊl] *noun* wool of very good quality

fingers ['fɪŋgəz] *noun* pieces of metal which project from and are attached to the cutter bar of a binder, mower or harvester. The knives pass across the fingers, cutting the grass or cereal crop as they do so

◊ **finger millet** ['fɪŋgə 'mɪlɪt] *noun* important grain crop of East and Central Africa, South India and Sri Lanka. It can be stored for long periods and so is useful as a reserve grain in case of famine

finish ['fɪnɪʃ] **1** *noun* the furrow left at the

edge of a ploughed field **2** *verb* to feed cattle *or* sheep at a rate of growth which increases the ratio of muscle to bone, and increases the proportion of fatty tissue in the carcass to a level at which the animal is considered to be fit for slaughter (market prices include prices for 'finished lambs', 'finished pigs', etc.)

◊ **finishing** ['fɪnɪʃɪŋ] *noun* action of feeding cattle *or* sheep until they are ready for slaughter; **finishing ration** = feed given to animals to prepare them for slaughter; *a finishing ration includes silage, beet pulp and by-products such as outsize carrots*

Finncattle [fɪn'kætl] *noun* breed of dairy cattle derived from three Finnish breeds; the animals are medium-sized and brown

◊ **Finnish Ayrshire** ['fɪnɪʃ 'eəʃə] *noun* breed of cattle found in northern Finland, similar to the Ayrshire; they are mainly reared for milk

fir (tree) [fɜː] *noun* common evergreen softwood tree *(Labies)*

◊ **fircone** ['fɜː 'kəʊn] *noun* hard casing of the seeds of a fir tree

fire ['faɪə] *noun* substance in the state of burning, which can be seen in the form of flames or a glow; *forest fires create enormous amounts of atmospheric pollution;* **fire climax** = condition by which an ecosystem is maintained by fire

◊ **fireblight** ['faɪəblaɪt] *noun* disease of apples and pears, which is characterized by dead flowers and branches; it is caused by the bacterium *Erwinia amylovora*

◊ **firebreak** ['faɪəbreɪk] *noun (in a forest)* area where no trees are planted, so that a forest fire cannot pass across and spread to other parts of the forest

COMMENT: in shifting cultivation, the practice of clearing vegetation by burning is widespread. One of the simplest forms involves burning off thick and dry secondary vegetation. Immediately after burning, a crop like maize is planted and matures before the secondary vegetation has recovered. Where fire clearance methods are used, the ash acts as a fertilizer

QUOTE ecologists accept that forest fires are, over the long term, essential for vigorous tree growth. Burning clears out undergrowth that might choke saplings, removes old dead wood and recycles nutrients into the soil
Nature

firm [fɜːm] **1** *adjective* solid; *the soil is firm and not too crumbly* **2** *verb* to make cultivated soil solid, before sowing; *a roller is used to firm the soil*

first calf heifer [fɜːst 'kɑːf 'hefə] *noun* a heifer which has borne its first calf

fish [fɪʃ] *noun* cold-blooded aquatic vertebrate, some species of which are eaten for food (fish are high in protein, phosphorus, iodine and vitamins A and D. White fish have very little fat); **fish farming** = breeding edible fish in special pools for sale as food

◊ **fishmeal** ['fɪʃmiːl] *noun* dried fish reduced to a powder, used as an animal feed or as a fertilizer (a natural phosphate)

fixation [fɪk'seɪʃn] *noun* act of fixing something; **nitrogen fixation** = process by which nitrogen in air is converted by bacteria in certain plants into nitrogen compounds, and when the plants die the nitrogen is released into the soil and acts as a fertilizer; **nitrogen-fixing plants** = plants, such as lucerne or beans, which form an association with bacteria which convert nitrogen from the air into nitrogen compounds which pass into the soil

fixed costs ['fɪkst 'kɒsts] *noun* costs (such as rent) which do not increase with the quantity of a product produced

flail [fleɪl] **1** *noun* **(a)** a wooden hand tool used for beating grain to separate the seeds from the waste parts; **flail forage harvester** = type of forage harvester which uses a high-speed flail rotor; the cut crop passes through a vertical chute and is discharged into a trailer; **flail mower** = mowing machine with a high-speed rotor fitted with swinging flails which cut the grass and leave it bruised in a fluffy swath **(b)** type of hedgecutter, with rapidly turning cutting arms **2** *verb* to beat grain with a flail

flaked maize ['fleɪkt meɪz] *noun* animal feedingstuff made from maize which has been treated with steam, rolled and dried; it is highly digestible, rich in starch and often given to pigs

flat deck piggery [flæt 'dek 'pɪgəri] *noun* piggery used for rearing weaned piglets from between two and eight weeks of age; it has a mesh floor, self-feed hoppers and controlled heating and ventilation

flat rate feeding ['flæt reɪt 'fiːdɪŋ] *noun* a

system of feeding concentrates to dairy cows, involving few changes to the level of concentrate input; it lasts from calving to turnout

flatten ['flætən] *verb* to make something flat; to make plants lie flat on the ground; *the stems are flattened by a roller-crusher; harvesting is difficult after the entire crop has been flattened by rain*

flatworm ['flætwɜ:m] *noun* type of parasitic worm with a flat body, such as a fluke

QUOTE killer flatworms from New Zealand which devour earthworms are invading Britain, causing serious alarm among scientists who believe soil fertility will be damaged
Guardian

flax [flæks] *noun Linum usitatissimum,* the linseed plant, of which the leaf fibres are obtained after 'retting': the plant is left in water to rot the tissues. The fibres are used in making linen

◊ **flaxseed** ['flækssi:d] *noun* seed from the flax plant, crushed to produce linseed oil

flea [fli:] *noun* small insect which lives as a parasite on animals, sucking their blood and causing disease, such as plague

◊ **flea beetle** ['fli: 'bi:tl] *noun* a small dark beetle which causes damage to Brassica seedlings, especially during hot dry weather between April and mid-May

fleece [fli:s] **1** *noun* coat of wool covering a sheep or goat **2** *verb* to shear (a sheep), to cut the fleece off an animal

◊ **fleeced** [fli:st] *adjective* covered with a coat of wool

COMMENT: the fleece is both the wool growing on the animal and the wool which has been cut off, usually in one piece, when shearing

flies [flaɪz] *see* FLY

flights *or* **flight feathers** [flaɪts *or* 'flaɪt 'feðəz] *noun* the main feathers on a bird's wing, properly called the 'primaries'

◊ **flightless bird** ['flaɪtləs 'bɜːd] *noun* bird (such as the ostrich) which cannot fly

flint [flɪnt] *noun* hard stone of nearly pure silica, found in lumps in chalk soils

COMMENT: sharp-edged flints of various sizes found in some calcareous soils are very wearing on farm implements and tractor tyres; they can also cause damage to machinery if they are picked up during harvesting. Flints are very durable and are used to form a weather-resistant facing to farm buildings and walls

flixweed ['flɪkswi:d] *noun* a common annual weed *(Descurainia sophia)*

float [fləut] *noun* another name for a Dutch harrow

flocculation [flɒkjuˈleɪʃn] *noun* grouping of small particles of soil together to form larger ones; *the flocculation of particles is very important in making clay soils easy to work*

flock [flɒk] *noun* large group of birds *or* of sheep or goats; **flock book** = record of the pedigree of a particular breed of sheep, kept by the breed society; **flock mating** = mating system which uses several males to mate with the females of a flock (NOTE: the word 'flock' is used for sheep, goats, and domesticated birds such as chickens or geese. The word used for cattle is 'herd')

◊ **flockmaster** ['flɒkmɑːstə] *noun* farm worker in charge of a flock of sheep

flood [flʌd] **1** *noun* large amount of water covering land which is normally dry (caused by melting snow, heavy rain, high tides, storms, etc.); *after the rainstorm there were floods in the valleys; the spring floods have washed away most of the topsoil;* **flood alleviation** *or* **flood control measures** = ways of controlling rivers which are liable to flood; **flood damage** = damage caused by floodwater; **flood irrigation** *or* **basin irrigation** = irrigation using water brought down by a river in flood (the floodwaters are led off into specially prepared basins) **2** *verb* to cover dry land with a large amount of water

◊ **floodplain** ['flʌd 'pleɪn] *noun* wide flat part of the bottom of a valley which is usually covered with water when the river floods

◊ **floodwater** ['flʌdwɔːtə] *noun* water which floods land

flora ['flɔːrə] *noun* (i) wild plants of a certain region; (ii) book *or* list describing the plants of a region; *compare* FAUNA

◊ **floral** ['flɔːrəl] *adjective* referring to plants *or* flowers

◊ **floret** ['flɒrɪt] *noun* little flower which forms parts of a larger composite flower head

flour ['flaʊə] *noun* soft fine powder made from ground cereal grains, and used for making bread

◊ **floury** ['flaʊri] *adjective* soft and powdery, like flour; **floury potatoes** = varieties of potato which, when cooked, turn easily into flour

COMMENT: flour is made by grinding grain, and removing impurities. White flour is flour which has had all bran and germ removed: the best quality of white flour is patent flour, which is very fine. Self-raising flour is white flour with baking powder added. Brown flour is not refined, and still contains the wheat germ and parts of the bran: if it contains all the grain, it is called 'wholemeal' and if it contains most of the grain it is 'wheatmeal'. Some expensive flours are 'stone-ground', that is they are made in the traditional way with millstones

flourish ['flʌrɪʃ] *verb* to live well and spread; *the colony of rabbits flourished in the absence of any predators; the island has a flourishing plant community*

flower ['flaʊə] *noun* usually brightly-coloured reproductive part of a plant, with an external calyx, petals, stamens and pistils which bear pollen; **flowers of sulphur** = powdered sulphur, sulphur which is used to dust on plants to prevent mildew

fluke [fluːk] *noun* parasitic flatworm which settles inside the liver (liver flukes), in the blood stream (Schistosoma) and other parts of the body; a serious disease of sheep and cattle

fluoride ['flʊəraɪd] *noun* any chemical compound of fluorine (usually found with sodium *or* potassium *or* tin)

◊ **fluorine** ['flʊəriːn] *noun* chemical element (a yellowish gas) (NOTE: chemical symbol is **F**; atomic number is **9**)

◊ **fluorosis** [fluːəˈrəʊsɪs] *noun* condition caused by excessive fluoride in drinking water or in eaten vegetable matter (it causes discoloration of the teeth and affects the milk yields of cattle)

COMMENT: fluorides such as hydrogen fluoride are emitted as pollutants from certain industrial processes and can affect plants (especially citrus fruit) by reducing chlorophyll. They also affect cattle by reducing milk yields

flush [flʌʃ] *noun* rapid growth of grass

◊ **flushing ewes** ['flʌʃɪŋ 'juːz] *noun* ewes brought into good condition prior to breeding, usually by improving their diet

fly [flaɪ] *noun* general term for a small insect with two wings, of the order *Diptera*. Some flies cause diseases of plants (the frit fly) and some harm animals (the gadfly); **fly blown** = fleece laden with maggot-fly eggs

◊ **flying flock** ['flaɪɪŋ 'flɒk] *noun* a flock of sheep imported onto a farm for a time, normally for less than a year, and then sold

FMA = FERTILIZER MANUFACTURERS ASSOCIATION

FMBA = FLOUR MILLING AND BAKING ASSOCIATION

foal [fəʊl] **1** *noun* a young horse (of either sex) in its first year; **mare in-foal** = pregnant mare **2** *verb* (*of a mare*) to produce an offspring

◊ **foaling** ['fəʊlɪŋ] *noun* parturition in a mare (giving birth to a foal)

fodder ['fɒdə] *noun* general term for food given to livestock, in particular straw and hay made from dried plants such as grass, clover; **fodder beet** = a root crop (bred from sugar beet and mangolds) usually grown after cereals and used to feed stock; **fodder crop** = a crop, such as kale, lucerne or hay, grown for use as animal feed; *see also* FORAGE CROP; **fodder radish** = brassica grown primarily for use as a green fodder crop; **fodder storage** = the storing of fodder for use in winter; buildings used for storing fodder may be of the simple Dutch barn type, and can be built cheaply using poles and a galvanized iron roof

FoE = FRIENDS OF THE EARTH

foehn [fɜːn] *noun* warm dry wind which blows down the lee side of a mountain (it is caused when moist air rises up the mountain on the windward side, loses its moisture as precipitation and then goes down the other side as a dry wind); *similar to* CHINOOK

foetus *or US* **fetus** ['fiːtəs] *noun* embryo

of an animal developing in the mother's womb

◊ **foetal** *or* *US* **fetal** ['fiːtəl] *adjective* referring to a foetus

foggage ['fɒgɪdʒ] *noun* (a) winter grazing of cattle on non-ryegrass swards (b) grass left standing to provide winter grazing for sheep and cattle

foggara [fɒ'gɑːrə] *noun* slightly sloping underground channels, used to bring irrigation water from aquifers near the foot of mountains to neighbouring plains, especially in the Sahara

foggia ['fɒdʒiə] *noun* pits or cellars used for grain storage or as cisterns for drinking water in south-eastern Italy

fold [fəʊld] **1** *noun* moveable enclosure, made of hurdles or of electric wire fencing, used to keep cattle or sheep in a certain place **2** *verb* to put sheep into a fold; **folded sheep** = sheep kept in movable folds, as a means of controlling their grazing

◊ **foldland** ['fəʊldlənd] *noun* formerly, an area of land allotted to each manor for the purpose of grazing the manor's sheep

foliage ['fəʊliɪdʒ] *noun* leaves (on plants); *in a forest, much of the rainfall is lost through evaporation from foliage surfaces*

◊ **foliar** ['fəʊliə] *adjective* referring to leaves; **foliar feed** *or* **foliar spray** = liquid nutrient used by farmers and gardeners to spray onto the leaves of plants which then absorb it

folic acid ['fɒlɪk 'æsɪd] *noun* vitamin of the B complex

follicle ['fɒlɪkl] *noun* tiny hole in the skin from which a hair grows; in birds, feathers grow from follicles

followers ['fɒləʊəz] *noun* (a) young cows in a dairy herd which are not yet in milk, and are being reared to replace the older stock (b) cattle put to graze a pasture after another group of animals has used it; *see also* LEADER

following crop ['fɒləʊɪŋ 'krɒp] *noun* crop sown by a tenant farmer before leaving the farm at the end of his tenancy; he is permitted to return and harvest the crop and remove it (NOTE: also known as **away-going crop** or **off-going crop**)

food [fuːd] *noun* things which are eaten (NOTE: for materials provided to be eaten by animals, use the word **feed**) **food balance** = the balance between food supplies and the demand for food from the population; **food chain** = series of organisms which pass energy from one to another as each provides food for the next (the first organism in the chain is a producer, all the rest are consumers); **food web** = series of food chains which are linked together in an ecosystem

COMMENT: two basic kinds of food chain exist: the grazing food chain and the detrital food chain in which both plant-eaters and detritus-eaters take part. In practice, food chains are interconnected, making up food webs

Food and Agriculture Organization (FAO) international organization based in Rome; an agency of the United Nations, it was established with the purpose of encouraging higher standards of nutrition and eradicating malnutrition and hunger

Food and Environmental Protection Act, 1986 legislation which brings the use of agrochemicals under statutory control (as opposed to the previous voluntary arrangement)

food poisoning ['fuːd pɔɪzənɪŋ] *noun* illness caused by eating food which is contaminated with bacterial or chemical agents, or by eating food which is itself poisonous (such as forms of fungi). The commonest form of food poisoning is probably that caused by the Salmonella group of pathogens, found in meat and animal products

foodstuffs ['fuːdstʌfs] *noun* different types of food (NOTE: when referring to animals, it is usual to use the word **feedingstuff**)

fool's parsley ['fuːlz 'pɑːsli] *noun* species of hemlock (*Aethusa cynapium*) which looks like parsley; it is not a common source of poisoning in animals

foot and mouth disease [fʊtn 'maʊθ dɪ'ziːz] *noun* a contagious viral disease of cattle, sheep, goats and pigs. The disease is not characterized by a high mortality level. Most of the affected animals simply lose weight and milk yields from affected dairy cattle decline. Diseased animals are slaughtered in the UK; the disease is notifiable

footbath ['futbɑːθ] *noun* (i) trough containing disinfectant through which sheep or cattle are driven to prevent or cure various diseases such as foot rot; (ii) also, a shallow container containing disinfectant in which a person walks to disinfect shoes or boots

footpath ['futpɑːθ] *noun* way along which people can walk on foot; *long-distance footpaths have been laid out through the mountain regions*

foot rot ['fut 'rɒt] *noun* a disease of the horny parts and the soft tissue of feet of sheep. It occurs particularly in wet marshy and badly-drained pastures, and is caused primarily by the organism *Fusiformis necrophorus* and sometimes *Fusiformis nodosus*. It makes sheep lame

forage ['fɒrɪdʒ] **1** *noun* crops grown for consumption by livestock; **forage box** = large movable container used mainly to transport forage from a silo to a trough; **forage feeding** = practice of cutting herbage from a sward or foliage from other crops for feeding fresh to animals; **forage maize** = maize grown for ensilage; **forage wagon** = mobile container with a pick-up attachment used for collecting and carrying cut forage **2** *verb* to look for food

◊ **forage harvester** *noun* a machine which cuts, chops and loads green crops such as lucerne into a trailer, to make silage. There are three main groups: (i) the single chop machines which use a flail to cut the crop and produce a chop length of 150mm and above. Two different types of this machine are in use: the in-line, directly behind the tractor, and the off-set which allows a field to be cut round and round. (ii) double chop machines cut the crop twice. (iii) precision chop machines are used for short cut material which gives better clamp fermentation

COMMENT: forage crops are highly digestible and palatable, and are either very quick growing or very high-yielding. They have the advantage that, whether grazed or harvested and stored, they provide feed at times when grass growth is poor. There are a number of different types of forage harvester: trailed harvesters can be power take-off or engine-driven. Self-propelled machines are becoming more

widely used. After the crop is cut, it is either chopped or lacerated and loaded into a trailer

foreleg ['fɔːleg] *noun* one of the two front legs of an animal (as opposed to the hind legs)

forest ['fɒrɪst] **1** *noun* **(a)** natural group of plants and animals, of which the dominant organisms are trees (a large area is a forest, a small area is a wood); **high forest** = forest made up of tall trees which block the light to the forest floor; **mixed forest** = forest with more than one species of tree; **the forest floor** = ground at the base of the trees in a forest **(b)** another name for what is really a plantation (trees and shrubs planted in rows, usually for commercial exploitation); *see also* AFFORESTATION **(c)** area in which a king has the right to keep deer for hunting **(d)** common land within a forest where deer live **2** *verb* to manage a forest, by cutting wood as necessary, and planting new trees

◊ **forested** ['fɒrɪstɪd] *adjective* (land) covered with forest

◊ **forester** ['fɒrɪstə] *noun* person who manages woods and plantations of trees

◊ **forestry** ['fɒrɪstri] *noun* management of forests, woodlands and plantations of trees

◊ **Forestry Commission** government agency responsible for the management of state-owned forests in the UK (90% of the state-owned forests in the UK are coniferous). A civil service review team was set up in April 1993 to consider the privatization of the 2 million acres owned by the government Forestry Commission. 410,000 acres have been sold since 1991.

QUOTE one of the most vital functions fulfilled by forests is the control of water runoff to rivers. In a well-forested watershed, 95% of the annual rainfall is trapped in the network of roots that underlies the forest floor. This water is then released slowly over the year, keeping streams and rivers flowing during the dry season. When the forest is removed, the rains rush down the denuded slopes, straight into the local streams and rivers, only 5% of the rainwater being absorbed into the soil
Ecologist

forge [fɔːdʒ] *noun* blacksmith's workshop

forget-me-not [fəˈgetmɪnɒt] *noun*

widespread weed *(Myosotis arvensis)* found in all soils, especially near woodland

fork [fɔːk] **1** *noun* common hand implement for turning over soil, and lifting out weeds. Forks have four square prongs each sharpened to a point. Forks with flat prongs are used for lifting potatoes; larger, five-pronged forks with round curved prongs, are used for spreading manure **2** *verb* to dig ground with a fork; **to fork a bed over** = to dig a whole bed using a fork; **to fork manure in** = to spread manure with a fork, turning the soil over to cover it

formula ['fɔːmjʊlə] *see* DENTAL FORMULA

formulation [fɔːmjuˈleɪʃn] *noun* the form in which a pesticide is sold for use; most pesticides are not soluble in water and have to be formulated so they can be mixed and applied as liquids; they are supplied as emulsions, wettable powders, etc. , which can be mixed with water

fortified ['fɔːtɪfaɪd] *adjective* with something added to make stronger; **fortified food** = food with vitamins or proteins added to make it more nutritional; **fortified wine** = wine (such as sherry or port) with extra alcohol added

forward ['fɔːwəd] *adjective (of crops)* earlier than usual *or* too early

◊ **forward creep grazing** ['fɔːwəd 'kriːp 'greɪzɪŋ] *noun* grazing method where grassland which has been allocated to ewes and lambs during the fattening period, is divided into paddocks, each separated by portable fencing; as one area is finished, the fencing is moved to allow the animals to move to the next

foul of the foot ['faʊl əv ðə 'fʊt] *noun* disease of cattle caused by damage to the cleft of the hoof and invasion by a germ *(Fusiformis necrophorus)*

four tooth sheep [fɔː tuːθ 'ʃiːp] *noun* sheep which is 18-21 months of age

Fourth World ['fɔːθ 'wɜːld] *noun* the poorest countries in the non-aligned Third World

four-wheel drive vehicle ['fɔː wiːl 'draɪv] *noun* vehicle in which the power is transmitted to all four wheels, as opposed to only one pair of wheels as it usual in cars;

four-wheel drive is very necessary in farm vehicles as it gives them greater power on difficult terrain

fowl [faʊl] *noun* the chicken, a bird *(Gallus domesticus)* raised for food; **fowl pest** = Newcastle disease, an acute contagious disease of fowls; affected birds suffer loss of appetite, diarrhoea and respiratory problems; mortality rates are high; **fowl pox** = a viral disease in which wart-like nodules appear on the comb, wattles, eyelids and openings of the nostrils of fowls

fox [fɒks] *noun* a carnivorous canine predator *(Vulpes vulpes)* with red fur and a large bushy tail

◊ **foxglove** ['fɒksglʌv] *noun* common weed *(Digitalis purpurea).* The plant is poisonous and harmful to animals

◊ **foxtail millet** ['fɒksteɪl 'mɪlɪt] *noun* Setaria italica, the first cereal to be cultivated in China; used for silage, hay, brewing and flour in many parts of the world; in Britain it is used as birdseed

fragmentation [frægmənˈteɪʃn] *noun* **farm fragmentation** = situation where the fields of a farm are scattered over an area, so that the holding is not made up of a single unit of land; it can be the result of inheritance practices where the land of a person who has died is split between all the children, of land reclamation schemes where the land is reclaimed piece by piece, or because old open-field systems have been kept

frame [freɪm] *noun* **(a) (cold) frame** = box construction, with a glass lid, used for raising or keeping plants out of doors but with a certain amount of protection against frost **(b)** main part of a plough, to which the ploughshare and mouldboard are attached

Fraxinus ['fræksɪnəs] *Latin name for* the ash tree

free [friː] *adjective* not attached *or* not controlled; **free-living animal** = animal which exists in its environment without being a parasite on another; **free-running sleeve** = loose sleeve fitted over shafts to stop clothing becoming entangled by riding on the shaft if contact is made, for example, on manure spreader beater drive shafts

freehold ['friːhəʊld] *noun* the absolute right to hold land or property for an unlimited time without paying rent; **freehold property** = property which the owner holds in freehold

◊ **freeholder** ['fri:həʊldə] *noun* person who holds a freehold property; *compare* LEASEHOLD, LEASEHOLDER

freerange eggs ['fri:reɪndʒ 'egz] *noun* eggs from hens that are allowed to run about in the open and eat more natural food (as opposed to battery hens)

freestone ['fri:stəʊn] *noun* varieties of peach, where the flesh does not cling to the stone (as opposed to clingstone varieties)

freeze drying [fri:z 'draɪɪŋ] *noun* method of preserving food by freezing rapidly and drying in a vacuum

FREGG = FREE RANGE EGG ASSOCIATION

French bean ['frenʃ 'bi:n] *noun* a common green vegetable *(Phaseolus vulgaris)* ; the beans grow on dwarf bush plants, and are grown for sale fresh or for processing as canned, frozen or dried vegetables. Some are harvested as a dried seed crop, for sale dried as haricot beans, or for processing (as for example into baked beans)

fresh [freʃ] *adjective* not processed (i. e. not canned *or* frozen); *fresh vegetables are more expensive in winter, because they have to be imported; beans are grown both for the fresh market and also for preserving;* **fresh fruit bunch (FFB)** = very large bunch of fruitlets of the oil palm tree

QUOTE poor drainage and frequent flooding hinders the harvesting and collection of FFB, resulting in lower yields
The Planter

friable ['fraɪəbl] *adjective* (soil) which is light *or* which crumbles easily into pieces

Friends of the Earth (FoE) *noun* pressure group formed to influence local and central governments on environmental matters

Friesian ['fri:zɪən] *noun* a breed of black and white dairy cattle, famous for its very high milk yields

COMMENT: there are three main types of Friesian recognized today: the Dutch Friesian, the British Friesian and the Holstein-Friesian, which is the North American type. The Friesian is the most important breed in British dairy herds

Friesland ['fri:zlənd] *noun* breed of sheep whose milk is used in the production of soft cheese and yoghurt

frit fly ['frɪt 'flaɪ] *noun* a small black fly *(Oscinella frit)* that attacks wheat, maize and oats

◊ **frits** [frɪts] *noun* trace elements fused with silica to form glass; this is crushed into small pieces for distribution on the soil

frog [frɒg] *noun* part of a plough to which the mouldboard and share are attached

frond [frɒnd] *noun* leaves of the fern group of plants *(Pteridophyta)*

frost [frɒst] *noun* freezing weather when the temperature is below the freezing point of water (it may lead to a deposit of crystals of ice on surfaces); *there was a frost last night;* **air frost** = condition where the air temperature is below 0°C, but not at ground level; **ground frost** = condition where the air temperature at ground level is below 0°C; **frost pocket** *or* **frost hollow** = low-lying area where cold air collects; crops which are susceptible to frost should not be planted in such areas; **frost-free region** = region where there are no frosts; **frost-hardy plant** = plant which can survive frost; **frost-tender plant** = plant which is killed by frost

fructose ['frʌktəʊz] *noun* fruit sugar, found in honey as well as in fruit; **high fructose corn syrup (HFCS)** = sweetener used in the soft drinks industry, extracted from maize (NOTE: also called **isoglucose**)

fruit [fru:t] **1** *noun* (i) ripe ovary of a plant and its contents of seeds; (ii) in general usage, the fleshy material round the fruit which is eaten as food; *a diet of fresh fruit and vegetables;* **fruit fly** = a fly which attacks fruit **2** *verb (of a tree)* to have fruit; *some varieties of apple fruit very early; the drought has damaged fruiting trees;* the **fruiting season** = time of year when a particular tree has fruit

◊ **fruitwaste** ['fru:tweɪst] *noun* residue left after juice has been extracted from fruit, used as an animal feed

◊ **fruitwood** ['fruːtwʊd] *noun* wood from a fruit tree (such as apple *or* cherry) which may be used to make furniture

COMMENT: fruit contains fructose which is a good source of vitamin C and some dietary fibre. Dried fruit have a higher sugar content but less vitamin C than fresh fruit

fuel ['fjuːəl] *noun* substance (such as wood *or* coal *or* gas *or* oil) which can be burnt to provide heat *or* power

◊ **fuelwood** ['fjuːlwʊd] *noun* wood which is grown to be used as fuel

QUOTE over 200,000 woodstoves have been sold in the UK, and the fuelwood market is estimated at 250, 000 tonnes per annum
Environment Now

QUOTE fuelwood and charcoal remain the principal sources of energy in most developing countries
Forestry Chronicle

fullering ['fʊlərɪŋ] *noun* making a groove on the lower surface of a horse's shoes, into which the heads of the nails will fit

full-mouthed [fʊl'maʊðd] *adjective* (animal) which has a complete set of permanent teeth

full-time farmer [fʊl'taɪm 'faːmə] *noun* farmer who derives his living from agriculture, as distinct from a part-time farmer; **full-time worker** = farmworker engaged full-time in work on a farm

fumigate ['fjuːmɪgeɪt] *verb* to kill germs *or* insects by using fumes

◊ **fumigation** [fjuːmɪ'geɪʃn] *noun* disinfection by means of gas or fumes which penetrate into cracks and holes, probably more efficient that spraying or scrubbing

◊ **fumigant** ['fjuːmɪgənt] *noun* chemical compound which is heated and becomes volatile, used to kill insects

fumitory ['fjuːmɪtəri] *noun* common weed *(Fumaria officinalis)* affecting cereal and clover crops

fungus ['fʌngəs] *noun* simple plant organism with thread-like cells (such as yeast, mushrooms, mould) and without green chlorophyll; **fungus disease** = disease caused by a fungus; **fungus poisoning** = poisoning by eating a poisonous fungus (NOTE: plural is **fungi** ['fʌngiː]. For other

terms referring to fungi, see words beginning with **myc-**)

◊ **fungal** ['fʌngəl] *adjective* referring to fungi

◊ **fungicidal** [fʌngɪ'saɪdl] *adjective* (paint) which kills fungi

◊ **fungicide** ['fʌngɪsaɪd] *adjective & noun* (chemical) used to kill fungi or restrict their growth

◊ **fungoid** ['fʌngɔɪd] *adjective* like a fungus

COMMENT: some fungi can become parasites of animals and cause diseases such as aspergillosis. Other fungi cause plant diseases, such as blight. Others, such as yeast, react with sugar to form alcohol. Fungicides are available as sprays or dusts for use on crops

funicle ['fjuːnɪkl] *noun* short stalk attaching a seed to the inside of the pod

fur [fɜː] *noun* **(a)** coat of hair, covering an animal; *the rabbit has a thick coat of winter fur* **(b)** skin and hair of an animal, used to make clothes

furlong ['fɜːlɒŋ] *noun* one eighth of a mile, or 220 yards. Originally a furlong was the length of a furrow in the common field

furrow ['fʌrəʊ] *noun* long trench and ridge cut in the soil by the mouldboard of a plough; **furrow press** = special type of very heavy ring roller attached to the plough, used to press the furrow slices

furze [fɜːz] *noun* common shrub *(Ulex europoeus),* found in wasteland; formerly often cut and used as fodder after chaffing. It contains a small amount of a poisonous alkaloid called ulexine, which is seldom present in dangerous quantities

futures ['fjuːtʃəz] *noun* stocks of produce which are bought or sold for shipping at some later date, and which may not even have been produced when they are on the market (as opposed to 'actuals', which are stocks of real produce, available for delivery now)

QUOTE Dutch traders point out that current London futures prices for April (about £190/t recently) would seem to be on the high side
Farmers Weekly

FUW = FARMERS' UNION OF WALES

FWAG = FARMING AND WILDLIFE ADVISORY GROUP a voluntary organization, established in 1969 to encourage understanding between farmers and conservationists

FWT = FARMING AND WILDLIFE TRUST a trust formed to fund the FWAG and also to provide advisors in various parts of the country

fym *or* **FYM** = FARMYARD MANURE

Gg

gadfly ['gædflaɪ] *noun* a fly that bites cattle, one of the genera *Tabanus* (the horsefly) or *Oestrus* (the bot fly), most common from late May onwards, and causing considerable trouble to cattle

gage [geɪdʒ] *noun* variety of plum, especially the greengage

Galician blond [gə'lɪsɪən 'blɒnd] *noun* breed of cattle from northern Spain; it is a triple-purpose breed, red in colour, with yellow horns

gall [gɔːl] *noun* hard growth on a plant caused by a parasitic insect

Galla ['gælə] *noun* breed of goat found in Kenya and Somalia, raised both for meat and for their skins; the animals are white

gallon ['gælən] *noun* **(a)** a measure of capacity, equivalent to eight pints or 4.55 litres; used both for liquids and for measuring dry goods, such as grain (NOTE: this is also called the 'imperial gallon') **(b)** *US* measure of capacity, equal to 3.78 litres; used only for liquids

Galloway ['gæləweɪ] *noun* a hardy breed of completely black hornless cattle. The coat is distinctive, being formed of long wavy hairs covering a soft undercoat. The Galloway originated in the wet and hilly south-west of Scotland. It is reared mainly for beef

Gallus ['gæləs] *Latin name for* the domestic chicken

galvanized iron ['gælvənaɪzd 'aɪən] *noun* iron which has been coated with zinc to prevent it rusting; galvanized iron sheeting is widely used for roofs

Galway ['gɔːlweɪ] *noun* breed of sheep found in the Irish Republic. The white-faced Galway is the only native Irish breed and is used to produce store lambs

game [geɪm] *noun* animals which are killed for sport (and food), and listed in the UK under the Game Act which defines the shooting seasons; **game birds** = wild birds which are classified as game, and which can be shot only during certain seasons (the most important in the UK are pheasant, partridge and grouse)

◊ **the Game Conservancy** [geɪm kən'sɜːvənsi] an organization concerned with the conservation of game species; it advises on shoots and woodland management. The long-term aim of the Conservancy is to increase the production of wild game by careful management of habitats. It claims that certain game birds such as the black grouse, the grey partridge and the capercaillie are at risk

◊ **gamekeeper** ['geɪmkiːpə] *noun* person working on a private estate who protects wild birds and animals bred to be hunted

COMMENT: game, such as pheasants and partridges, is an important asset on some farms, and letting land for sport shooting is a source of high income

gamete ['gæmiːt] *noun* reproductive cell of an animal or plant, a germ cell which can develop into a spermatozoon or ovum

gamma rays ['gæmə 'reɪz] *noun* rays which are shorter than X-rays, given off by radioactive substances and used in food irradiation

gander ['gændə] *noun* male goose

gang [gæŋ] *noun* group of workers working together (such as a gang of sheep shearers)

Ganga tiri ['gæŋgə 'tɪri] *noun* breed of cow found in India

gangrene ['gæŋgriːn] *noun* serious rot affecting potato tubers; caused by fungi, it spreads in storage

◊ **gangrenous mastitis** ['gæŋgrɪnəs mæ'staɪtɪs] *noun* form of the mastitis disease affecting cattle. It may begin as staphylococcal mastitis. The udder becomes blue and cold

gantry ['gæntri] *noun* new type of farm machine consisting of a long steel beam with implement carriers. The engine and cab are at one end of the beam, and the drive wheel is at the other end; currently, gantries are 12m wide

> QUOTE changing over to a field gantry system won't mean changing over to prairie-like fields and selling all the tractors. But it could bring substantial savings, particularly in cultivation costs
> *Farmers Weekly*

gapes [geɪps] *noun* disease affecting the breathing function of poultry, caused by small worms in the windpipe

garbanzo [gɑːˈbænzəʊ] *noun* variety of chickpea grown in South America

garden ['gɑːdən] *noun* **(a)** land cultivated as a hobby *or* for pleasure, rather than to produce an income; **flower garden** = garden where only flowers are grown; **kitchen garden** = garden with herbs and small vegetables, ready for use in the kitchen **(b)** **market garden** = farm near a town, providing fresh vegetables and salad crops for sale in the town's market

◊ **gardener** [gɑːdnə] *noun* person who looks after a garden

◊ **gardening** ['gɑːdnɪŋ] *noun* horticulture *or* looking after a garden; **market gardening** = growing fresh vegetables and salad crops for sale in a nearby town

gari ['gæri] *noun* coarse meal made from fermented cassava, an important food in West Africa; *compare* FARINHA

garlic ['gɑːlɪk] *noun* plant *(Allium sativum)* with a strong-smelling pungent root used as a flavouring in cooking. The bulb consists of a series of wedge-shaped cloves, surrounded by a white fibrous skin

garrigue [gəˈriːg] *noun* dense undergrowth of aromatic shrubs found in Mediterranean regions accompanying evergreen and cork oak

Gasconne ['gæskɒn] *noun* breed of beef cattle from the Gascony area of south-west France; the animals are silver-grey in colour with medium length horns

gastr- *or* **gastro-** [gæstr *or* 'gæstrəʊ] *prefix* referring to the stomach

◊ **gastric** ['gæstrɪk] *adjective* referring to the stomach; **gastric acid** = hydrochloric acid secreted into the stomach by acid-forming cells; **gastric juices** = mixture of hydrochloric acid, pepsin, intrinsic factor and mucus secreted by the cells of the lining membrane of the stomach to help the digestion of food

◊ **gastro-enteritis** [gæstrəʊentəˈraɪtɪs] *noun* viral infection which causes inflammation of the membranes lining the stomach and intestines; **parasitic gastro-enteritis (PGE)** = infection of the stomach caused by roundworms, especially *Osteragia;* **transmissible gastro-enteritis (TGE)** = very infectious form of the disease, which affects young pigs

gatherers ['gæðərəz] *noun* lowest order of economic activity in which people collect food and materials (practised by a few primitive groups only)

GATT = GENERAL AGREEMENT ON TARIFFS AND TRADE international organization aiming to reduce restrictions on trade between countries. Replaced in 1995 by the World Trade Organization (WTO)

gaur [gauə] *noun* the Indian bison; a large wild ox which is dark brown in colour with long curved horns

Gayal ['gaɪəl] *noun* ox similar to the gaur, but domesticated in north-west India to produce meat and hides

GCT = GAME CONSERVANCY TRUST

geese [giːs] *see* GOOSE

geest [geɪst] *noun* infertile sandy lowland region of North and East Germany, covered with heath

gelatin ['dʒelətɪn] *noun* protein which is soluble in water, and which is extracted from animal bones and horns; it is used in laboratories as a culture medium

geld [geld] *verb* to castrate (especially a horse)

◊ **gelding** ['geldɪŋ] *noun* a castrated horse

Gelbvieh ['gelbviː] *noun* breed of dairy cattle from Bavaria in south Germany; the colour varies from cream to yellow (NOTE: also called **German Yellow)**

gene [dʒiːn] *noun* unit of DNA on a chromosome which governs the synthesis of one protein, usually an enzyme, and determines a particular characteristic of an organism; **gene bank** = collection of seeds from wild plants, which may be found to be useful in the future for breeding new varieties; **gene pool** = all the best quality animals and plants which can be used for further breeding; *see* GENETIC

COMMENT: genes are either dominant, when the characteristic is always passed on to the offspring, or recessive, when the characteristic only appears if both parents have contributed a recessive gene

genera ['dʒenərə] *see* GENUS

generic [dʒə'nerɪk] *adjective* referring to a genus; **generic name** = the name of a genus

COMMENT: organisms are usually identified by using their generic and specific names, e. g. *Homo sapiens* (man) and *Felix catus* (domestic cat). The generic name is written or printed with a capital letter. Both names are usually given in italics, or are underlined if written or typed

genetic [dʒə'netɪk] *adjective* referring to the genes; *breeders of new crop plants depend on genetic materials from wild forms of maize and wheat;* **genetic code** = information which determines the synthesis of a cell, is held in the DNA of a cell and is passed on when the cell divides; **genetic engineering** = techniques used to change the genetic composition of an organism so that certain characteristics can be created artificially; **genetic improvement** = the improvement of an animal or plant by breeding

◊ **genetics** [dʒə'netɪks] *noun* study of the way the characteristics of an organism are inherited through the genes

COMMENT: comparisons of today's farm animals with those of the past show considerable differences in appearance and productivity. Today's dairy cattle have no horns, and produce two or three times as much milk as their ancestors in the 19th century. This is in part due to genetic improvement of livestock by selection of superior animals for breeding

QUOTE the emergent field of genetic engineering by which science devises new variations of life forms, does not render wild genes useless. This new science must be based on existing genetic material and makes such material more valuable

Brundtland Report

QUOTE Scientists will be able to transfer from a species a gene for crop disease resistance, or for a toxin which can destroy a pest, or one which confers immunity to illness

Guardian

QUOTE Genetic engineering offers the agrochemical industry prospects not seen since the pesticides boom of the 1950s. Scientists are on the brink of rapid growth in the development and use of designer insects and pest-resistant plants

Report of the Royal Commission on Environmental Pollution (1989)

genotype ['dʒenətaɪp] *noun* genetic composition of an organism

Gentile di Puglia [dʒen'tiːleɪ dɪ 'puːljə] *noun* breed of Italian sheep found in the Foggia region. A fine-wool merino breed used in a transhumance system

genus ['dʒiːnəs] *noun* **(a)** the second major rank in classification of living things, a group of closely-related species, the members of which are more closely related to each other than they are to members of other genera (several genera form a family) (NOTE: plural is **genera) (b)** name of the former Milk Marketing Board farm services company which is separate from Milk Marque

Gerber test ['dʒɜːbə] *noun* a test to determine the butterfat content of milk

germ [dʒɜːm] *noun* **(a)** part of an organism which develops into a new organism; **germ cell** = gamete *or* cell which is capable of developing into a spermatozoon or ovum; **germ plasm** = the protoplasm of a germ cell, containing the genes **(b)** central part of a seed, formed of the embryo; it contains valuable nutrients; *see also* WHEATGERM **(c)** microbe (such as a virus *or* bacterium) which causes a disease; *germs cannot be seen by the naked eye*

◊ **germicide** ['dʒɜːmɪsaɪd] *noun* substance used to kill germs

German Red Pied ['dʒɜːmən red 'paɪd] *noun* breed of cattle from north-west Germany; mainly raised for meat, the animals are red and white in colour

◊ **German Yellow** ['dʒɜːmən 'jeləʊ] *noun* breed of dairy cattle from Bavaria in south Germany; the colour varies from cream to yellow (NOTE: also called **Gelbvieh**)

germinate ['dʒɜːmɪneɪt] *verb (of a plant seed or spore)* to start to grow

◊ **germination** [dʒɜːmɪ'neɪʃn] *noun* beginning of the growth of a seed, resulting from moisture and a high enough temperature; **germination percentage** = the number of seeds which germinate, taken from a representative sample of 100 seeds

gestation [dʒes'teɪʃn] *noun* period when a female animal has living young in her uterus; **gestation period** = the time between conception and birth. This varies with different animals: for cows it is 283 days, for mares 340 days, for ewes and goats 144 - 150 days and for sows 114 days

get-away cage ['getəweɪ 'keɪdʒ] *noun* a poultry cage which is designed to provide more space for each bird. It separates the roosting area from the laying area and provides means of feeding and drinking

Gezira [gə'zaɪrə] *noun* large-scale irrigation and farming area in the Sudan; with over 100,000 tenant farmers producing cotton, sorghum, groundnuts, wheat, fodder crops and vegetables

ghee [giː] *noun* butter made from buffalo milk, clarified to look like oil. It is used widely in the Indian subcontinent as a cooking fat. Vegetable ghee can be made from groundnut and palm oil

gherkin ['gɜːkɪn] *noun* small cucumber grown for pickling

gibberellin [dʒɪbə'relɪn] *noun* plant growth hormone which helps flower formation and accelerates germination

gid [gɪd] *noun* brain disease of young sheep which also occurs in cattle. Caused by ingestion of tapeworm eggs voided by dogs and foxes. Blindness is an early symptom

gilt [gɪlt] *noun* a young female pig

gimmer ['gɪmə] *noun* female sheep after its first shearing

QUOTE all ewe lambs and gimmers should be vaccinated against clostridial disease
Farmers Weekly

ginger ['dʒɪndʒə] *noun* spice plant *(Zinziber officinale)* grown in many parts of the tropics. It grows from one to three feet high and consists of an enlarged underground stem or rhizome, with erect stems with narrow leaves above the soil. The stem is used in oriental cooking, and powdered ginger prepared from the dried rhizome is used in particular in confectionery

gir [gɜːr] *noun* Indian breed of cattle, raised for milk. The animals are red or grey, with horns and a pronounced hump

gizzard ['gɪzəd] *noun* the muscular stomach of a bird, in which food is ground up, helped by the grit and small stones which collect there

glanders ['glændəz] *noun* serious contagious disease of horses, no longer present in Britain, but still found in Asia and Africa

glasshouse ['glɑːshaʊs] *noun* greenhouse, a building made of glass, used for raising plants. The most important glasshouse crop in England is tomatoes. When heated it is also called a 'hothouse'

glaucous ['glɔːkəs] *adjective* green colour with a white bloom, giving a plant a green-blue appearance (as in some varieties of cabbage)

gley [gleɪ] *noun* type of soil which shows signs of gleying

◊ **gleying** ['gleɪɪŋ] *noun* series of properties of soil which indicate poor drainage and lack of oxygen (anaerobism); the signs are a blue-grey colour, rusty patches and standing surface water

glidewort ['glaɪdwət] *noun* another name for the common hemp nettle

globe [gləʊb] *noun* ball-shaped vegetable, such as the globe artichoke, or a variety of mangel

Global Environment Facility

['gləʊbəl ɪn'vaɪrənmənt fæsɪlɪti] *noun* organization set up in 1991 to tackle environmental problems that go beyond country boundaries. It is funded by the World Bank

Gloucester ['glɒstə] *noun* **(a)** a hard British cheese; **Double Gloucester** = rich orange coloured British cheese **(b)** rare breed of cattle, mahogany in colour, with a white strip passing down the back, over the tail, down the hind quarters and along the belly; its milk was originally used in the production of Double Gloucester cheese **(c) Gloucester Old Spot** = breed of pig from the South West of England (Wiltshire, Somerset and Gloucester); it is large, with clearly defined black spots on a white coat; now a rare breed

glucose ['gluːkəʊz] *noun* monosaccharide sugar, used by plants and animals as a source of energy

glucosinolate [gluːkəʊ'sɪnəʊleɪt] *noun* compound left in rape meal after the oil has been extracted (NOTE: also called **glucos)**

COMMENT: the animals convert the compound to toxin after eating it; although glucosinolates can be removed by processing, plant breeders are trying to breed new varieties of rape that are low in glucos, and therefore avoid the extra production cost

glume [gluːm] *noun* small leaf or scale enclosing a grass spikelet; most grasses have two glumes

◊ **glume blotch** ['gluːm 'blɒtʃ] *noun* *Leptosphaeria nodorum,* a fungal disease of wheat

gluten ['gluːtən] *noun* the protein in a seed, such as wheat or maize; the nitrogenous part of flour which remains when the starch is extracted; **corn gluten** *or* **maize gluten** = type of animal feed made from the residues after starch has been extracted from grain

COMMENT: the gluten is what makes dough elastic and bread soft; millet and rice do not contain gluten and so cannot be used for making bread

glyphosate ['glaɪfəseɪt] *noun* major translocated herbicide used in agriculture

goad [gəʊd] *noun* a spiked stick used to prod cattle

goat [gəʊt] *noun* a hardy ruminant, usually

with horns (NOTE: males are **bucks,** females are **does,** and the young are **kids)**

◊ **goatling** ['gəʊtlɪŋ] *noun* female goat between the ages of one and two years, which has not yet borne a kid

COMMENT: In Europe goats are important for milk production; goat's milk has a higher protein and butterfat content than cow's milk, and is used especially for making cheese. Elsewhere goats are reared for meat. They are useful as browsers and will eat materials which are not normally eaten by cattle

QUOTE goat numbers in the UK are between 100, 000 and 150, 000, and are likely to expand at twice the rate of the sheep population
Farmers Weekly

Golden Guernsey ['gəʊldən 'gɜːnzi] *noun* breed of goat

Golden Triangle ['gəʊldən 'traɪæŋgl] *noun* an area of South East Asia on the borders of Burma, Thailand, Laos and China, believed to be the area where rice was first domesticated

goose [guːs] *noun* large heavy bird, between a duck and a swan in size. Possibly this was one of the first wild birds to be domesticated. Geese are raised especially for table birds at Christmas. In France, goose livers are used to make pâté de foie gras (NOTE: the males are **ganders,** the young are **goslings)**

◊ **gooseberry** ['gʊzbəri] *noun* soft fruit, usually green in colour, from a small prickly bush; gooseberries are rarely eaten raw, but are usually cooked or preserved

◊ **goosegrass** ['guːsgrɑːs] *see* CLEAVERS

gosling ['gɒzlɪŋ] *noun* young goose

Gossypium [gɒ'sɪpiəm] *Latin name for* cotton

gourd [gʊəd] *noun* fruit of a trailing or climbing plant. Many varieties are cultivated either as ornamental plants or to provide dried bottle-like containers which can be used as utensils such as water carriers

gout fly ['gaʊt 'flaɪ] *noun* small fly whose larvae hatch and feed on shoots and ears of cereals; mainly barley is affected

QUOTE severe and widespread attacks of gout fly, an old and almost forgotten pest of cereals, have hit early-sown crops of winter wheat and winter barley in Suffolk. Damage has been more severe than ever seen before. The base of tillers swell to abnormal size, giving the plants an onion-like appearance. Many insecticides currently applied to winter cereals are likely to provide satisfactory control

Farmers Weekly

government agencies ['gʌvnmənt 'eɪdʒənsɪz] *noun* organizations which provide specialist advice for farmers (such as the ADAS, set up by the British Ministry of Agriculture, Fisheries and Food)

◊ **government assistance** ['gʌvnmənt ə'sɪstəns] *noun* financial aid in the form of grants and subsidies provided by governments to help farmers

gr = GRAIN, GRAM

grade [greɪd] **1** *noun* category of something which is classified according to quality or size **2** *verb* to divide produce into different categories, according to its quality or size; *eggs are graded into classes A, B, and C;* **graded seed** = seed such as sugar beet which is formed of a cluster of seeds and can be separated out by rubbing (NOTE: also called **rubbed seed)**

COMMENT: agricultural land is classified into five grades. Grade 1 is land with very minor or no physical limitations to agricultural use. Grade 2 has some minor limitations in soil texture, depth or drainage. Grade 3 has moderate limitations due to soil, relief or climate, it has no potential for horticulture, but can produce good crops of cereals, roots and grass. Grade 4 has severe limitations and is basically used for pasture. Grade 5 is of little agricultural value, mainly for rough grazing

grader ['greɪdə] *noun* machine which grades fruit or vegetables, according to size

◊ **grading up** ['greɪdɪŋ 'ʌp] *noun* selective breeding process, using the males of one breed to mate with females of another breed for at least four generations. The result will be that the female breed will disappear and be replaced by that of the males

graft [grɑːft] **1** *noun* piece of plant *or* animal

tissue transplanted onto another plant *or* animal and growing there **2** *verb* to transplant a piece of tissue from one plant *or* animal to another

COMMENT: many cultivated plants are grafted. The piece of tissue from the original plant (or scion) is placed on a cut made in the outer bark of the host plant (the stock) so that a bond takes place. The aim is usually to ensure that the hardy qualities of the stock are able to benefit the weaker cultivated scion. The scion tissue may be a short shoot or simply a bud. The main methods of grafting are: the crown graft, where a branch of a tree is cut across at right angles, slits made in the bark around the edge of the stump, and shoots inserted into the slits; and the v-graft, where the stem of the stock is trimmed to a point, and the stem of the cutting is split to allow it to be fitted over the point of the stock

grain [greɪn] *noun* **(a)** cereal crop, such as corn, of which the seeds are dried and eaten; **grain drill** = machine used for sowing cereals in rows; **grain drier** = machine which dries moist grain, before storage; the grain is subjected to hot or warm air; **grain lifters** = attachments to the cutter bar of a combine harvester, which lift the stems of cereal crops which have been beaten down by bad weather, and so allow the crop to be cut and gathered; **grain pan** = part of a combine where the threshed grain collects and is shaken through to the bottom of the machine; **grain reserves** = amount of grain held in a store by a country, which is estimated to be above the country's requirements for one year; **grain rolled** = cereal rolled or crushed between two rotating cylinders for feeding to livestock; **grain spear** = instrument for measuring the temperature and moisture of stored grain (it consists of a thermometer and hygrometer at the end of a long rod which is pushed into the grain); **grain storage** = keeping grain until it is sold or used. Most grain is stored on the farm until it is sold, and is kept in bins or in bulk on the floor of the granary. The system of storage depends on whether the grain is to be used for feedingstuff on the farm, or is to be sold; **grain tank** = storage area at the top of a combine, in which threshed grain is kept; when the tank is full, the grain is transferred to a trailer; **grain weevil** = reddish-brown weevil; it lays eggs in stored grain; the larvae feed inside the grain, where they also pupate (NOTE: in this sense, **grain** does not have a plural) **(b)**

single seed from a cereal plant; *see also* COARSE GRAINS, FINE GRAINS, HEAVY GRAINS, LIGHT GRAINS **(c)** measure of weight equal to .0648 grams (NOTE: in this sense, when used with numbers, **grain** is usually written **gr**)

gram [græm] *noun (in India and Pakistan)* the chick pea, the most important pulse grown in the Indian subcontinent; **green gram** = mung, an Indian pulse used as a vegetable or also to produce bean sprouts; **red gram** = the pigeon pea, used in India to make dhal

Gramineae *or* **Graminales** [græ'mɪnɪi: *or* græmɪ'nɑːliːz] *noun* the grasses, a very large family of plants including cereals such as wheat, maize, etc.

graminicide [græ'mɪnɪsaɪd] *noun* herbicide which kills grasses

Granadilla [grænə'dɪlə] *noun Passiflora edulis,* the passion fruit, a climbing plant with purple juicy fruit. It is native to Brazil

granary ['grænəri] *noun* place where threshed grain is stored

granule ['grænjuːl] *noun* small artificially made particle; fertilizers are produced in granule form, which is easier to handle and distribute than powder

◊ **granular** ['grænjʊlə] *adjective* in the form of granules

grape [greɪp] *noun* fruit of woody perennial vines *(Vitis)*

◊ **grapevine** ['greɪpvaɪn] *noun* the vine on which grapes grow

COMMENT: grapes are grown in most areas of the world that have a Mediterranean climate, and even in temperate areas like southern England and central Germany. They are eaten as fruit, dried to make currants and raisins, or crushed to make grape juice and wine

grapefruit ['greɪpfruːt] *noun* citrus fruit of a tree *(Citrus paradisi)* similar to the orange. The fruit is lemon-yellow when ripe, about twice the size of an orange, and very juicy

grass [grɑːs] *noun* flowering monocotyledon of which there are a great many genera, including wheat, barley, rice, oats; an important food for herbivores and humans; **cows at grass** = cows which are grazing in a field; **grass sickness** = sudden and usually fatal illness affecting sheep and cattle. Symptoms include depression, inflamed membranes, discharge from nostrils. No effective treatment; **grass staggers** = hypomagnesaemia, a condition caused by lack of magnesium in the blood stream; animals shiver and stagger. Cattle may be affected shortly after being turned out onto spring pastures after having wintered indoors; *see also* GRAMINEAE

◊ **grassland** ['grɑːslænd] *noun* land covered by grasses, especially wide open spaces such as the prairies of North America or the pampas of South America

COMMENT: grass is the most important crop in the UK. It occupies about two-thirds of the total crop area. Grasslands can be divided into the following types: *Rough Mountain and Hill grazing:* not of great value, the plants being mainly fescues, bents, nardus and molinia grasses; *permanent pastures:* these are never ploughed, and the quality depends on the percentage of perennial ryegrass; *leys:* these are temporary grasslands which are sown to grass for a limited period (usually one to five years). The year in which the seed mixture is sown is known as the 'seeding year'. At the end of the first year there is the first year harvest. Sowing the seeds mixture with a cover crop is known as 'undersowing'. 'Direct sowing' is sowing on bare ground without a cover crop. The main species used in grasslands are: *Grasses:* perennial ryegrass, cocksfoot, Timothy, Italian ryegrass and meadow fescue; *clovers:* red clover, white clover; *other legumes:* lucerne, sainfoin; *herbs:* yarrow, chicory, rib grass, burnet. Farmers depend on reliable seed firms to supply them with standard seed mixtures. Varieties and strains of herbage plants have different growth characteristics and the choice of mixtures will depend on the purposes of the ley

QUOTE the first problem for the pasture agronomist is to find a grass that can produce dense ground cover quickly. Legumes must not compete too much with the grass because cows prefer to eat grasses

New Scientist

gravity feed ['grævɪti 'fiːd] *noun* system where pellets, seeds or granules, fall from a hopper into a distribution channel

graze [greɪz] *verb* **(a)** *(of animals)* to feed on grass; *sheep grazing on the hillside will keep the grass short* **(b)** *(of a farmer)* to put animals in a field to eat grass

◊ **grazier** ['greɪzɪə] *noun* farmer who looks after grazing animals

◊ **grazing** ['greɪzɪŋ] *noun* **(a)** action of feeding animals on growing grass; **grazing cycle** = length of time between the beginning of one grazing period and the next; **grazing food chain** = food chain in which the energy in vegetation is eaten by animals, digested, then passed into the soil as dung and so taken up again by plants which are eaten by animals, etc.; **grazing pressure** = the number of animals of a specified class per unit weight of herbage at a point of time **(b)** land covered with grass, suitable for animals to feed on; *there is good grazing on the mountain pastures*; **grazing systems** = different methods of pasture management; *see* FORWARD CREEP GRAZING, PADDOCK GRAZING, ROTATIONAL GRAZING, STRIP GRAZING, ZERO GRAZING

greaseband ['griːsbænd] *noun* strip of paper covered with a sticky substance, wrapped round the trunk of a tree to prevent pests from climbing up the tree

greasy pig disease ['griːsi pɪg dɪ'ziːz] *noun* a bacterial disease which causes skin abrasions. Can rapidly affect an entire litter

Great Plains ['greɪt 'pleɪnz] *noun* an area of North America, stretching from Texas in the south through Kansas, Oklahoma, and the Dakotas into Canada. Millions of tons of wheat are grown each year, much of it for export. The Great Plains are also used as grazing land

green [griːn] **1** *adjective* **(a)** the colour like the colour of grass or leaves (the green colour in plants is provided by chlorophyll); **green gram** = mung, an Indian pulse used as a vegetable or also to produce bean sprouts; **green manure** = rapid-growing green vegetation (such as mustard plants or rape) which is grown and ploughed into the soil to rot and act as manure; **green manuring** = growing green crops and ploughing them in to increase the organic content of the soil **(b)** **green village** = a nucleated village settlement, built round a village green **(c)** immature; **green wood** = new shoots on a tree, which have not ripened fully **(d)** referring to an interest

in ecological and environmental problems; **green consumerism** = movement to encourage people to buy food and other products which are environmentally good (such as organic foods, lead-free petrol, etc.); **green currencies** *or* **green rates** = fixed exchange rates for currencies used for agricultural payments in the EU; **green pound** = the fixed sterling exchange rate as used for agricultural payments in sterling between the UK and other members of the EU; **green top milk** = untreated milk, identified by the green tops to the bottles. (Sales to the public are banned in the UK) **2** *noun* *(in the UK)* **village green** = grass-covered area in the centre of a village, around which cottages are built

Green Belt ['griːn 'belt] *noun* area of agricultural land *or* woodland *or* parkland which surrounds an urban area

> COMMENT: Green Belt land is protected, and building is restricted and often completely forbidden. The aim of setting up a Green Belt is to prevent urban sprawl and reduce city pollution

greenfly ['griːnflaɪ] *noun* type of aphid, a small insect which sucks sap from plants and can multiply very rapidly

> COMMENT: greenfly attack young shoots which have a softer texture; various species of greenfly feed on cereal crops in May and June. Greenfly can carry virus diseases from infected plants to clean ones

greengage ['griːngeɪdʒ] *noun* variety of cooking plum, which is hard and green

greenhouse ['griːnhaʊs] *noun* building made mostly of glass, used to raise and protect plants

> COMMENT: greenhouses are used in temperate areas to grow plants which cannot be grown out of doors; either to bring the plants on early (raising seedlings to be planted out later) or to grow plants out of season (tomatoes can be grown in greenhouses during the winter months). A cold greenhouse (i. e. a greenhouse without any heating) can be used for protection of more or less hardy plants during the winter or for growing plants in late spring and summer; a heated greenhouse will be necessary to raise tender plants during the winter

greenhouse effect ['griːnhaʊs ɪ'fekt] *noun* effect produced by the accumulation of carbon dioxide crystals and water vapour in the upper atmosphere, which insulates the earth and raises the atmospheric temperature by preventing heat loss; **greenhouse gases** = gases (carbon dioxide, methane, CFCs and nitrogen oxides) which are produced by burning fossil fuels and which rise into the atmosphere, preventing heat loss

COMMENT: carbon dioxide particles allow the sun's radiation to pass through and reach the earth, but prevent heat from radiating back into the atmosphere. This results in a rise in the earth's atmospheric temperature, as if the atmosphere were a greenhouse protecting the earth. Even a small rise of less than one degree Celsius in the atmospheric temperature could have serious effects on the climate of the earth as a whole. The temperature over the poles would rise, melting the polar ice caps and sea water would become warmer, and expand, causing sea levels to rise everywhere. This would cause permanent flooding in many parts of world. In addition, temperate areas in Asia and America would experience hotter and drier conditions, causing crop failures. Carbon dioxide is largely formed from burning fossil fuels; other gases contribute to the greenhouse effect, for instance, methane is increasingly produced by rotting vegetation in swamps, from paddy fields, from termites' excreta and even from the stomachs of cows. There is an opposite effect to the greenhouse, which is the cooling of the earth's atmosphere caused by the dust veil, a mass of particles of dust which circulates in the upper atmosphere and prevents the sun's radiation from passing though to the earth's surface

QUOTE in the next 40 or 50 years, the greenhouse gases in the atmosphere will double and the world will get warmer by between 1 and 4.5°C
Guardian

QUOTE in general, a temperature rise will decrease the time available for the specific phases of cereal crops from leaf growth through to reproduction. Yields will probably decrease as temperatures increase. But more CO_2 will put yields up and if temperature and CO_2 were the only factors, output from most crops would not change much
Farmers Weekly

greenhouse mealy bug ['gri:nhaus

'mi:li bʌg] *noun (Planococcus citri)* a horticultural pest, a distant relative of the aphid. It may spoil the appearance of some glasshouse crops, particularly orchids

greening ['gri:nɪŋ] *noun* turning green (as of potatoes, if left too long in the light)

Green Revolution ['gri:n revə'lu:ʃn] *noun* development of new forms of cereal plants such as wheat and rice and the use of more powerful fertilizers, which give much higher yields and increase the food production especially in tropical countries

QUOTE in Asia, the Green Revolution, which featured high-input, high-yield agricultural technology, heralded a remarkable increase in grain production to the point that India and Pakistan became self-sufficient having formerly been large importers
Appropriate Technology

greens [gri:nz] *noun* green vegetables, such as cabbages; **winter greens** = hardy varieties of Brassica which are grown for use during the winter

Greyface ['greifeis] *noun* crossbred sheep resulting from a Border Leicester ram and a Blackface ewe. The ewes are mated with Suffolk rams to produce good quality lambs

grey leaf ['grei 'li:f] *noun* disease of cereals caused by manganese deficiency

grid [grid] *noun* **cattle grid** = type of grill made of parallel bars, covering a hole dug in the road; it prevents stock from crossing the grid and leaving their pasture, but allows vehicles and humans to pass

grind [graind] *verb* to crush or roll something until it is reduced to powder; *wheat grain is ground in mills to make various grades of flour*

grist [grist] *noun* (i) corn for grinding; (ii) malt crushed for brewing

grit [grit] *noun* small particles of various substances fed to poultry

COMMENT: there are two different kinds of grit: hard insoluble grit, such as flint and gravel which the fowl has to take into its gizzard to do the grinding of its feed; and the soluble grit, such as oyster-shell and limestone, which contains lime and which the birds need for bone formation and, later, for the formation of egg shells

grits ['grɪts] *noun US* type of porridge made from buckwheat flour

Groningen Whiteheaded ['grɒnɪŋən waɪt'hedɪd] *noun* dual-purpose breed of cattle developed in the Netherlands; the body is black but the head is white

groom [gru:m] **1** *noun* person who looks after horses **2** *verb* to look after animals, especially horses, by brushing cleaning and combing

ground [graʊnd] *noun* **(a)** surface layer of soil *or* earth; **ground frost** = condition where the air temperature at ground level is below 0°C; **ground water** = water which stays in the top layers of soil or in porous rocks and can collect pollution (as opposed to surface water which drains away rapidly, and leaches pollutants from the surface) **(b)** area; **breeding grounds** = area where birds *or* animals come each year to breed; **wintering grounds** = area where birds come each year to spend the winter

groundnut ['graʊnnʌt] *noun Arachis hypogaea* the peanut, a grain legume, and one of the main oilseeds; **groundnut cake** = the residue left after oil extraction from groundnuts, a valuable protein concentrate for livestock

COMMENT: groundnuts (or peanuts) are used in the production of vegetable oil for cooking, in salad dressings and in the making of margarine. Poorer quality oils are used to make soap. Eaten raw or roasted or cooked in various ways, groundnuts figure prominently in West African cooking. In the USA, much of the crop is made into peanut butter. The USA, Argentina, Nigeria, Sudan and Indonesia are major exporters of groundnuts, while Canada and Western Europe are the main importing countries

groundsel ['graʊndsəl] *noun* common weed *(Senecio vulgaris)* which affects most crops; also called 'birdseed'

grouse [graʊs] *noun* small game bird; there are two main species in Europe: the rare **black grouse** *(Lyrurus tetrix)* and the Scottish **red grouse** *(Lagopus scoticus)*

grow [grəʊ] *verb* **(a)** *(of a plant)* to exist and flourish; *bananas grow only in warm humid conditions; tomatoes grow well in greenhouses* **(b)** *(of plants and animals)* to become taller or bigger; *the tree grows a new ring each year; the plant grew three centimetres in one year;* **growing point** = point on the stem of a plant where growth occurs (often at the tip of the stem or branch); **growing season** = period of the year when a plant grows; the growing season is especially important in temperate latitudes where the length of the growing season is usually limited by low temperatures; *alpine plants have a short growing season* **(c)** *(of people)* to cultivate (plants); *farmers here grow two crops in a year; he grows peas for the local canning factory; growing early vegetables is difficult because of the cold winters*

◊ **growth** [grəʊθ] *noun* **(a)** increase in size; amount by which something increases in size; *the disease stunts the conifers' growth; the growth in the goat population since 1960; the rings show the annual growth of the tree;* **growth promoter** = substance which makes a plant *or* animal grow; **growth regulator** = a chemical used to control the growth of certain plants; mainly used for weed control in cereals and grassland; **growth retardant** = substance used to slow or stop the growth of a plant; **growth rings** = rings seen in the cross-section of the trunk of a felled tree, each one added during a single year; *see also* ANNUAL RINGS; **growth stages** *or* **stages of growth** = the different stages of development of a crop; growth is measured as an increase in weight or area. Chemical fertilizers and other chemicals must be applied to the crop at the correct time to ensure maximum effectiveness **(b)** shoot which has grown from a plant; *the cordon should be pruned by cutting back all growths over one metre long* **(c)** type of plant which grows in a certain area (for example vines growing in different areas of France, coffee growing in different areas of Colombia, etc.)

grub [grʌb] **1** *noun* small caterpillar or larva **2** *verb* **to grub out** *or* **to grub up** = to remove a plant from the ground with its roots; *miles of hedgerows have been grubbed up to make larger fields*

grunt [grʌnt] **1** *noun* noise made by a pig **2** *verb (of a pig)* to make a noise; *compare* BLEAT, LOW, NEIGH

guano ['gwɑ:nəʊ] *noun* mass of accumulated bird droppings, found especially on small islands in the sea, gathered and used as fertilizer (it is a natural phosphate)

guaranteed prices [gærən'tiːd 'praɪsɪs] *noun* a feature of national agricultural policy in which the producers of a certain commodity are guaranteed a certain minimum price for their produce

guava ['gwɑːvə] *noun* common tree (*Psidium guajava*) originating in West Africa but now grown in tropical areas all over the world; it is easily grown and produces yellow-skinned, pink-fleshed fruits which are excellent for eating

guayale [gwæ'jɑːli] *noun* shrub containing sap which is extracted from mature plants and used as a rubber substitute; it is grown in West Africa

Guernsey ['gɜːnzi] *noun* breed of dairy cattle; the coat is fawn, with distinct patches of white. The Guernsey is similar to the Jersey, but larger. The breed is noted for the rich colour of the milk, the high butterfat content and the good milk yields obtained

guild [gɪld] *noun* group of plants *or* animals of the same species which live in the same type of environment

guinea ['gɪni] *noun* former British coin (equivalent to the present £1. 08) which is still used in quoting prices at livestock sales (NOTE: it is abbreviated in prices to **gn: 3,400gns were paid for the Longhorn bull)**

◊ **guinea corn** ['gɪni 'kɔːn] *noun* (*in West Africa)* sorghum

◊ **guinea fowl** ['gɪni 'faʊl] *noun* table bird, found wild in savanna regions of Africa. They are now raised for their meat

which has a delicate flavour similar to that of game birds

gully ['gʌli] *noun* **(a)** deep channel formed by soil erosion, and which cannot be removed by cultivation **(b)** small channel for water (such as an artificial channel dug at the edge of a field or a natural channel in rock)

gum arabic [gʌm 'ærəbɪk] *noun* resin secreted from acacia trees

COMMENT: gums look like resins and have the same origin. They exist in liquid state in the trunks and branches of trees. Gum arabic is produced from various species of *Acacia*. It is exported from Northern Africa and the Sudan, and is used in confectionery, pharmacy and stationery

gun [gʌn] *see* BALLING GUN, DOSING GUN

Gunter's chain ['gʌntəz 'tʃeɪn] *noun* chain originally used by surveyors to measure land

gur [guːə] *noun* (*in the Indian subcontinent and parts of Africa)* sugar made from sugar cane, but without spinning to extract it. Originally gur was made by tapping the sweet sap of certain species of palm

gut [gʌt] *noun* **(a)** the large intestine **(b)** the digestive processes; *at calving, a cow's appetite decreases and her gut is slow*

Guzerat ['guːzəræt] *noun* American Brahman breed of cattle

gypsum ['dʒɪpsəm] *noun* calcium sulphate, sometimes used to improve salt-affected soils

Hh

H *chemical symbol for* hydrogen

ha = HECTARE

habit ['hæbɪt] *noun* way in which a plant grows; *a bush with an erect habit* or *with a creeping habit*

habitat ['hæbɪtæt] *noun* area of the environment in which an organism lives

QUOTE throughout India, elephant habitat is suffering encroachment by coffee and tea plantations
BBC Wildlife

hacienda [hæsi'endə] *noun* large agricultural estate found mainly in Spanish-speaking countries

hack [hæk] **1** *noun* **(a)** riding horse **(b)** horse let out to hire **2** *verb* to ride a horse; especially to ride a horse to a show (as opposed to taking the horse in a box)

hackles ['hækəlz] *noun* long feathers on the neck of a domestic cock

hackney ['hækni] *noun* type of horse used both for riding and as a draught animal

haemoglobin (Hb) [hi:mə'gləubɪn] *noun* red respiratory pigment containing iron in red blood cells, which gives blood its red colour; *see also* CARBOXY-HAEMOGLOBIN

COMMENT: haemoglobin absorbs oxygen in the lungs and carries it in the blood to the tissues. Haemoglobin is also attracted to carbon monoxide and easily absorbs it instead of oxygen, causing carbon monoxide poisoning

Hagberg (test) ['hægbɜ:g] *noun* test used to determine the milling quality of wheat; **Hagberg falling number** = the falling time in seconds in the Hagberg test; *a top quality wheat with a specific weight of 79-80, Hagberg 350 and protein of 12% see also* ALPHA AMYLASE

COMMENT: the test measures the falling time of wheat, using ground wheat in suspension in water. A good milling wheat has a high falling time, and wheat with low falling times is not normally used in milling. Hagberg test kits are available for farmers to make their own tests on samples of wheat

QUOTE Hagbergs have generally been over 250, and up to 300 in many cases
Farmers Weekly

hair balls ['heə 'bɔ:lz] *noun* balls of hair which collect in the stomach making digestion difficult. Can cause fits and convulsions in very young calves and sight may be slightly impaired

hairworm ['heəwɜ:m] *noun* very thin worm of the species *Capellaria*, which infests poultry

hake bar ['heɪk 'ba:] *noun* attachment which links a trailed plough to the tractor

half-breed ['ha:fbri:d] *noun* animal of mixed breed, mainly applied to crossbred sheep

half-hardy [ha:f'ha:di] *adjective* (garden plants) which can stand a certain amount of cold

Half long ['ha:f 'lɒŋ] *noun* breed of sheep. Cheviot ram crossed with Blackface ewe

half-standard [ha:f'stændəd] *noun* type of fruit tree, with a stem trunk which is shorter than for a full standard (about 1.2m from the ground to the first branches)

Hallikar ['hælɪka:] *noun* breed of cattle from south east India. Dark grey in colour, essentially a draught breed

halo- ['hæləu] *prefix* meaning salt

◊ **halomorphic soil** [hæləu'mɔːfɪk sɔɪl] *noun* soil which contains large amounts of salt

◊ **halophyte** ['hæləfaɪt] *noun* plant which lives in salty soil

halo blight ['heɪləʊ 'blaɪt] *noun* disease which affects the pods of peas and beans, making them brown and withered

halter ['hɔːltə] *noun* rope with a noose for holding and leading horses or cattle

ham [hæm] *noun* **(a)** the thigh of a pig (i.e., its back leg) **(b)** meat from this part of the pig, usually cured in brine and dried in smoke

hamlet ['hæmlət] *noun* group of farmhouses, cottages and farm buildings in a rural area

hammer mill ['hæmə 'mɪl] *noun* machine used in the preparation of animal feed by grinding cereals into meal

> COMMENT: a typical hammer mill has a high-speed shaft with a grinding rotor at one end and a fan at the other. Eight flails are attached to the rotor which beat the grain into meal

Hampshire ['hæmpʃə] *noun* American breed of black-haired pig with white markings; it is similar to the British Saddleback which has black skin and a white saddle

◊ **Hampshire Down** ['hæmpʃə 'daʊn] *noun* short stocky early-maturing sheep, originating from Berkshire ewe flocks and Southdown rams

hand [hænd] *noun* **(a)** measure used to show the height of a horse; one hand is 10.16cm, and the measurement is taken from the ground to the withers of the horse **(b) by hand** = work which is done using the hands, as opposed to machines; **hand collection** *or* **hand picking** = picking of fruit (such as bananas or peaches) by hand; **hand pulling** = pulling weeds or plants out of the ground by hand; **hand feeding** *or* **hand rearing** = bringing up orphaned animals by feeding them with a bottle

hank [hæŋk] *noun* wool which has been spun into a thread and coiled into a loop for convenience; a hank is 560 yards long

harden off ['hɑːdən 'ɒf] *verb* to make plants which have been raised in a greenhouse become gradually more used to the natural temperature outdoors; *after seedlings have been grown in the greenhouse, they need to be hardened off before planting outside in the open ground*

hardjo ['hɑːdjəʊ] *see* LEPTOSPIRA

hardpan ['hɑːdpæn] *noun* hard cement-like layer in the soil or subsoil, which can be very harmful as it prevents good drainage and stops root development

◊ **hard wheat** ['hɑːd 'wiːt] *noun* wheat with a hard grain rich in gluten

◊ **hardwood** ['hɑːdwʊd] *noun* slow-growing broad-leaved tree (such as oak or teak) which produces a hard wood with a fine grain; **hardwoods** = forests of hardwood trees; *compare* SOFTWOOD

hardy ['hɑːdi] *adjective* (plant *or* animal) which can withstand the cold, especially severe winter conditions; **frost-hardy plant** = plant which is able to withstand frost; *see also* HALF-HARDY

hare [heə] *noun* long-eared furry animal; a rodent with hind legs longer than forelegs. Hares can be a serious pest when they eat growing crops, especially young cereals

haricot bean ['hærɪkəʊ 'biːn] *noun* dry ripe seed of the French bean

harmattan [hɑːˈmætən] *noun* hot dry winter wind which blows from the north-east and causes dust storms in the Sahara

harrow ['hærəʊ] **1** *noun* farm implement, made of a series of spikes or tines, which is dragged across the surface of ploughed soil to prepare seed beds and to destroy weeds. Harrows are also used to cover seeds with soil. There are a number of different types, including chain, disc, Dutch, power and spike tooth harrows **2** *verb* to level the surface of ploughed soil, covering seeds which have been sown in furrows

harvest ['hɑːvɪst] **1** *noun* (i) action of gathering a crop; (ii) time when a crop is gathered; (iii) the crop which is gathered; *we think this year's rice harvest will be a good one* **2** *verb* to gather a crop which is ripe; *they are harvesting the rice crop; clearcutting is one of the methods of harvesting timber*

◊ **harvester** ['hɑːvɪstə] *noun* machine which harvests a crop; most crops are now harvested by machines such as combine harvesters or sugar beet harvesters; *see also* FORAGE HARVESTER

haryana [hæˈrjɑːnə] *noun* breed of cattle found in the Indian sub-continent

hatch [hætʃ] **1** *noun* brood of chicks **2** *verb* *(of an animal in an egg)* to become mature and break out of the egg

◊ **hatchery** ['hætʃəri] *noun* unit which specializes in the hatching of eggs

COMMENT: a hen will naturally sit on her eggs until they hatch. Incubators are used to hatch large numbers of eggs by artificial means

haulm [hɔːm] *noun* stalks and stems of peas, beans and potatoes; **haulm roller** = roller found on potato harvesters and grading machinery; **haulm silage** = silage made from the stems and leaves of peas and beans left after harvest

haunch [hɔːnʃ] *noun* hind leg of an animal, especially a deer

hawthorn ['hɔːθɔːn] *noun* small tree *(Crataegus monogyna)* with spiny shoots, used for hedges round grazing areas

hay [heɪ] *noun* grass which has been cut and dried and conserved for fodder; **hay bale** = hay which has been compressed into a square, rectangular or round bale, so that it can be handled and stored more easily; **hay baler** = machine which gathers cut hay and makes it into bales; **hay net** = coarse meshed net bag which is filled with hay and hung up for horses to feed from; **hay quality** = the nutritional value of hay, which can depend on the weather conditions and the time taken to dry; **hay rack** = wooden frame containing hay, which is placed where livestock can feed from it; **hay seed** = grass seed obtained from hay

◊ **hayage** ['heɪɪdʒ] *noun* hay for silage, cut and compressed in plastic bags so that it stays green without any fungus being able to spread; *compare* HAYLAGE

◊ **haycock** ['heɪkɒk] *noun* formerly, a conical heap of raked hay

◊ **haylage** ['heɪlɪdʒ] *noun* dry hay, which has been harvested and is kept in a silo; *compare* HAYAGE

◊ **hayloader** ['heɪləudə] *noun* implement for loading hay from the field into a trailer

◊ **haymaking** ['heɪmeɪkɪŋ] *noun* conserving a grass crop for use as fodder

◊ **hay rake** ['heɪ 'reɪk] *noun* implement used for raking hay prior to collection or baling. Tractor rakes are wide, usually wider than farm gateways and are constructed so that they can be reduced to a narrow width. Long-handled wooden rakes are still used to gather hay by hand

◊ **haystack** ['heɪstæk] *noun* heap of hay, built in the open air and protected by thatching; not used very often nowadays

◊ **hay-sweep** ['heɪswiːp] *noun* implement used to collect hay from swaths and carry it to a stack

COMMENT: hay is cut before the grass flowers; at this stage in its growth it is a nutritious fodder. If it is mowed after it has flowered it is called straw, and is of much less use as a food, and so is used for bedding. Haymaking normally needs three to four days of fine weather in early season and two to three days when the humidity falls as temperatures rise. The critical period for hay occurs when the crop is partly dried in the field. The object should be to dry the crop as much as possible without too much exposure to sun, and the least possible movement after the crop is partially dry. Field-dried hay is normally baled, and further barn drying is common

hazel ['heɪzəl] *noun* a nut-bearing tree *(Corylus avellana);* the Kentish cob or filbert is a commercially grown hazel

Hb = HAEMOGLOBIN

HCH *see* BHC, LINDANE

HDC = HORTICULTURE DEVELOPMENT COUNCIL

HDRA = HENRY DOUBLEDAY RESEARCH ASSOCIATION

head corn ['hed 'kɔːn] *noun* largest grains in a sample of cereal

◊ **header** ['hedə] *noun* machine which removes the seed heads from plants; *see also* STRIPPER-HEADER

◊ **headland** ['hedlænd] *noun* land at the edges of a field where the plough is turned; it is the last part of the field to be ploughed

◊ **headrail** ['hedreɪl] *noun* rail across the front of a cubicle, to which a halter can be attached

heart [hɑːt] **1** *noun* **(a)** central growing point of a plant, with very tender leaves; *a lettuce with a good heart* **(b)** soil which is in good heart = soil which is fertile and produces large yields of crops **2** *verb (of a* *plant)* to develop a heart; *the lettuces are hearting up well*

◊ **heart rot** ['hɑːtrɒt] *noun* disease of sugar beet and mangolds, caused by boron deficiency; a dry rot spreads from the crown downwards and attacks the roots. The growing point is killed, and replaced by a mass of small deformed leaves

◊ **heartwood** ['hɑːtwʊd] *noun* the central hard dead wood in a tree trunk, as opposed to the sapwood

heat [hiːt] *noun* **(a)** warmth which is produced by electricity or hot air or warm pipes; **to sow lettuces under heat** = to sow lettuce seed in a heated greenhouse; **heatsealing** = method of closing plastic food containers; air is removed from a plastic bag with the food inside and the bag is then pressed by a hot plate which melts the plastic and seals the contents in the vacuum **(b)** the oestrus period, when a female will allow mating; **an animal on heat** = female animal in the oestrus period, when she will accept a male

heath [hiːθ] *noun* area of dry sandy acid soil, with low shrubs such as heather and gorse growing on it

◊ **heathland** ['hiːθlænd] *noun* wide area of heath, largely waste ground, though it may provide some grazing

heather ['heðə] *noun* plant *(Calluna vulgaris)* found on acid soils, common in upland areas; used by game birds such as grouse, for cover and food

COMMENT: the main competing uses for heather moorland in the UK are grouse shooting, sheep grazing, afforestation, recreational use and landscape conservation

heaves [hiːvz] *noun* condition of horses, where spores from mouldy hay block the animals' lungs, making breathing difficult

heavy grains (HG) ['hevi 'greɪnz] *noun* cereal crops such as maize, rye and wheat (as opposed to the light grains, barley and oats)

heavy soils ['hevi 'sɔɪlz] *noun* soils with a high clay content, which need more tractor power when ploughing and cultivating

Hebridean sheep [hebrɪ'diːən 'ʃiːp] *noun* a rare breed of small black sheep of Scandinavian origin; the fleece is jet-black in colour, and the animals have one pair of horns curling downwards and another pair almost upright

hectare ['hekteə] *noun* area of land measuring 100 by 100 metres, i.e. 10,000 square metres or 2.47 acres (NOTE: usually written **ha** after figures: **2,500ha)**

QUOTE the UK is well-placed to grow up to 200,000ha (0.5m acres) of linseed to replace imports and provide exports to other EU countries
Farmers Weekly

hecto- ['hektəʊ] *prefix* meaning one hundred

◊ **hectolitre** *or US* **hectoliter** ['hektəliːtə] *noun* one hundred litres (NOTE: usually written **hl** after figures: **10hl)**

hedge [hedʒ] *noun* row of bushes planted and kept trimmed to provide a barrier along a field boundary

◊ **hedgecutter** *or* **hedgetrimmer** ['hedʒkʌtə *or* 'hedʒtrɪmə] *noun* implement attached to a tractor, used to trim hedges; smaller hand-held units are available

◊ **hedgelaying** ['hedʒleɪɪŋ] *noun* traditional method of cultivating hedges, where tall saplings are partly cut through near the base, and then bent over so that they lie horizontally and make a thick barrier

◊ **hedgerow** ['hedʒrəʊ] *noun* line of bushes forming a hedge

◊ **hedging** ['hedʒɪŋ] *noun* art of cultivating hedges

COMMENT: you can judge the age of a hedge by counting the species in a 30-yard stretch and multiplying by 110, adding 30; the more species there are, the older the hedge is. About three-quarters of the farms in England and Wales have hedges, with an estimated total length of 500,000km. In the last years, 8,000km of hedges have been removed and about 4,000km have been planted

heel in ['hiːl 'ɪn] *verb* to place plants in a trench and cover with soil until needed for permanent planting

heft [heft] *noun* a group of mountain sheep which graze the same area in which they were born, although not kept in by fences

heifer ['hefə] *noun* female cow which has not calved or has calved for the first time

QUOTE high heifer prices coupled with the cost of milk quota makes expanding milk production an expensive business
Farmers Weekly

hemlock ['hemlɒk] *noun* **(a)** North American softwood tree **(b)** poisonous plant *(Conium maculatum)*; **hemlock poisoning** = poisoning of young cattle by eating fresh shoots of the hemlock; sheep and goats are believed to be resistant to the poison

hemp [hemp] *noun* plant *(Cannabis* sp) used to make rope; also producing an addictive drug

◊ **hemp nettle** ['hemp 'netl] *noun* common weed *(Galeopsis tetrahit)* which affects spring cereals and vegetables; also known as 'glidewort' and 'holyrope'

hen [hen] *noun* **(a)** female of the common domestic fowl **(b)** also any female bird, such as a hen pheasant; *see also* FAT HEN

◊ **henhouse** ['henhaus] *noun* small wooden building for keeping hens

herb [hɜːb] *noun* **(a)** plant which has no perennial stem above the ground during the winter period **(b)** plant which can be used as a medicine *or* to give a certain taste to food *or* to give a certain scent

◊ **herb-** [hɜːb] *prefix* referring to plants *or* vegetation

◊ **herbaceous** [hɜːˈbeɪʃəs] *adjective* (plant) without perennial stems above the ground; **herbaceous border** = bed of herbaceous plants growing along the edge of a lawn

◊ **herbage** ['hɜːbɪdʒ] *noun* the parts of a sward above ground, seen as an accumulation of plant material with characteristics of mass and nutritive value; **herbage allowance** = weight of herbage per unit of live weight at a point in time; **herbage consumed** = mass of herbage once it has been consumed by grazing animals; **herbage residual** = herbage remaining after defoliation

◊ **herbarium** [hɜːˈbeəriəm] *noun* collection of plant specimens

◊ **herbicide** ['hɜːbɪsaɪd] *noun* chemical which kills plants, especially weeds; **contact herbicide** = herbicide which kills plants by contact with the leaves, but may not kill a plant with large roots; **residual herbicide** = herbicide applied to the surface of the soil which acts through the roots of the plants; not only growing weeds are killed but also new plants as they germinate; **selective herbicide** = herbicide which only kills certain types of plant, used for weed control in growing crops; **systemic herbicide** = herbicide which is absorbed into the plant's sap system through its leaves (NOTE: also called **translocated**)

◊ **herbivore** *or* **herbivorous animal** ['hɜːbɪvɔː *or* hɜːˈbɪvərəs] *noun* animal which eats plants, such as cattle or sheep, as opposed to flesh-eating animals; *compare* CARNIVORE, OMNIVORE

herd [hɜːd] **1** *noun* a number of animals, such as cattle, kept together on a farm or looked after by a farmer; *they have a herd of beef cattle; dairy herds have been reduced since the introduction of the milk quota system;* **herd book** = the record of animals kept by breeding societies in which only the offspring of registered animals can be recorded; *(in New Zealand)* **herd tester** = person who tests a dairy herd for butterfat content (NOTE: the word 'herd' is usually used with cattle; for sheep, goats, and birds such as hens or geese, the word to use is 'flock') **2** *verb* to tend a herd of animals; to crowd animals together; *livestock herding is the main source of revenue for the people of the area*

◊ **herdsman** ['hɜːdzmən] *noun* male farm worker who looks after a herd of livestock

◊ **herdsperson** ['hɜːdzpɜːsən] *noun* farm worker who looks after a herd of livestock

Herdwick ['hɜːdwɪk] *noun* mountain breed of sheep, which are able to survive in bitter winter conditions; a native of the Lake District. The rams have horns. The age of a Herdwick can be judged by its colour: lambs are born almost black, but the animals get lighter in colour as they get older

hereditary [həˈredɪtəri] *adjective* which is transmitted from parents to offspring

◊ **heredity** [həˈredɪti] *noun* occurrence of physical *or* mental characteristics in offspring which are inherited from their parents

Hereford ['herɪfəd] *noun* breed of large, hardy cattle; they are deep red in colour, with a white head and chest. They are an early-maturing breed, and are important for beef production

herringbone parlour ['herɪŋbəʊn 'pɑːlə] *noun* type of milking parlour with no stalls and where the operator works from a central pit. The cattle stand side by side at an angle on each side of the central pit

heterotrophic organism *or* **heterotroph** [hetərəʊ'trofɪk *or* 'hetərətrɒf] *noun* organism (animal *or* parasite *or* fungus, etc.) which obtains its energy from carbon, by breaking down organic matter in the ecosystem and eating other organisms; *compare* AUTOTROPH

Hevea ['hiːviːə] *Latin name for the* rubber tree

HFCS = HIGH FRUCTOSE CORN SYRUP

HFRO = HILL FARMING RESEARCH ORGANISATION

Hg *chemical symbol for* mercury

HG = HEAVY GRAINS

HGCA = HOME GROWN CEREALS AUTHORITY

Hickstead ['hɪksted] *noun* site of the all-England jumping course in West Sussex

HIDB = HIGHLANDS AND ISLANDS DEVELOPMENT BOARD

hide [haɪd] *noun* animal skin, which is important commercially both in its raw state and as leather

high fructose corn syrup (HFCS) ['haɪ frʌktəʊz 'kɔːn sɪrʌp] *noun* sweetener used in the soft drinks industry, extracted from maize; also called 'isoglucose'

Highland ['haɪlənd] *noun* hardy breed of beef cattle, with a long shaggy hair hiding a dense undercoat; the breed is small, with very long horns, and matures slowly; it is native to the Highlands and Western Isles of Scotland

◊ **highlands** ['haɪləndz] *noun* area of mountains; *vegetation in the highlands or highland vegetation is mainly grass, heather and herbs*

high temperature short time method (HTST) ['haɪ 'temprtʃə ʃɔːt 'taɪm] *noun* usual method of pasteurizing milk, where the milk is heated to 72°C for 15 seconds and then rapidly cooled

high-yielding ['haɪ 'jiːldɪŋ] *adjective* which yields a high crop; *they have started to sow high-yielding varieties (HYV) of seed with very good results*

hill [hɪl] *noun* higher ground than the surrounding areas, but not as high as a mountain; **hill drainage** = small open channels about ten to sixty metres apart, dug to drain hilly grazing areas; **hill farms** = farms in mountainous country, with 95% or more of their land classified as rough grazing, depending mainly on breeding ewe flocks; **hill grazing** = grassland used for sheep and cattle grazing in hilly and mountainous areas. The main plants grown are fescues, bents, nardus and molinia grasses, as well as cotton grass, heather and gorse; **hill land** = land on hills, mountains or moors; the Hill Livestock Compensatory allowance order classifies such marginal land in upland and hilly areas and makes payments to compensate for farming in these more difficult conditions

Hill Radnor ['hɪl 'rædnə] *see* RADNOR

hilum ['haɪləm] *noun* the point where a seed is attached to a pod; when the seed is ripe and has been separated from the pod, a black scar can be seen on the seed

Himalayan nettle [hɪmə'leɪən 'netl] *see* NILGIRI NETTLE

hind [haɪnd] **1** *adjective* referring to the back part of an animal; **hind legs** = the back legs of an animal **2** *noun* a female deer

◊ **hindquarters** ['haɪndkwɔːtəz] *noun* back part of an animal, including the haunches and hind legs

hinge [hɪnʃ] *noun* soil which is left uncut by a plough when it has failed to cut a full furrow

hirsel ['hɔːsəl] *noun* **(a)** heft of sheep **(b)** piece of ground and flock looked after by one shepherd

Hisex Brown *or* **Hisex White** ['haɪseks] *noun* Czechoslovakian hybrid breeds of laying fowl

hitch [hɪtʃ] **1** *noun* the mechanism for connecting implements to tractors; *also called* DRAWBAR **2** *verb* to attach an implement or trailer to a tractor

hive [haɪv] *noun* box in which bees are kept; **hive-bee** = the domesticated bee *(Apis mellifera)*

hl = HECTOLITRE

HLCA = HILL LIVESTOCK COMPENSATORY ALLOWANCE

HMSO = HER MAJESTY'S STATIONERY OFFICE

HMV = HIGH MOBILITY VEHICLE

hock [hɒk] *noun* hind leg joint of mammals, between the knee and the fetlock

hoe [həʊ] **1** *noun* implement pulled by a tractor to turn the soil between rows of crops, and so to control weeds; **hand hoe** = garden implement, with a small sharp blade, used to break up the surface of the soil or to cut off weeds **2** *verb* to cultivate land with a hoe

COMMENT: there are several types of hand hoe, including the Dutch hoe, where the blade is more or less straight and is pushed by the operator, the draw hoe has the blade set at right angles to the handle and is used for drawing drills at seed-sowing time. The Canterbury hoe does not have a blade, but is like a three-pronged fork, with the prongs set a right angles to the handle

hog [hɒg] *noun* **(a)** castrated male pig **(b)** *US* any pig

hogg [hɒg] *noun* young sheep before the first shearing

◊ **hogget** ['hɒgɪt] *noun* sheep roughly six to twelve months old

hold [həʊld] *verb (of an animal)* to conceive after artificial insemination; *twenty-five ewes were AI'd and sixteen of them held*

◊ **holding** ['həʊldɪŋ] *noun* land and buildings held by a freehold or leasehold occupier

◊ **holdover** ['həʊldəʊvə] *noun* situation where a tenant farmer uses buildings and crops on a farm, after leaving the farm at the end of his tenancy; it may, for example, involve harvesting crops later in the year after the tenancy has expired

hollow ['hɒləʊ] *noun* place which is lower than the rest of the surface; *frost occurs often in low-lying areas, called frost hollows*

holly ['hɒli] *noun* evergreen tree *(Ilex aquifolium)* producing hard white timber

Holstein ['hɒlsteɪn] *noun* Friesian cattle imported into Canada from Holland at the end of the 19th century, now a breed of dairy cattle, black and white in colour; also called Canadian Holstein or Holstein-Friesian in the UK

holyrope ['həʊlɪrəʊp] *noun* popular name for the hemp nettle *(Galeopsis tetrahit)*

home farm ['həʊm 'fɑːm] *noun* farm on a large estate, usually farmed by the owner

◊ **Home Grown Cereals Authority (HGCA)** organization established to improve the production and marketing of home-grown wheat, barley, maize, oats and rye; the authority promotes research into cereal growing; it is also engaged in intervention buying and selling of cereals

◊ **homestead** ['həʊmsted] *noun* farmhouse with dependent buildings and the land which surrounds it

homogenized milk [hə'mɒdʒənaɪzd 'mɪlk] *noun* milk which is made more digestible by breaking up the fat droplets into smaller particles which are evenly distributed through the liquid

homograft ['həʊməʊgrɑːft] *noun* graft of tissue from one specimen to another of the same species

honey ['hʌni] *noun* sweet yellow fluid collected by bees from nectar in flowers, and stored in a beehive

◊ **honey bee** ['hʌni 'biː] *noun* common hive-bee

◊ **honeycomb** ['hʌnɪkəʊm] *noun* construction of wax, made by bees for storing honey

◊ **honey fungus** ['hʌni 'fʌŋgəs] *noun* fungus which primarily attacks trees and shrubs. Roots become infected by rhizomorphs. In the spring the foliage wilts and turns yellow

hoof [huːf] *noun* horny casing of the foot of

a horse, a cow or other animals (NOTE: the plural is either **hoofs** or **hooves)**

◊ **hoof and horn meal** ['huːf nd 'hɔːn miːl] *noun* fertilizer made from animal hooves and horns

hookworm ['hʊkwɜːm] *noun* type of roundworm infesting animals

hoose [huːs] *noun* popular name for lungworms

hop [hɒp] *noun* climbing perennial plant *(Humulus lupulus)*, cultivated for its seed cones which are used to give a bitter flavour to beer; **hop bine** = the new shoot of a hop plant, which has to be made to coil round climbing strings; **Hop Marketing Board** = organization set up in 1932 to control overproduction of hops and to set quotas for growers (now replaced by English Hops Ltd.); **hop mildew** = fungal disease of hops

hopper ['hɒpə] *noun* container with a hole at the bottom, for holding seed or fertilizer granules; the seed drops from the hole onto the ground or into channels taking it to drills

Hordeum ['hɔːdiəm] *Latin name for* barley

horizon [hə'raɪzən] *noun* layer of soil which is of a different colour *or* texture from the rest: the topsoil containing humus, the subsoil containing minerals leached from the topsoil, etc.

COMMENT: in general, the topsoil is called the 'A' horizon, the subsoil the 'B' horizon, and the underlying rock the 'C' horizon

hormone ['hɔːməʊn] *noun* substance produced in animals and plants in one part of the body which has a particular effect in another part of the body; **plant hormones** = hormones (such as auxin) which particularly affect plant growth (they are more accurately called 'plant growth substances'); **juvenile hormone** = hormone in an insect larva which regulates its development into an adult; **hormone rooting powder** *or* **auxin** = substance which if placed on a shoot of a plant will encourage the formation of roots; **hormone weedkiller** = substance which is absorbed by a plant and stimulates overgrowth, so starving the plant to death

◊ **hormonal** [hɔː'məʊnəl] *adjective* referring to hormones; **hormonal deficiency** = lack of necessary hormones

horn [hɔːn] *noun* hard growth which is formed on the tops of the heads of animals such as cattle, deer, goats and sheep; *see also* DISBUD, DEHORN

◊ **horned** [hɔːnd] *adjective* with horns; *a horned variety of sheep*

◊ **hornless** ['hɔːnləs] *adjective* without horns; *a hornless breed of cattle*

hornbeam ['hɔːnbiːm] *noun Carpinus betulus,* tree which produces a very hard wood, formerly used in making wheels for farm carts

horse [hɔːs] *noun* hoofed animal with a flowing mane and tail, used on farms as a working animal, now mainly replaced by tractors; **horse bean** = a broad bean used as a fodder

COMMENT: the main groups of horses are: the Heavy Draught Class, such as the *shire horse;* the Light Draught Class, such as the *Cleveland Bay;* the Saddle and Harness Class, such as the *hackney;* the Pony Class, such as the *Shetland*

horsebox ['hɔːsbɒks] *noun* closed vehicle used for transporting horses

◊ **horsebreaker** ['hɔːsbreɪkə] *noun* person who breaks in (i.e. trains) a horse

◊ **horsefly** ['hɔːsflaɪ] *noun* general name for many bloodsucking Tabanid flies (NOTE: also called **cleg)**

◊ **horseshoe** ['hɔːsʃuː] *noun* iron shoe nailed to the hard part of a horse's hoof

◊ **horsetail** ['hɔːsteɪl] *noun* poisonous weed found in grassland

horticulture ['hɔːtɪkʌltʃə] *noun* gardening, the growing of fruit, flowers and vegetables for food or decoration

◊ **horticultural** [hɔːtɪ'kʌltʃərəl] *adjective* referring to horticulture; *allotment holders show their vegetables at the local horticultural show*

◊ **horticulturist** [hɔːtɪ'kʌltʃərɪst] *noun* gardener, person who specializes in horticulture

Hosier system ['həʊsiə 'sɪstəm] *noun* system of dairy cattle management, where the milking of cows is done in the field using a milking pail

host [həʊst] *noun* plant *or* animal on which

a parasite lives; **intermediate host** = host in which a parasite lives for a time before passing on to another host; **definitive host** = host in which a parasite settles and breeds

hothouse ['hɒthaʊs] *noun* heated greenhouse

house [haʊs] **1** *noun* building in which livestock is kept, such as a poultry house **2** *verb* to keep livestock in a building; *the animals are housed in clean cubicles*

◊ **housing** ['haʊzɪŋ] *noun* series of buildings for livestock

HRI = HORTICULTURAL RESEARCH INTERNATIONAL

HSE = HEALTH AND SAFETY EXECUTIVE British government organization responsible for checking the conditions of work of workers, including farmworkers

HTA = HORTICULTURAL TRADES ASSOCIATION

HTST method = HIGH TEMPERATURE SHORT TIME METHOD

huerta ['hweətə] *noun* small highly-cultivated lowland area in the Spanish provinces of Valencia, Murcia and Granada; the area is capable of producing two or more crops per year with irrigation

hull [hʌl] *noun* (i) outer covering of a cereal seed (hulls form bran); (ii) the pod of peas or beans

◊ **huller** ['hʌlə] *noun* kind of threshing machine which removes seeds from their husks

humate ['hju:meɪt] *noun* salt which comes from humus

humid ['hju:mɪd] *adjective* (air) which is damp *or* which contains moisture vapour; *decomposition of organic matter is rapid in hot and humid conditions*

◊ **humidify** [hju:'mɪdɪfaɪ] *noun* to add water vapour to air to make it more humid

◊ **humidity** [hju:'mɪdɪti] *noun* measurement of how much water vapour is contained in the air; **absolute humidity** = ratio of the mass of water vapour in a given quantity of air to the amount of air; **relative humidity** = amount of water vapour in a

given quantity of air, shown as a proportion of the amount needed to saturate the air

humify ['hju:mɪfaɪ] *verb* to break down rotting organic waste to form humus

◊ **humification** [hju:mɪfɪ'keɪʃn] *noun* breaking down of rotting organic waste to form humus

hump [hʌmp] *noun* rounded flesh on the back or shoulders of an animal, such as a camel, or certain breeds of cattle

humus ['hju:məs] *noun* **(a)** well decomposed organic matter in soil, which makes the soil dark and binds it together **(b)** dark organic residue left after sewage has been treated in sewage works

hundredweight ['hʌndrədweɪt] *noun* measure of weight of dry goods, such as grain (NOTE: abbreviated after numbers to **cwt: 5cwt**)

> COMMENT: the British hundredweight is equivalent to 50.8kg, and the US hundredweight is equivalent to 45.4kg

hungry soil ['hʌngri 'sɔɪl] *noun* soil which lacks nutrients, and so needs large amounts of fertilizer to produce good crops; **nitrogen-hungry plants** = plants which need a lot of nitrogen

hurdle ['hɜ:dl] *noun* portable rectangular wooden frame used for temporary fencing for sheep

husbanding ['hʌzbəndɪŋ] *noun* using and keeping carefully; *a policy of husbanding scarce natural resources*

◊ **husbandry** ['hʌzbəndri] *noun* farming *or* looking after animals and crops; *a new system of intensive cattle husbandry*

> QUOTE there is an added risk of the spread of infection under intensive farming methods, both by the stock being housed under more concentrated conditions and by the greater use of artificial feeding in this method of husbandry
>
> *Farmers Weekly*

husk [hʌsk] **1** *noun* **(a)** dry outer covering of cereal grains, which has little food value, and which is removed during threshing **(b)** parasitic form of bronchitis, caused by lungworms; found mainly in cattle **2** *verb* to remove the husk from seeds (NOTE: **dehusk** means the same)

hybrid ['haɪbrɪd] *noun* **(a)** cross between two varieties *or* species of plant or animal; **hybrid vigour** = increase in size *or* rate of growth *or* fertility *or* resistance to disease found in offspring of a cross between two species **(b)** the first generation of a cross between two species or two which differ in one or more genes

◊ **hybridization** [haɪbrɪdaɪ'zeɪʃn] *noun* production of hybrids

◊ **hybridize** ['haɪbrɪdaɪz] *verb* to produce hybrids by crossing varieties of plants or animals

hydrated lime (Ca(OH)$_2$) [haɪ'dreɪtɪd 'laɪm] *noun* substance produced when burnt lime is wetted; also called 'slaked lime', it is a mixture of calcium oxide and water, used to improve soil quality; the lime is in powder form, having been burnt to break it down from large lumps

hydrochloric acid [haɪdrə'klɒrɪk 'æsɪd] *noun* an inorganic acid which forms in the stomach and is part of the gastric juices

hydrocool [haɪdrə'kuːl] *verb* to cool fresh fruit to prevent it from deteriorating during transport to the retail shop or market

hydrogen ['haɪdrədʒən] *noun* chemical element, a gas which combines with oxygen to form water, and with other elements to form acids, and is present in all animal tissue (NOTE: chemical symbol is **H**; atomic number is **1)**

◊ **hydromorphic soil** [haɪdrə'mɔːfɪk 'sɔɪl] *noun* waterlogged soil found in bogs and marshes

◊ **hydrophobia** [haɪdrə'fəubiə] *noun* another name for rabies

◊ **hydrophyte** ['haɪdrəfaɪt] *noun* plant which lives in water *or* in marshy conditions

◊ **hydroponics** [haɪdrə'pɒnɪks] *noun* science of growing plants without soil but in sand *or* other granular material using a liquid solution of nutrients to feed them

hydrops uteri ['haɪdrɒps 'jutəri] *noun* disease of livestock caused by excessive amount of fluid in the pregnant womb, womb may need to be drained

hygrometer [haɪ'grɒmɪtə] *noun* device for measuring humidity (as for example, the humidity in stored grain)

hyperphosphate [haɪpə'fɒsfeɪt] *noun* soft rock phosphate obtained from North America

hyphae ['haɪfiː] *noun* long thin filaments which make up a typical fungus

hypomagnesaemia [haɪpəumægni'ziːmiə] *noun* condition due to lack of magnesium in the bloodstream; animals shiver and stagger. Cattle may be affected shortly after being turned out onto spring pastures after having wintered indoors. Also called 'grass staggers'

HYV = HIGH-YIELDING VARIETY

Ii

I *chemical symbol for* iodine

IACR = INSTITUTE OF ARABLE CROPS RESEARCH

IACS = INTEGRATED ADMINISTRATION AND CONTROL SYSTEM farmers claiming arable aid payments, beef premiums and suckler cow premiums must complete IACS forms. This asks for maps of field-by-field cropping and other details

IAH = INSTITUTE FOR ANIMAL HEALTH

IBAP = INTERVENTION BOARD FOR AGRICULTURAL PRODUCE

ICA = INTERNATIONAL COFFEE AGREEMENT agreement between countries to stabilize the price of coffee

ICC = INDEPENDENT CROP CONSULTANT

ICCA = INTERNATIONAL COCOA AGREEMENT agreement between countries to stabilize the price of cocoa

ICCO = INTERNATIONAL COCOA ORGANIZATION international organization set up to stabilize the international market in cocoa by holding buffer stocks to offset seasonal differences in production quantities

-icide [ɪ'saɪd] *suffix* meaning 'which kills'; **pesticide** = substance which kills pests

ICO = INTERNATIONAL COFFEE ORGANIZATION international organization set up to stabilize the international market in coffee by holding buffer stocks to offset seasonal differences in production quantities

ICRISAT = INTERNATIONAL CROPS RESEARCH INSTITUTE FOR THE SEMI-ARID TROPICS

IFA = IRISH FARMERS ASSOCIATION

IFR = INSTITUTE OF FOOD RESEARCH

IGAP = INSTITUTE FOR GRASSLAND AND ANIMAL PRODUCTION

IGER = INSTITUTE OF GRASSLAND AND ENVIRONMENTAL RESEARCH

IHR = INSTITUTE OF HORTICULTURE RESEARCH

IITA = INTERNATIONAL INSTITUTE OF TROPICAL AGRICULTURE

ILCA = INTERNATIONAL LIVESTOCK CENTRE FOR AFRICA

Ile de France ['iːldəfrɑːns] *noun* large French breed of sheep; rams are kept to provide crossbred lambs for meat production

illuviation [ɪluːviˈeɪʃn] *noun* deposition of particles and chemicals leached out from the topsoil into the subsoil

ILRAD = INTERNATIONAL LABORATORY FOR RESEARCH ON ANIMAL DISEASES

imbalance [ɪmˈbæləns] *noun* lack of balance *or* situation where one substance *or* species is dominant; *lack of vitamins A and E creates hormonal imbalances in farm animals*

immature [ɪməˈtjʊə] *adjective* not mature *or* not fully grown

immune [ɪˈmjuːn] *adjective* protected against an infection *or* allergic disease; *this strain is not immune to virus*

◊ **immunity** [ɪˈmjuːnɪti] *noun* **(a)** *(in animals)* ability to resist attacks of a disease because antibodies are produced; **acquired immunity** = immunity which a body acquires and which is not congenital; **natural immunity** = immunity from disease inherited by newborn offspring from the mother **(b)** *(in plants)* resistance to disease through a protective covering on leaves *or* the formation of protoplasts *or* development of inactive forms of viruses

◊ **immunization** [ɪmjuːnaɪˈzeɪʃn] *noun* making an animal immune to an infection, either by injecting an antiserum (passive immunization) or by giving the body the disease in such a small dose that the body does not develop the disease, but produces antibodies to counteract it

◊ **immunize** [ˈɪmjuːnaɪz] *verb* to inject an animal with a substance to make it immune to an infection

impermeable [ɪmˈpɜːmiəbl] *adjective* (substance) which does not allow a liquid *or* gas to pass through; (membrane) which allows a liquid to pass through, but not solid particles suspended in the liquid; *rocks which are impermeable to water*

implement 1 *noun* [ˈɪmplɪmənt] tool, piece of equipment used for a certain job; **agricultural implements** = tools, such as billhooks, scythes, which are used on farms; **garden implements** = implements such as forks and spades which are used in the garden **2** *verb* [ˈɪmplɪment] to put legislation into action

import 1 *noun* [ˈɪmpɔːt] crops or produce which are bought abroad and brought into the country; **import quotas** = limits set to the amount of a type of produce which can be imported **2** *verb* [ɪmˈpɔːt] to buy crops or produce in foreign countries and bring them back into the home country; *they import dates from North Africa*

◊ **importer** [ɪmˈpɔːtə] *noun* person or company which imports produce; *a grain importer*

impoverish [ɪmˈpɒvərɪʃ] *verb* to make poor; *overfarming has impoverished the soil*

◊ **impoverished** [ɪmˈpɒvərɪʃt] *adjective* made poor; *if impoverished soil is left fallow for some years, nutrients may build up in the soil again*

◊ **impoverishment** [ɪmˈpɒvərɪʃmənt] *noun* making poor; *overexploitation led to the impoverishment of the soil*

impregnate [ˈɪmpregneɪt] *verb* to fertilize a female, by introducing male spermatozoa into the female's body so that they link with the female's ova

improve [ɪmˈpruːv] *verb* to make better; **improved varieties** = new species of plants

which are stronger, or more productive than old species

◊ **improvement** [ɪmˈpruːvmənt] *noun* making better; **improvement grant** = grant available to a farmer to improve the standard of the farm; grants are available for a variety of purposes, e.g. for the eradication of bracken on pasture land, for draining, for providing a water supply to fields

in- [ɪn] *prefix* used to refer to a pregnant female animal; **in-calf, in-pig, in-foal** = cow, sow or mare which is going to have young; **in-lay** = period when a hen is laying eggs; **in-milk** = lactation period of a cow; **in season** = period when a female animal is ready for mating

inactive [ɪnˈæktɪv] *adjective* which does not act

◊ **inactivate** [ɪnˈæktɪveɪt] *verb* to make something inactive; *biopesticides are easily inactivated in sunlight*

inbreeding [ˈɪnbriːdɪŋ] *noun* breeding between a closely related male and female, who have the same parents or grandparents, which strengthens the dominant trait in the breed, but also increases the possibility of congenital defects; **inbreeding depression** = reduction of characteristic strong traits by inbreeding; *compare* OUTBREEDING

◊ **inbred** [ɪnˈbred] *adjective* suffering from inbreeding

COMMENT: inbreeding sorts out some of the best qualities in stock and has been used to establish uniform flocks or herds carrying distinctive traits

incisor [ɪnˈsaɪzə] *noun* cutting tooth at the front of the animal's mouth

incorporate [ɪnˈkɔːpəreɪt] *verb* (i) to apply chemicals, such as slug pellets, by spreading them in rows at the same time as the seed is sown, as opposed to broadcasting; (ii) to plough back straw or green manure into the soil

◊ **incorporation** [ɪnkɔːpəˈreɪʃn] *noun* introducing chopped straw, green manure, etc., into the soil as it is being ploughed

QUOTE having incorporated straw now for ten years, he finds that it has improved soil texture
Arable Farming

incubation [ɪŋkjuː'beɪʃn] *noun* hatching of eggs either by a sitting bird or by artificial means in an incubator; **incubation period** = (i) the length of time take for a chick to hatch; (ii) length of time between the infection of an animal by a disease and the appearance of the first symptoms

◊ **incubator** ['ɪŋkjuːbeɪtə] *noun* special unit providing artificial heat used to hatch eggs; available as small trays, or as large rooms for large-scale producers

indefinite inflorescence [ɪn'defɪnɪt ɪnfləˈresəns] *noun* inflorescence where the stems bearing the flowers continue to grow

Indian buffalo ['ɪndiən 'bʌfələʊ] *noun* domesticated water buffalo used as a draught animal in India and Sri Lanka; the animals are grey, black or dark brown in colour, with large horns. Buffalo milk is used for making butter

◊ **Indian corn** ['ɪndiən 'kɔːn] *see* MAIZE

◊ **Indian game** ['ɪndiən 'geɪm] *noun* breed of table poultry, often black with yellow legs

indigenous [ɪn'dɪdʒənəs] *adjective* which is native to a place; *oaks are indigenous to the British Isles; the government encourages the planting of indigenous hardwoods where possible*

indigestible [ɪndɪ'dʒestɪbl] *adjective* (food) which cannot be digested, such as roughage

indigo ['ɪndɪgəʊ] *noun* blue dye extracted from species of the *Indigofera* bush

infect [ɪn'fekt] *verb* to contaminate with disease-producing microorganisms *or* toxins; to transmit infection; **infected area** = place where animals must be kept in isolation as a result of a notice issued by the Animals Inspector when an animal is suspected or known to have a notifiable disease

◊ **infection** [ɪn'fekʃn] *noun* entry of microbes into the body, which then multiply in the body

◊ **infectious** [ɪn'fekʃəs] *adjective* (disease) which is caused by microbes and can be transmitted to other animals by direct means

◊ **infectious bovine rhinotracheitis** [ɪn'fekʃəs 'bəʊvaɪn raɪnəʊtrækɪ'aɪtɪs] *noun* virus disease of cattle which affects the reproductive, nervous, respiratory or digestive systems. Milk yield is depressed as adults run a high fever

◊ **infective** [ɪn'fektɪv] *adjective* (disease) caused by a microbe, which can be caught from another animal but which cannot always be directly transmitted

◊ **infectivity** [ɪnfek'tɪvɪti] *noun* being infective

infertile [ɪn'fɜːtaɪl] *adjective* (i) not able to bear fruit; (ii) *(of animal)* not able to produce young; (iii) *(of soil)* barren *or* not able to produce good crops

◊ **infertility** [ɪnfə'tɪlɪti] *noun* not being fertile *or* not being able to reproduce

COMMENT: an infertile soil is one which is deficient in plant nutrients. The fertility of a soil at any one time is partly due to its natural makeup, and partly to its condition, which is largely dependent on the management of the soil in recent times. Application of fertilizers can raise soil fertility; bad management can decrease it. Infertility in cattle can be a serious problem, and may be caused by disease such as contagious abortion, or by lack of or imbalance in minerals in the diet

infest [ɪn'fest] *verb (of parasites)* to be present in large numbers; *pine forests have been infested with beetles; infested plants should be dug up and burnt*

◊ **infestation** [ɪnfes'teɪʃn] *noun* having large numbers of parasites; *the crop showed serious infestation with greenfly*

in-field ['ɪnfiːld] *noun* formerly, the field nearest the farmstead, regularly manured and cultivated; in-fields are still preserved on some hill farms

infiltration [ɪnfɪl'treɪʃn] *noun* (i) passing of waste water into the soil *or* into a drainage system; (ii) irrigation system where water passes through many small channels to reach the fields; **infiltration capacity** = rate at which water is absorbed by soil

inflorescence [ɪnflə'resəns] *noun* a flower or group of small flowers, together with the stems called spikelets

COMMENT: there are two types of inflorescence: indefinite inflorescence is where the branches bearing the flowers continue to grow; where the main stem ends in a single flower and then stops growing is known as definite inflorescence

in-ground valuation ['ɪngraʊnd vælju'eɪʃən] *noun* value of tillages or cultivations including direct costs of seed, fertilizer and spray

in-going payment ['ɪngəʊɪŋ 'peɪmənt] *noun* sum of money paid by a new tenant for the value of the improvements made by the former tenant of a holding

inherit [ɪn'herɪt] *verb (of offspring)* to receive characteristics from a parent's genes

inhibit [ɪn'hɪbɪt] *verb* to make it difficult for something to happen; *lack of nutrients may inhibit plant growth*

◊ **inhibitor** [ɪn'hɪbɪtə] *noun* **growth inhibitor** = substance which stops a plant growing; **nitrification inhibitor** = chemical substance which prevents ammonium nitrate from forming in slurry

inject [ɪn'dʒekt] *verb* to put a liquid into an animal's body under pressure, by using a hollow needle inserted into the tissues; *the cow was injected with antibiotics*

◊ **injection** [ɪn'dʒekʃn] *noun* **(a)** act of injecting a liquid into the body; **intramuscular injection** = injection of liquid into a muscle (as for a slow release of a drug); **intravenous injection** = injection of liquid into a vein (as for fast release of a drug); **hypodermic injection** *or* **subcutaneous injection** = injection of a liquid beneath the skin (as for pain-killing drugs) **(b)** liquid introduced into the body; *the vet gave the bull an injection of antibiotics*

inoculate [ɪ'nɒkjuleɪt] *verb* to introduce vaccine into an animal's body in order to make the body create antibodies, so making the animal immune to the disease; *the animals are inoculated against pneumonia*

◊ **inoculation** [ɪnɒkju'leɪʃn] *noun* action of inoculating

inorganic [ɪnɔː'gænɪk] *adjective* (substance) which does not come from animal or vegetable sources and does not contain carbon; *inorganic substances include acids, alkalis and metals;* **inorganic fertilizer** = artificial synthesized fertilizer (as opposed to manure, compost and other

organic fertilizers which are produced from bones, blood and other parts of formerly living matter); **inorganic pesticide** *or* **fungicide** *or* **herbicide** = pesticide *or* fungicide *or* herbicide made from inorganic substances such as sulphur

inputs ['ɪnpʊts] *noun* substances put into the soil (such as fertilizers which are applied by a farmer)

ins [ɪnz] *noun* term used to describe the points where the plough enters the ground when leaving the headland (the points where the plough is lifted out of the soil are the outs)

insect ['ɪnsekt] *noun* small animal with six legs and a body in three parts; **insect bites** = stings caused by insects which puncture the skin and in so doing introduce irritants

◊ **insecticide** [ɪn'sektɪsaɪd] *noun* natural or synthetic substance which kills insects

COMMENT: insects form the class Insecta. The body of an insect is divided into three distinct parts: the head, the thorax and the abdomen. The six legs are attached to the thorax. There are two antennae on the head. Most insect bites are simply irritating, but some people can be extremely sensitive to the bites of certain types of insect (such as bee stings). Other insect bites can be more serious, as insects can carry the bacteria which produce typhus, sleeping sickness, malaria, filariasis, etc. Natural insecticides (i.e. those produced from plant extracts) are less harmful to the environment than the synthetic insecticides which, though effective, are often persistent and kill not only insects but also other larger animals when they get into the food chain. In agriculture, most pesticides are chemically based: they are either chlorinated hydrocarbons, organophosphorous compounds or carbamate compounds. Insecticides are used in a number of ways, including spraying and dusting, or in granular forms as seed dressings. In the form of a gas, insecticides are used to fumigate greenhouses and granaries

QUOTE the perennial problem of locust infestation in North Africa can be contained by comprehensive spraying of insecticides. The widespread banning of the use of dieldrin, a potent agent against the locust, is adding to costs, since other more benign chemicals require more frequent application
Middle East Agribusiness

inseminate [ɪnˈsemɪneɪt] *verb* to impregnate, by introducing male spermatozoa into the female's body so that they link with the female's ova

◊ **insemination** [ɪnsemɪˈneɪʃn] *noun* the introduction of sperm into the vagina; **artificial insemination (AI)** = way of breeding livestock by fertilizing females with sperm from specially selected males; AI is important in cattle breeding. Milk Marque operates a number of AI centres

inspector [ɪnˈspektə] *noun* official whose job is to examine animals *or* soil *or* buildings, etc., to see if they are in conformity with government regulations; **Animals Inspector** = official whose job is to inspect animals to see if they have notifiable diseases, are being kept in acceptable conditions, etc.

instant [ˈɪnstənt] *adjective* (food) which can be prepared very quickly; **instant coffee** = coffee powder which can be made into a drink by adding boiling water

intake [ˈɪnteɪk] *noun* (a) amount of a substance taken into an organism (either eaten or absorbed); *a study of food intake among grassland animals; the bird's daily intake of insects is more than half its own weight* (b) hill pasture which has been improved and fenced in

intensity [ɪnˈtensɪti] *noun* degree to which land is used; **low-intensity land** = land on which crops are not intensively cultivated

◊ **intensification** [ɪntensɪfɪˈkeɪʃn] *noun* doing something in an intensive way; *intensification of farming has contributed to soil erosion*

intensive [ɪnˈtensɪv] *adjective* **intensive agriculture** = system of farming characterized by a high level of inputs, especially capital, labour and fertilizer, per unit area of land; **intensive beef production** = production of a young lean beef animal in a period of less than a year; **intensive cultivation** *or* **intensive farming** = (farming) in which as much use is made of the land as possible (growing crops close together *or* growing several crops in a year *or* using large amounts of fertilizer); **intensive livestock production** = specialized system of livestock production where the livestock are housed indoors; this system can be started up at any time of the year; disease

hazards are those related to diet and permanent housing for the whole of the animal's life

◊ **intensively** [ɪnˈtensɪvli] *adverb* in an intensive way (NOTE: the opposite is **extensive, extensification)**

> QUOTE the diet of animals kept in intensive units is usually quite different from what the animals would eat if foraging for themselves. Grazing animals have the ability to seek out certain plants which contribute to their health
> *Ecologist*

inter- [ˈɪntə] *prefix* meaning between

interbreed [ɪntəˈbriːd] *verb* to breed together (individuals from the same species can interbreed; those from different species cannot)

> COMMENT: interbreeding of close relatives can sometimes give a concentration of desirable traits. This was much used by breed pioneers, but it can also increase the frequency of inherited physical defects and breeding plans based on interbreeding are now rare

intercrop [ˈɪntəkrɒp] **1** *noun* crop which is grown between the rows of other crops, such as barley and mustard or pigeon pea and black gram **2** *verb* to grow crops between the rows of other crops

> QUOTE in intercropping systems, two or more crops differing in height, canopy spread, rooting behaviour and growth habit, are grown simultaneously on the same piece of land in separate rows in such a way that they accommodate each other with least competition
> *Indian Farming*

interculture [ˈɪntəkʌltʃə] *noun* the practice of mixed cropping, where two or more different crops are grown together on the same area of land

intermediate [ɪntəˈmiːdiət] *adjective* between two extremes; **intermediate host** = host on which a parasite lives for a time before passing on to another; **intermediate technology** = technology which is between the advanced electronic technology of industrialized countries and the primitive technology in underdeveloped countries

intermuscular fat [ɪntəˈmʌskjələ ˈfæt]

noun fat between muscles (as opposed to intramuscular fat)

internal laying [ɪn'tɜːnəl 'leɪɪŋ] *noun* condition in hens caused by a fault in the oviduct, which results in the yolks not being passed along the oviduct for covering with membranes and shell

international [ɪntə'næʃnəl] *adjective* between countries; **International Union for the Conservation of Nature (IUCN)** = international organization which coordinates the protection of species throughout the world; **International Cocoa Organization (ICCO)** = international organization set up to stabilize the international market in cocoa by holding buffer stocks to offset seasonal differences in production quantities; **International Coffee Organization (ICO)** = international organization set up to stabilize the international market in coffee by holding buffer stocks to offset seasonal differences in production quantities; **International Crops Research Institute for the Semi-Arid Tropics (ICRISAT)** = organization established in 1972 at Hyderabad, India; **International Institute of Tropical Agriculture (IITA)** = organization established at Ibadan, Nigeria in 1965; **International Livestock Centre for Africa (ILCA)** = organization established in 1974 at Addis Ababa, Ethiopia; **International Laboratory for Research on Animal Diseases (ILRAD)** = organization established at Nairobi, Kenya in 1974; **International Rice Research Institute (IRRI)** = organization established at Los Banos, Philippines in 1959; **International Sugar Organization (ISO)** = international organization formed of sugar-exporting countries; **International Wheat Council (IWC)** = group of wheat-exporting countries; **International Wool Secretariat (IWS)** = group which represents countries which export wool

interrelay cropping [ɪntə'riːleɪ 'krɒpɪŋ] *noun* cropping system in which the crops are grown in quick succession, so much so, that the succeeding crop is sown in the standing one, some time before it is harvested

intersow ['ɪntəsəʊ] *verb* to sow seed between rows of existing plants

COMMENT: studies conducted at the Punjab Agricultural University suggest the possibility of intersowing wheat in the furrows between the consecutive potato

ridges at the time of earthing up; summer mungbean (green gram) can then be intersown in the standing wheat crop a few days before harvest, using the space released by the potatoes

interveinal [ɪntə'veɪnəl] *adjective* between the veins; **interveinal yellowing** = condition of plants caused by magnesium deficiency, where the surface of the leaves turns yellow and the veins stay green

intervention [ɪntə'venʃn] *noun* acting to make a change; **Intervention Board** = body set up in 1972 to implement the regulations of the Common Agricultural Policy in the UK; the Board is mainly concerned with compensatory payments; **intervention buying** = a feature of the Common Agricultural Policy, whereby governments or their agents offer to buy surplus agricultural produce at a predetermined price; it is subject to a minimum quality standard; **intervention price** = price at which the EU will buy farm produce which farmers cannot sell, in order to store it; **intervention stocks** = the various amounts of commodities held by the EU in store; **to sell into intervention** = to sell to a government agency at an intervention price because the market price is too low

COMMENT: the intervention price is the price at which the national intervention agencies are obliged to buy up agricultural commodities offered to them. There are intervention prices on products such as wheat, barley, beef and pigmeat. The application of the system of intervention prices has led to the accumulation of vast stocks of commodities, some of which are sold on the world markets at very low prices

QUOTE the threat of having to sell into intervention loomed closer this week as grain prices eased again despite lower than expected harvest estimates
Farmers Weekly

intestine [ɪn'testɪn] *noun* **the intestines** = the tract which goes from the stomach to the anus, in which food passes and is digested

◊ **intestinal** [ɪntɪs'taɪnəl] *adjective* referring to the intestine; **intestinal diseases** = diseases and conditions, such as anthrax, dysentery, parasites, enteritis, swine fever, which affect the intestines of animals

intramuscular ['ɪntrə'mʌsjələ] *adjective*

which is inside the muscle (as intramuscular fat in meat)

introduce [ɪntrə'djuːs] *verb* to bring something into a new place

◊ **introduction** [ɪntrə'dʌkʃn] *noun* **(a)** bringing something to a new place; *before the introduction of grey squirrels, the red squirrel was widespread* **(b)** plant *or* animal which has been introduced to a new area; *it is not an indigenous species but a 19th century introduction*

invertebrate [ɪn'vɜːtɪbrət] *adjective* & *noun* (animal) that has no backbone; **invertebrate pests** = pests such as grain mites and storage insects such as saw-toothed beetles and the grain weevil, which cause considerable damage to crops in tropical or warm temperate areas (NOTE: the opposite is **vertebrate)**

iodine ['aɪədiːn] *noun* chemical element which is essential to the body, especially to the functioning of the thyroid gland, found in seaweed (NOTE: chemical symbol is **I**; atomic number is **53**)

◊ **iodophor** [aɪ'ɒdəfɔː] *noun* disinfectant used to disinfect teats of cows to prevent mastitis

ion ['aɪən] *noun* atom which has an electric charge (ions with a positive charge are called cations, and those with a negative charge are anions); **ion exchange** = exchanging of ions between a solid and a solution

IPU = ISOPROTURON

Irish Moiled ['aɪrɪʃ 'mɔɪld] *noun* a rare breed of medium-sized dual-purpose cattle; the animals have a distinctive white back strip

iron ['aɪən] *noun* metallic element essential to biological life and an essential part of human diet (found in liver, eggs, etc.) (NOTE: chemical symbol is **Fe**; atomic number is **26**)

COMMENT: iron is an essential part of the red pigment in red blood cells. Lack of iron in haemoglobin results in iron-deficiency anaemia. Its role in the physiology of plants appears to be associated with specific enzymatic reactions and the production of chlorophyll

irongrass *or* **ironweed** ['aɪəngrɑːs *or*

'aɪənwiːd] *noun* popular names for the knotgrass

irradiate [ɪ'reɪdɪeɪt] *verb* (i) to subject something to radiation; (ii) to treat food with radiation to prevent it going bad

◊ **irradiation** [ɪreɪdɪ'eɪʃn] *noun* **(a)** spread from a centre, as nerve impulses **(b)** use of rays to kill bacteria in food

COMMENT: food is irradiated with gamma rays from isotopes which kill bacteria. It is not certain, however, that irradiated food is safe for humans to eat, as the effects of irradiation on food are not known. In some countries irradiation is only permitted as a treatment of certain foods

IRRI = INTERNATIONAL RICE RESEARCH INSTITUTE

irrigate ['ɪrɪgeɪt] *verb* to supply land with water

◊ **irrigation** [ɪrɪ'geɪʃn] *noun* the artificial supplying and application of water to land with growing crops; *new areas of land must be brought under irrigation to meet the rising demand for food;* **basin irrigation** = irrigation technique where water is trapped in basins surrounded by low mud walls (as in rice fields); **flood irrigation** = irrigation system where water is allowed to flood over the land; **furrow irrigation** = irrigation technique where water is allowed to flow along furrows

◊ **irrigator** ['ɪrɪgeɪtə] *noun* device for irrigating, such as the Baars irrigator

COMMENT: irrigation can be carried out using powered rotary sprinklers, rain guns, spray lines or by channelling water along underground pipes or small irrigation canals from reservoirs or rivers. Irrigation water can be more effectively used than the equivalent amount of rainfall, because a regular supply is ensured. Basin or flood irrigation is a primitive form of irrigation, where flood waters from rivers are led to prepared basins. Perennial irrigation allows the land to be irrigated at any time. This may be by primitive means such as shadufs, or by distributing water from barrages by canal and ditches. It is usual to measure irrigation water in millimetres: 1mm on one hectare equals 10m^3 or ten tonnes. Irrigation is not necessarily always advantageous to the land, as it can cause salinization of the soil. This happens when the soil becomes waterlogged so that salts in the soil rise to the surface. At the surface, the irrigated water rapidly evaporates, leaving the salts behind in the form of a saline crust. Irrigation also has the further disadvantage of increasing the spread of disease. Water insects easily spread through irrigation canals and reservoirs. In the United Kingdom the greatest need for irrigation is in the east, where the lower rainfall and higher potential evaporation and transpiration means that irrigation is beneficial nine years out of ten. In the UK, potatoes, sugar beet, horticultural crops and grassland are the main irrigated crops

QUOTE in 1950 about 2 million hectares of the Aral Sea zone were irrigated. These days, irrigation covers some 7 million hectares and provides the Soviet Union with 95% of its cotton

Guardian

QUOTE the most significant progression in agricultural irrigation development has been the move away from flood irrigation to spray type with sprinklers

Arab World Agribusiness

isinglass ['aɪzɪŋglɑːs] *noun* pure soluble gelatin, used to make alcoholic drinks clear; formerly used to preserve eggs

iso- ['aɪsəʊ] *prefix* meaning equal

◊ **isobar** ['aɪsəbɑː] *noun* line on a map linking points of equal barometric pressure at the same time

◊ **isoglucose** [aɪsə'gluːkəʊz] *noun* sweetener used in the soft drinks industry, extracted from maize (NOTE: also called **high fructose corn syrup**)

◊ **isohyet** [aɪsəʊ'haɪət] *noun* line on a map linking points of equal rainfall

isolate ['aɪsəleɪt] *verb* to make something stand alone; *to isolate infected animals*

◊ **isolation** [aɪsə'leɪʃn] *noun* keeping infected animals away from others

isoleucine [aɪsəʊ'luːsiːn] *noun* essential amino acid

◊ **isotherm** ['aɪsəʊθɜːm] *noun* line on a map linking points of equal temperature

◊ **Isoproturon** [aɪsə'prɒtjʊrɒn] *noun* a selective weedkiller used on certain cereals, especially against blackgrass

Italian ryegrass [ɪ'tælɪən 'raɪgrɑːs] *noun* short lived ryegrass *(Lolium multiflorum);* it is sown in spring, and is very quick to establish; it produces good growth in its seeding year and early graze the following year; it is commonly used for short duration leys

itch [ɪtʃ] *noun* form of mange; **itch mite** = an arachnid *(Sarcoptes scabiei)* which burrows into the animal's skin, causing itching

IUCN = INTERNATIONAL UNION FOR THE CONSERVATION OF NATURE

ivy ['aɪvi] *noun* climbing evergreen plant *(Hedera helix)*

◊ **ivy-leaved speedwell** ['aɪvɪliːvd 'spiːdwel] *noun* widespread weed *(Veronica hederifolia)* which affects most autumn sown crops; *also called* BIRD'S-EYE, EYE-BRIGHT

IWC = INTERNATIONAL WHEAT COUNCIL

IWS = INTERNATIONAL WOOL SECRETARIAT

Jj

J = JOULE

Jack bean ['dʒæk 'biːn] *noun* tropical legume *(Canavalia ensiformis)* grown as a fodder crop

Jacob ['dʒeɪkəb] *noun* rare breed of sheep with multi-coloured fleece; it is medium-sized and multi-horned, with a white coat and brown or black patches

jaggery ['dʒægəri] *noun* *(Indian subcontinent & parts of Africa)* non-centrifugal sugar from sugar cane, originally used to refer to sugar tapped from palm trees

Jamaica Hope [dʒə'meɪkə 'həʊp] *noun* breed of dairy cattle developed by the Jamaican government from Jersey cows and Sahiwal Indian bulls; the colour is fawn or brown

Japanese millet ['dʒæpəniːz 'mɪlɪt] *noun* grain crop *(Echinochloa frumentacea)* cultivated in Japan and Korea for human consumption; grown in the USA as a forage crop

Java cotton ['dʒævə 'kɒtən] *noun* silky fibres which cover the seeds of the kapok tree *(Ceiba pentandra)* used for filling pillows and cushions

Jersey ['dʒɜːzi] *noun* important breed of dairy cattle, originally from the island of Jersey; Jersey cows are smaller than most other breeds and produce high yields of high butterfat content milk. The cattle are variously coloured from light fawn to red and almost black

Jerusalem artichoke [dʒə'ruːsələm 'aːtɪtʃəʊk] *see* ARTICHOKE

jetting ['dʒetɪŋ] *noun* **(a)** a method of applying insecticide under pressure, used on sheep; **jetting gun** = gun used to apply insecticide **(b)** method of cleaning out blocked field drains using high pressure water jets

Jhum [jʌm] *noun* form of slash-and-burn shifting cultivation practised in Nagaland and Arunachal Pradesh, India

Johne's disease ['dʒəʊnz dɪ'ziːz] *noun* serious infectious inflammation of the intestines, particularly in cattle; affected animals rapidly become extremely thin

joint [dʒɔɪnt] *noun* **(a)** part of an organism where two sections are joined; **leaf joint** = point on the stem of a plant where a new shoot may grow **(b)** piece of meat ready for cooking, usually containing a bone; *a joint of beef; a bacon joint*

◊ **joint-ill** ['dʒɔɪnt 'ɪl] *noun* disease of young livestock, especially newborn calves, kids, and lambs; it causes abscesses at the navel and swellings in some joints; *also known as* NAVEL-ILL

jojoba [hə'həʊbə] *noun* *Simmondsia chinensis*, a perennial plant whose seeds yield an oil which is liquid wax, grown in the USA

joule [dʒuːl] *noun* SI unit of energy (NOTE: usually written **J** with figures: **25J**)

> COMMENT: one joule is the amount of energy used to move one kilogram the distance of one metre, using the force of one newton. 4.184 joules equals one calorie, and has now replaced the calorie in the SI system

juice [dʒuːs] *noun* liquid inside a fruit or vegetable; liquid inside cooked meat; **juice extractor** = device for extracting juice from a fruit or vegetable

Jumna pari ['dʒʌmnə 'paːri] *noun* breed of goat found in India; a large animal with long lop ears, it is reared for its milk

June agricultural census [dʒuːn 'sensəs] *noun* in Great Britain, the annual farm returns on 4th June each year provide the basis for a census of agriculture carried out by the MAFF; the census covers crop areas, numbers of livestock, production and yields, number and size of holdings, numbers of workers, farm machinery, prices and incomes

◊ **June drop** ['dʒuːn 'drɒp] *noun* the fall of small fruit from apples and pears, which takes place in summer as the fruit begin to grow; some fruit fall off, allowing the remaining fruit to mature more easily

juniper ['dʒuːnɪpə] *noun* coniferous tree of the genus *Juniperus;* the berries are used for flavouring, especially in making gin

jute [dʒuːt] *noun* coarse fibre from a plant *(Corchorus* sp), used to make sacks, coarse cloth and cheap twine

COMMENT: the main producers of jute are Bangladesh (which produces over 50% of the total world production), India and Brazil

juvenile ['dʒuːvənaɪl] *noun* stage in the development of an animal, before it becomes adult (used especially of insects)

Kk

K *symbol for* potassium

k *abbreviation for* KILO **kg** = kilogram; **kJ** = kilojoule

Vitamin K [vɪtəmɪn 'keɪ] *noun* vitamin found in green vegetables like spinach and cabbage, which helps the clotting of blood and is needed to activate prothrombin

kaffir corn ['kæfə 'kɔːn] *noun* variety of grain sorghum, rather similar to maize; grown for food and animal forage in tropical Africa

kainite ['keɪnaɪt] *noun* potash fertilizer, made of a mixture of potassium and sodium salts, with sometimes magnesium salts added, used mainly on sugar beet and similar crops

kale [keɪl] *noun* type of brassica, sometimes used as a green vegetable for human consumption, but mainly grown as animal forage (NOTE: also called **bore cole**)

COMMENT: kale can be fed to animals in the field, or made into silage for use during winter. The main types of kale are the marrowstem, which produces heavy crops but is not winter hardy; the thousand-headed, which is hardier; and dwarf thousand-head, which produces a large number of new shoots late in the winter. Other hybrid varieties are also available. Kale is the commonest of green crops other than grass. The highest feeding value is in the leaf rather than the stem

Kambing Katjang ['kæmbɪŋ 'kætʃæŋ] *noun* breed of goat found in Malaysia and Indonesia; the colour is usually black. The animals are horned and bearded, and are reared for meat

kampong ['kæmpɒŋ] *noun* rural settlement in Malaysia, with houses surrounded by fruit trees and vegetable gardens

Kankrej ['kæŋkreɪ] *noun* Indian breed of cattle; a heavy draught animal with a well-developed hump. It is grey, with long horns

Kano Brown ['kænəʊ 'braʊn] *noun* West African breed of goat

kapok ['keɪpɒk] *noun* silky fibres which cover the seeds of the kapok tree *(Ceiba pentandra)* used for filling pillows and cushions; *also called* JAVA COTTON

Karakul lamb ['kærəkʊl 'læm] *noun* Asiatic breed of broad-tailed sheep; the newborn lambs are slaughtered for their glossy black fur

karoo [kə'ruː] *noun* wide elevated plateau lands in South Africa, with clayey soil; the karoo is waterless and arid during the dry season

karraya gum [kæ'ræjə 'gʌm] *noun* gum from the Indian tree *(Sterculia urens)*, used industrially

karri ['kæri] *noun* Australian tree, one of the blue gums; its wood is red, and is used in the building industry

karst [kɑːst] *noun* terrain typical of limestone country, with an uneven surface and holes and cracks due to weathering, usually with underground drainage and cave systems

Kashmir ['kæʃmɪə] *noun* breed of goat with a white or black and white coat; the breed is valued for the fineness of its long hair. Found in Tibet, Iran, Turkey and Russia

kauri gum ['kaʊri 'gʌm] *noun* resin obtained from a coniferous tree found in New Zealand

ked [ked] *noun* the sheep tick; a blood-sucking fly *(Melophagus ovinus)* which is a parasite of sheep and causes extreme irritation

keep [kiːp] **1** *noun* grass or fodder crops for

the grazing of livestock **2** *verb (of produce)* to remain in good condition after harvest; *Conference pears will keep until spring*

◊ **keeper** ['ki:pə] *noun* **(a)** person who looks after deer, pheasants or other animals and birds which are reared to be hunted **(b)** fruit which keeps well

keiserite ['keizərait] *noun* magnesium sulphate powder, used as a fertilizer where magnesium deficiency is evident, especially in light sandy soil

kemp [kemp] *noun* very coarse fibre in fleece, covered with a thick sheath and shed each year; **kemp-free mohair** = mohair which does not have any kemp

kenaf [kə'næf] *noun Hibiscus cannabinus,* a fibre-producing plant similar to jute

kenana [kə'nænə] *noun* Zebu breed of cattle, grey in colour

kennel ['kenəl] *noun* **(a)** small hut for a dog **(b)** **kennels** = commercial establishment where dogs are reared or kept for their owners **(c)** **cow kennels** = wooden building with stalls for cows

Kent [kent] *see* ROMNEY

◊ **Kentish cob** ['kentiʃ 'kɒb] *noun* commercially grown variety of hazel nut

kernel ['kɜːnəl] *noun* soft inside part of a nut, the part which is edible

Kerry ['keri] *noun* rare breed of dairy cattle which is native to Eire; the animals are black, sometimes with white patches on the udder, and are small with upturned horns

◊ **Kerry Hill** ['keri 'hil] *noun* breed of small hill sheep originating in the Kerry hills of Powys in Wales. It has a soft white fleece and speckled face and legs. The ewes are crossed with Down rams for lamb production

kg = KILOGRAM **kg/ha** = kilograms per hectare

khesari [ke'sɑːri] *noun* high-yielding vetch, grown as a catch crop with linseed, lentils, peas or mustard in paddy fallows in India

Khillari [ˈkɪlæri] *noun* Indian breed of cattle; the animals are grey or white in colour, with long pointed horns. They are

mainly found in Maharashtra State, and are primarily used as draught animals

Kholmogor [həʊlmə'gɔː] *noun* breed of dairy cattle found in Russia; the animals are mainly coloured black and white

kibbled ['kɪbəld] *adjective* coarsely ground, as in kibbled maize

kibbutz [kɪ'bʊts] *noun* form of settlement in Israel, based on the collective farming principles (NOTE: plural is **kibbutzim**)

kid [kɪd] *noun* young goat of either sex, up to one year old; **kid-snatching** = taking a new-born kid away from its mother to prevent her from licking it and so passing on CAE

◊ **kidding pen** ['kɪdɪŋ 'pen] *noun* pen in which a doe is kept when giving birth to kids

kidney bean ['kɪdni 'biːn] *noun* climbing French bean, with red seeds, used as a vegetable

killing age ['kɪlɪŋ 'eɪdʒ] *noun* age of an animal *or* bird when it is slaughtered; **killing out percentage** = deadweight of an animal expressed as a percentage of its live weight

kilo ['kiːləʊ] *abbreviation for* KILOGRAM

kilo- ['kɪləʊ] *prefix* meaning one thousand

◊ **kilocalorie** [kɪləʊ'kæləri] *noun* SI unit of measurement of heat (= 1,000 calories) (NOTE: when used with numbers **kilocalories** is usually written **Cal.** Note that it is now more usual to use the term **joule**)

◊ **kilogram** *or* **kilo** ['kɪləgræm] *noun* base SI unit of measurement of weight (= 1,000 grams); *two kilos of sugar; he weighs 62 kilos (62 kg)* (NOTE: when used with numbers **kilos** is usually written **kg**)

◊ **kilojoule** ['kɪlədʒuːl] *noun* SI measurement of energy or heat (= 1,000 joules) (NOTE: with figures usually written **kJ**)

kindling ['kɪndlɪŋ] *noun* birth of a litter, especially rabbits

kip [kɪp] *noun* hide of a young animal, used for leather

kitchen garden ['kɪtʃən 'gɑːdən] *noun* garden, near a house, where fruit, vegetables and salad crops are grown for use by the household

kiwi (fruit) ['kɪwi] *noun* subtropical woody climbing plant *(Actinidia chinensis)* which bears brownish oval fruit with a green juicy flesh; the plant has been developed in New Zealand, and is now grown in many subtropical regions; also called 'Yang Tao' or 'Chinese gooseberry'

kJ = KILOJOULE

knacker ['nækə] *noun* person who slaughters casualty animals, particularly horses

knapweed ['næpwiːd] *noun* a perennial weed *(Centaurea nigra)*

knee cap ['niː 'kæp] *noun* felt protector for the knees of horses, used especially when transporting them as a protection against damage caused when slipping. Also used on young horses when jumping

knife [naɪf] *noun* attachment on the cutter bar of a mower or combine harvester; there are two types: one with smooth sections which need frequent sharpening, and the other with serrated sections which need no sharpening

knotgrass ['nɒtgrɑːs] *noun* common weed *(Polygonum aviculare)* which affects spring cereals, sugar beet and vegetable crops; its spreading habit prevents other slower-growing plants from growing; *also known as* IRONWEED, IRONGRASS, PIGWEED, WIREWEED

knotter ['nɒtə] *noun* the mechanism on a baler which ties the bales; it has three basic parts: billhook, retainer disc and the knife

kohlrabi [kəʊl'rɑːbi] *noun* variety of cabbage with a swollen stem, used as a fodder crop, and also sometimes eaten as a vegetable; the leaves may be green or purple; also known as 'turnip-rooted cabbage'

kolkhoz [kɒl'hɒz] *noun* formerly, a large-scale collective farm in the Soviet Union; the land was leased to the workers by the state

kondo ['kɒndəʊ] *noun* coarse cereal grain grown in India

kraal [krɑːl] *noun* an African village surrounded by a thorn fence, which protects both people and livestock

Kuri ['kuːri] *noun* breed of cattle found in the Lake Chad region of West Africa; the animals are white or grey in colour, with very large dark horns. Mainly reared for meat and milk

kurrat ['kʌrət] *noun* leek-like plant, grown mainly in Egypt

kwashiorkor [kwɒʃi'ɔːkɔː] *noun* malnutrition of small children, mostly in tropical countries, causing anaemia, wasting of the body and swollen liver; it is caused by protein deficiency in the diet, especially where cassava is the staple foodstuff, since the protein content of cassava is almost nil

Kyloes ['kaɪləʊz] *noun* breed of small long-horned shaggy Highland cattle

LI

label ['leɪbəl] **1** *noun* piece of paper attached to produce, showing the price and other details; **quality label** = label which shows that the produce is of good quality, or that it meets government requirements **2** *verb* to stick a label onto something; *certified seed is tested, approved and then labelled*

COMMENT: government regulations cover the labelling of food; it should show not only the price and weight, but also where it comes from, the quality grade, and a sell-by date

lablab ['læblæb] *noun* legume *(Dolichos lablab)* grown in Egypt, India and South-East Asia; the pods and seed are used as a vegetable, and the plant can also be grown as a fodder crop

labourer ['leɪbərə] *noun* person who does heavy work; **agricultural labourer** = person who does heavy work on a farm, formerly a rural worker with no land, and sometimes still a worker with a special skill, such as ditching or hedging; **casual labourer** = worker who can be hired for a short period, usually at harvest time

QUOTE 35 people are leaving the land every day. That is, 250 every week or 1000 a month, 12,000 people a year are leaving the land
NFU president March 1992. British Farming

lactate [læk'teɪt] *verb* to produce milk

◊ **lactation** [læk'teɪʃn] *noun* (i) production of milk; (ii) suckling, the period during which young are nourished with milk from the mother's mammary glands

COMMENT: lactation is stimulated by the production of the hormone prolactin by the pituitary gland. In cows, goats and sheep kept for milk production, the lactation period is made longer by regular milking. For a dairy cow, the period is ten months, followed by a two-month rest before calving again

QUOTE feed requirements of cows in winter depend on whether they are dry or lactating. A suckler cow rearing one calf will require over 1.5 times as much feed as a dry cow
Practical Farmer

lactic acid ['læktɪk 'æsɪd] *noun* an organic acid formed by the action of bacteria on lactose, causing milk to go sour; used in the production of soured cream, cheese and yoghurt. Lactic acid is also produced in silage by bacteria present in the green crop acting on the carbohydrates and converting it from sugars present in the crop or added to it. Lactic acid and bacteria are also added to silage to assist the preservation process

lactose ['læktəuz] *noun* milk sugar, the sugar found in milk

Lacaune [læ'kəun] *noun* breed of sheep found mainly in Aveyron, France and used mainly for milk production

ladder farm ['lædə 'fɑːm] *noun* farm with a series of small long narrow fields

ladybird ['leɪdɪbɜːd] *noun* beetle of the *Coccinellidae* family, which is useful to the farmer because it feeds on aphids which would damage plants if they were not destroyed (NOTE: US English is **ladybug**)

lagoon [lə'guːn] *noun* pool of water or other liquid; *slurry can be stored in lagoons*

LAI = LEAF AREA INDEX

laid crop ['leɪd 'krɒp] *noun* crop which has been flattened by rain and wind

◊ **laid hedge** ['leɪd 'hedʒ] *noun* hedge that has been made by bending over each stem and weaving it between stakes driven into the ground

◊ **laid wool** ['leɪd 'wʊl] *noun* wool discoloured by the use of salves containing tar

lairage ['leərɪdʒ] *noun* shed or outdoor

enclosure for the temporary housing of animals, as on the way to market, or when they are being transported for export

lamb [læm] **1** *noun* **(a)** young sheep under six months of age **(b)** meat from a lamb (meat from an older sheep is called 'mutton', though this is not common) **2** *verb* to give birth to lambs; *most ewes lamb without difficulty, but some may need help*

◊ **lamb dysentery** ['lææm 'dɪsəntrɪ] *noun* bacterial disease which enters the lamb from the pasture. The bacteria infects the land for a very long time. It can be avoided by practising sound hygiene at lambing

◊ **lambing** ['læmɪŋ] *noun* the action of giving birth to lambs; **lambing pen** = pen in which a ewe is kept when giving birth to lambs; **lambing percentage** = the number of live lambs born per hundred ewes; **lambing season** = period of the year when a flock of ewes produces lambs, usually between December and January; the object is to produce lambs for the market when the price is highest, usually between February and May; **lambing sickness** = bacterial disease of sheep picked up from the soil, can cause rapid death

lamina ['læmɪnə] *noun* **(a)** the blade of a leaf **(b)** the sensitive membrane lying beneath the outer horny layer of an animal's hoof

◊ **laminitis** [læmɪ'naɪtɪs] *noun* inflammation of the lamina in a hoof, causing swelling, and often leading to deformed hooves; possibly caused by too much grain feed

land [lænd] *noun* **(a)** solid part of the earth's surface (as opposed to the sea); *the family have worked the land for generations;* **back to the land** = encouragement given to people who once lived in the country and moved to urban areas to find work, to return to the country; **land agent** = person employed to run a farm or an estate on behalf of the owner; **land consolidation** = the bringing together of small scattered plots of land into large single units, often helped by government grants, as in France and the Netherlands; **land reclamation** = the process of bringing back areas of desert, heath, marsh or other unproductive land to being productive for agricultural purposes **(b)** section of a field, divided from other sections by a shallow furrow (term used in systematic ploughing)

land capability ['lænd keɪpə'bɪlɪtɪ] *noun*

an estimate of the potential of land for agriculture, made on purely physical environmental factors, such as climate and soil. In 1993 the total area of agricultural land in the U.K. was 18,530,000 hectares occupying 77% of the total land area and only 17% of this was grade 1; *compare* LAND SUITABILITY

land classification ['lænd klæsɪfɪ'keɪʃn] *noun* classification of land into categories, according to its value for a broad land use type

COMMENT: in England and Wales, the MAFF classification map has five main grades, between Grade 1 (completely suitable for agriculture) and Grade 5 (land with severe limitations, because of its soil, relief or climate). The Soil Survey of England and Wales also has a scheme which divides land into seven classes; these range from Class I (land with no limitations, or only very minor limitations to use) to Class VII (land with severe limitations, which restrict its use to rough grazing, forestry or recreation). Classes II to VII are then divided into subclasses according to soil, wetness, climate, gradient and erosion limitations

land drainage ['lænd 'dreɪnɪdʒ] *noun* removing surplus water from land; **Land Drainage Service** = former division of the Agricultural Development and Advisory Service, with the responsibility to advise farmers on land drainage

COMMENT: if surplus water is prevented from moving through the soil and subsoil, it soon fills all the pore spaces in the soil and this will kill or stunt the growing crops. Well-drained land is better aerated, and crops are less likely to be damaged by root-destroying fungi; also aerated soil warms up more quickly in spring. Plants form deeper and more extensive roots systems, grassland is firmer, and disease risk from parasites is reduced. The main methods of draining land are underground pipe drains, mole drains and ditches

landfill site ['lændfɪl 'saɪt] *noun* area of excavated land (such as an old gravel pit) in which refuse is placed

landlord ['lændlɔːd] *noun* owner of land or building who lets it to a tenant for an agreed rent

landowner ['lændəʊnə] *noun* person who

owns land freehold, and may let it to a tenant, or may farm it himself

landrace ['lændreɪs] *noun* **(a)** native species of plant *or* animal which has not been cultivated **(b) Landrace** = breed of pig, native to Scandinavia; the animals have small heads, lop ears and long bodies; when crossed with the Large White it produces vigorous hybrid sows; almost all indoor and outdoor hybrid sows in the UK are 50% landrace

landscape ['lændskeɪp] **1** *noun* the general shape and appearance of an area of land **2** *verb* to change the appearance of a garden or park by planting trees, creating little hills, making lakes, etc.; **landscape gardener** = gardener who creates a new appearance for a garden

> COMMENT: many farmers find themselves as landscape managers, required to maintain the countryside in an aesthetically and environmentally pleasing condition for the predominantly urban population to enjoy

> QUOTE cheap food has revolutionised the British landscape. After the revolution comes the great terror. A leaderless army of unplanned golf courses, caravan parks, chalet villages, hypermarkets and sprawling infill is descending on the countryside and suburbanising it. Farmers should be a bulwark of this terror, mobilising the public in support
> *The Times*

Land Service ['lænd 'sɜːvɪs] *noun* a former division of the Agricultural Development and Advisory Service with the responsibility for advising farmers on farm management and resource planning

landside ['lændsaɪd] *noun* the part of the plough which takes the sideways thrust as the furrow is turned

land suitability ['lænd suːtə'bɪlɪti] *noun* the suitability of land for a certain agricultural purpose

> COMMENT: land suitability is similar to land capability, but defines its usefulness for a particular purpose; suitability tends to emphasize the positive value of land, while capability emphasizes its limitations

land tenure ['lænd 'tenjə] *noun* the way in which land is owned and possessed; this

may be by an individual owning the freehold, by a tenancy agreement between freeholder and tenant, or by a form of community ownership. In the U.K. two forms predominate, owner-occupation and agricultural tenancy, although other forms have local importance, such as crofting in the Highlands of Scotland. In June 1993 there were 70,614 owner-occupied holdings, 25,015 tenanted holdings and 28,635 mixed tenure holdings

land use ['lænd 'juːs] *noun* way in which the land is used by humans; *they are carrying out a study of land use in northern areas;* **land use classification** = the classification of land according to the way it is used

> COMMENT: in the UK, the main uses of land are classified as: crops and fallow, temporary grass, permanent grass, rough grazing, other land, urban land, forestry and woodland, and miscellaneous

lanolin ['lænəlɪn] *noun* fat which comes from the wool of sheep, used in the preparation of ointments

larch [lɑːtʃ] *noun* a coniferous European softwood tree *(Larix decidua)* a fast-growing tree used as a timber crop

Large Black ['lɑːdʒ 'blæk] *noun* dual-purpose hardy breed of pig; the animals are black with long lop ears. Formerly it was a useful outdoor breed because it could stand extreme weather, but is now less common. A Large Black sow when mated with a white boar produces hardy blue-and-white hybrids

◊ **Large White** ['lɑːdʒ 'waɪt] *noun* important commercial breed of pig; the animals are white with pricked up ears; it is the most popular UK boar breed and forms a part in almost all indoor hybrid sows (usually 50%) and hybrid boars. It has a large mature size and has been improved to produce good fecundity, growth and carcass characteristics

larva ['lɑːvə] *noun* stage (caterpillar *or* grub) in the development of an insect, after the egg has hatched but before it becomes adult (NOTE: plural is **larvae**)

◊ **larval** ['lɑːvəl] *adjective* referring to larvae; **larval stage** = early stage in the development of an insect after it has hatched from an egg

lasso [lə'suː] **1** *noun* rope with a noose at

the end, used to catch cattle **2** *verb* to catch cattle, using a lasso

LATB = LOCAL AGRICULTURAL TRAINING BOARD

lateral ['lætərəl] *noun* shoot which branches off from the leader or main branch of a tree or shrub

laterite ['lætərait] *noun* substance which when it dries out forms a hard material which cannot be wetted again; it is highly weathered and consists mainly of iron and aluminium oxides; *in some countries houses are built of laterite*

◊ **lateritic** [lætə'rıtık] *adjective* (soil) which contains laterite

◊ **laterization** [lætərai'zeiʃn] *noun* process of weathering soil into hard laterite

◊ **laterize** ['lætəraiz] *verb* to create laterite by exposing tropical soil to weather

COMMENT: when tropical rainforests are cleared, the soil beneath rapidly turns to laterite as nutrients are leached out by rain. Laterized land cannot be cultivated and such areas turn to desert

latex ['leiteks] *noun* (i) white fluid from a plant such as the poppy; (ii) white substance in the fluid from the rubber tree, which is treated and processed to make rubber

latifundium [læti'fʌndiəm] *noun* large landed estate, common in South America; latifundia are often split into small units which are rented out to peasants (NOTE: plural is **latifundia)**

latosol ['lætəsɒl] *noun* tropical soil in savanna and rainforest areas; its colour is most commonly red, though brown and yellow occur. The soil is rich in hydroxides of iron, aluminium and manganese, and is heavily leached

laurics *or* **lauric oils** ['lɔːrıks *or* 'lɔːrık 'ɔilz] *noun* oils from palm seed and coconut

lavender ['lævəndə] *noun* shrub (*Lavandula officinalis*) with small lilac-coloured flowers and narrow leaves, cultivated for perfume; once cut, the bloom is either dried or distilled: this is the traditional process for the extraction of lavender oil

laver ['lɑːvə] *noun* Welsh name for a variety of seaweed which is edible

laxative ['læksətiv] *adjective & noun* substance which encourage movements of the bowel; *succulent food such as root crops have a laxative effect*

Laxton's Superb ['lækstənz suː'pɜːb] *noun* variety of dessert apple formerly grown commercially in the UK

lay [lei] **1** *noun* hen in lay = bird which is laying eggs **2** *verb* (*of a bird*) to produce an egg

layer ['leiə] **1** *noun* (a) bird that is in lay; **layers' ailments** = disorders of fowls in lay, especially birds that are in heavy production; disorders of layers include egg binding, internal laying and layer's cramp; **layer's cramp** = condition found in pullets after the first few weeks of their laying life: the bird appears weak, but the trouble usually disappears after a few days **(b)** stem of a plant which has made roots where it touches the ground **2** *verb* to propagate a plant by bending a stem down until it touches the soil and letting it form roots there

◊ **layering** ['leiəriŋ] *noun* (a) method of propagation where the stem of a plant is bent until it touches the soil, and is fixed down on the soil surface until roots form; **air layering** = method of propagation where a stem is partially cut, then surrounded with damp moss, the whole being tied securely; roots will grow from the cut at the point where it is in contact with the moss **(b)** bending over the half-cut stems of hedge plants, and weaving them around stakes set in the ground, to form a new hedge

laying ['leiiŋ] *noun* the action of producing eggs; **laying cage** = specially built cage for laying hens; the cages are arranged in tiers and each cage should allow the birds to stand comfortably, allow the eggs roll forward and permit access to food and water, easy cleaning and easy handling of the birds; **laying period** = period during which a hen will continue to lay eggs; this begins at 22 weeks of age and normally lasts for 50 weeks; **internal laying** = condition caused by a fault in the oviduct, which results in the yolks not being passed along the oviduct for covering with membranes and shell

lazy-bed ['leizibed] *noun* small arable plot

used for growing potatoes, cereals and other crops, found in the West Highlands of Scotland; if the soil is thin, seed potatoes are placed on the surface of the soil and covered with turf

lea [liː] *noun* open ground left fallow or under grass

leach [liːtʃ] *verb* to wash something out of soil by passing water through it; *excess chemical fertilizers on the surface of the soil leach into rivers and cause pollution; nitrates have leached into ground water and contaminated the water supply*

◊ **leachate** ['liːtʃeɪt] *noun* (i) substance which is washed out of the soil; (ii) liquid which forms at the bottom of a landfill site

◊ **leaching** ['liːtʃɪŋ] *noun* process by which plant nutrients such as NPK are lost from the soil by water percolating through soil in humid climates; soils gradually become mineral-deficient and sour; **leaching field** = area round a septic tank with pipes which allow the sewage to drain away underground

lead [led] *noun* very heavy soft metallic element, which is poisonous in compounds; **lead poisoning** = the poisoning of animals which have eaten or licked paint; it is common in cattle and can be fatal (NOTE: chemical symbol is **Pb**; atomic number is **82**)

leader ['liːdə] *noun* animal which goes first, which leads the flock or herd; **leader-follower system** = system of grazing where priority is given to a group of animals (the leaders) and the crop is later grazed by a second group of animals (the followers); so first-year heifers might be followed by second-year heifers

leaf [liːf] *noun* green, usually flat, part of a plant, growing from a stem, whose purpose is to activate photosynthesis, the means by which the plant gets energy from sunlight; **leaf area index (LAI)** = the area of green leaf per unit area of ground; **leaf blotch** = disease of cereals *(Rhynchosporium secalis)* where dark grey lesions with dark brown margins occur on the leaves; **leaf burn** = damage done to leaves by severe weather conditions or herbicides; **leaf roll** = viral disease of potatoes, transmitted by aphids; the leaves roll up and become dry, and the crop yield is affected; **leaf spot** = fungal disease of brassicas, where the leaves

develop brown and black patches; **leaf stripe** = disease of barley and oats *(Pyrenophora graminea)* where the young leaves show pale stripes and seedlings often die

lean meat ['liːn 'miːt] *noun* meat with very little fat; *animals are bred to produce lean meat; venison is a very lean form of meat*

lease ['liːs] **1** *noun* legal contract for letting a building or piece of land for a period of time against payment of a rent **2** *verb* **(a)** to let or rent land *or* machinery for a period; *he leases workshops on his farm to small craft studios; we lease large pieces of equipment such as combines* **(b)** to use land *or* machinery for a time and pay a fee to the landlord *or* lessee; *he leases his farm from insurance company; all our farm equipment is leased*

◊ **leaseback** ['liːsbæk] *noun* arrangement where property is sold and then taken back by the former owner on a lease

◊ **leasehold** ['liːshəʊld] *noun & adjective & adverb* possessing property on a lease, for a fixed time; *to purchase a property leasehold; the property is for sale leasehold*

◊ **leaseholder** ['liːshəʊldə] *noun* person who holds a property on a lease

◊ **leasing** ['liːsɪŋ] *noun* which leases *or* working under a lease; *all the farm's tractors are owned, but the combines are leased*

leather ['leðə] *noun* the skin of an animal, tanned and prepared for use

◊ **leatherjacket** ['leðədʒækɪt] *noun* the larva of the cranefly *(Tipuda paludosa)* which hatch from eggs laid on the ground and feed on the young crop in spring; when grass is ploughed for cereal crops, the larvae feed on the seedling wheat, damaging the plants at or just below ground level

leek [liːk] *noun* hardy winter vegetable *(Allium ameloprasum)* with a mild onion taste; to produce high-quality leeks, the lower parts of the stems need to be blanched. The stems are used in soups and stews

Leghorn [leg'hɔːn] *noun* excellent laying breed of hen; a hardy bird, coloured black, brown and white. Leghorns produce good-sized white eggs

legume ['legjuːm] *noun* **(a)** any member of

the plant family *Leguminosae* which produce seeds in pods; **grass and legume associations are common in European pastureland (b)** dry seed from a single carpel, which splits into two halves (such as a pea)

◊ **Leguminosae** [legu:mɪn'əʊsɪ] *noun* family of plants (including pea and bean plants) which produce seeds in legume pods

◊ **leguminous** [lɪ'gju:mɪnəs] *adjective* (plant) which produces seeds in pods

COMMENT: there are many species of legume, and some are particularly valuable because they have root nodules that contain nitrogen-fixing bacteria. Such legumes have special value in maintaining soil fertility and are used in crop rotation. Peas, beans, peanuts, alfalfa, clover and vetch are all legumes

QUOTE the cowpea is a leguminous crop grown in tropical or semi-tropical climates for fodder, pulse, vegetable and green manuring
Indian Farming

Lehmann system ['leɪmən 'sɪstəm] *noun* system of pig breeding developed in Germany, where bulk food such as potatoes and fodder beet are fed after a basic ration of meal

Leicester longwool ['lestə 'lɒŋwʊl] *noun* breed of large hornless white-faced sheep, used a lot by Robert Bakewell, but now rare; *see also* BORDER LEICESTER, BLUE-FACED

lemma ['lemə] *noun* the outer bract which encloses the flowers of grass

lemon ['lemən] *noun* yellow edible fruit of an evergreen citrus tree *(Citrus limon);* lemons have a very tart flavour and are used in flavouring and in making drinks

lentil ['lentɪl] *noun* important pulse crop, growing in the warmer temperate and subtropical regions of the world. India is the largest producer of lentils. The dried seeds are often processed into dhal

Leptospira hardjo (LH *or* **lep hardjo)** ['leptəʊspaɪrə 'hɑːdjəʊ] *noun* bacterium which infects cattle and humans, causing leptospirosis and Wiels disease

◊ **leptospirosis** [leptəʊspaɪ'rəʊsɪs] *noun* disease of cattle caused by bacteria, which causes abortions and low milk yields; it can be carried by sheep or in running water

lesion ['liːʒn] *noun* open wound on the surface of a plant *or* on the skin of an animal, caused by disease or physical damage

less favoured area ['les feɪvəd 'eərɪə] *noun* former name for land in mountainous and hilly areas, which is capable of improvement and use as breeding and rearing land for sheep and cattle; now called Disadvantaged or Severely Disadvantaged Areas; the EU now recognizes such areas and gives financial help to farmers in them

let-down ['letdaʊn] *noun* **let-down of milk** = the release of milk from the mammary gland

COMMENT: the hormone oxytoxin activates the release of milk; the let-down lasts between seven and ten minutes, when the extraction of milk from the udder is easiest

lettuce ['letɪs] *noun* salad vegetable *(Lactuca sativa)* which comes in a variety of forms and leaf textures: the commonest are cos lettuce, cabbage lettuce, crisphead and loose-leaved lettuces; cabbage lettuces have roundish heads, while cos lettuces have longer leaves and are more upright

leucine ['luːsiːn] *noun* one of the amino acids

leucosis *or* **leukosis** [luːˈkəʊsɪs] *see* ENZOOTIC BOVINE LEUCOSIS

levada ['levɑːdə] *noun* irrigation canal leading to small irrigated plots on terraces

level ['levəl] **1** *adjective* flat *or* horizontal; *the mountains give way to miles of level grassy plain* **2** *noun* **(a)** amount; *forest will gradually die if pollution levels remain constant; present levels of water pollution are not acceptable* **(b)** flat low-lying area of usually marshy land, often reclaimed by artificial drainage in parts of Fen Country in Eastern England round the Wash

QUOTE atmospheric levels of carbon dioxide are increasing at about 0.35 per cent a year
New Scientist

levy ['levɪ] *noun* money which is demanded and collected by a government; **import levy** = tax on farm produce which is imported into the EU

ley [leɪ] *noun* (a) land which has been sown to grass for a time (b) field in which crops are grown in rotation with periods when the field is under pasture; **ley farming** = farming system in which fields are left to pasture in rotation

COMMENT: strictly speaking, ley farming is a system where a farm or group of fields is cropped completely with leys which are reseeded at regular intervals; alternatively, any cropping system which involves the use of leys is called ley farming. Ley farming is an essential part of organic farming: pasture land is fertilized by the animals which graze on it, and then is ploughed for crop growing. When the land has been exhausted by the crops, it is put back to pasture to recover

LH = LEPTOSPIRA HARDJO

lice [laɪs] *see* LOUSE

licence ['laɪsəns] *noun* permit, official document which allows someone to do something; **abstraction licence** = a licence issued by a Water Board to allow abstraction of water from a river or lake for domestic or commercial use. The licence is needed for irrigation; **movement licence** = licence which is needed in order to move animals from areas of infectious disease; restriction on the movement of animals is a measure used to prevent the spread of disease

lichee [laɪˈtʃiː] *see* LITCHI

lichen ['laɪkən] *noun* primitive plant which grows on the surface of stones *or* trunks of trees and can survive in arctic climates

COMMENT: lichens are formed of two organisms: a fungus which provides the outer shell and algae which provide chlorophyll and give the plant its colour. Lichens are very sensitive to pollution, especially sulphur dioxide, and act as indicators for atmospheric pollution. They also provide food for many arctic animals

lie [laɪ] *noun* place where an animal lies down; *livestock benefit from a dry lie at pasture*

LIFE = LOW INPUT FARMING AND ENVIRONMENT

lift [lɪft] *verb* to harvest root crops, such as potatoes, by digging them out of the ground; potatoes can be lifted from the soil and, using a spinner or an elevator digger, left in rows for hand-picking

◊ **liftings** ['lɪftɪŋz] *noun* crops which have been lifted

◊ **lifting unit** *or* **lifter** ['lɪftɪŋ 'juːnɪt *or* 'lɪftə] *noun* pair of wheels or a triangular-shaped share, used on a harvester to lift the roots and pass them to the main elevator

COMMENT: The roots are lifted by being squeezed out of the ground in between the two wheels. The distance between the two wheels or shares can be adjusted to suit the size of the crop. The wheels should be set quite close together at the bottom when harvesting small roots. The wheels run at an angle to each other so that their rims lie close together when in the soil and farther apart at the top

light grains ['laɪt 'greɪnz] *noun* cereals such as barley and oats

◊ **light soil** ['laɪt 'sɔɪl] *noun* soil consisting mainly of large particles which are loosely held together because of the relatively large pore space; light soil is usually easier to cultivate than heavy soil, but may dry out too quickly

◊ **Light Sussex** ['laɪt 'sʌsɪks] *noun* dual-purpose breed of poultry; the Light Sussex is one of the several varieties of the Sussex breed; the birds are white, with black stripes to the feathers of the neck and black feathers on the wings and tail

lignin ['lɪgnɪn] *noun* complex woody tissue deposited in plants and resulting in reduced levels of digestibility

◊ **lignify** ['lɪgnɪfaɪ] *verb* to become hard and woody; *plants are less digestible as they become lignified*

Lim *(informal)* = LIMOUSIN

Lima bean ['liːmə 'biːn] *noun* broad bean

limb [lɪm] *noun* leg of an animal

lime [laɪm] **1** *noun* (a) calcium compounds used to spread on soil to increase the pH level and correct acidity; lime is usually applied as simple chalk or limestone; also as quicklime or slaked lime (b) common European hardwood tree *(Tulia* sp) (c) *Citrus aurantifolia,* a citrus fruit tree, with green fruit similar to, but smaller than, a lemon **2** *verb* to treat acid soil by spreading lime on it

> QUOTE pulses are generally susceptible to soils which are acidic in nature and need liming if the soil pH is below 6
> *Indian Farming*

limestone ['laɪmstəʊn] *noun* common sedimentary rock

> COMMENT: limestone is formed of calcium minerals and often contains fossilized shells of sea animals. It is an important source of various types of lime

Limousin [lɪmu'zæn] *noun* French breed of beef cattle, developed on the uplands around Limoges in central France; the breed is relatively hardy; cattle are red, with large bodies. Limousin bulls are used on dairy cattle producing a good crossbred calf

linch pin ['lɪntʃ 'pɪn] *noun* pin used to lock an implement onto the three-point linkage at the rear of a tractor

Lincoln longwool ['lɪnkən 'lɒŋwʊl] *noun* rare breed of sheep, with white faces and long shiny wool; the animals are vary large and slow to mature; the breed is now found mainly in Lincolnshire

◊ **Lincoln red** ['lɪnkən 'red] *noun* breed of cattle bred from the shorthorn; the animals are deep red in colour; originally the breed was dual-purpose, but now is mainly used for crossing with dairy cows to produce beef calves

lindane ['lɪndeɪn] *noun* organochlorine pesticide, which is harmful to some animals such as bees and fish. Lindane is used in Britain as a farm insecticide and as a chemical for treating wood. It is banned in the USA, and its use is being examined in the UK (NOTE: also called **HCH**)

line [laɪn] *noun* **blood lines** = general term used to describe relationships between animals, the pedigree lines in a flock

◊ **linebreeding** ['laɪnbriːdɪŋ] *noun* mating of related animals with offspring

> COMMENT: the purpose of linebreeding is to try to preserve in succeeding generations the mix of genes responsible for a particularly excellent individual specimen

liner ['laɪnə] *noun* rubber inner tube of a teat cup in a milking machine

ling [lɪŋ] *noun* variety of heather *(Calluna vulgaris)*

link [lɪŋk] *noun* **(a)** measurement, forming one loop of a chain (one-hundredth of a surveying chain, or 7.92 inches) **(b)** something which joins two things; **critical link** = organism in a food chain that is responsible for taking up and storing nutrients, which are then passed on down the chain

◊ **linkage** ['lɪŋkɪdʒ] *noun* association of genes that appear to be inherited together

Linnaean system [lɪ'niːən 'sɪstəm] *adjective* scientific system of naming organisms worked out by the Swedish scientist Carolus Linnaeus (1707-1778)

> COMMENT: the Linnaean system (or binomial classification) gives each organism a name made up of two main parts. The first is a generic name referring to the genus to which the organism belongs, and the second is a specific name which refers to the particular species. Organisms are usually identified by using both their generic and specific names, e.g. *Homo sapiens* (man) and *Felix catus* (domestic cat). The generic name is written or printed with a capital letter. Both names are usually given in italics, or are underlined if written or typed

linseed ['lɪnsiːd] *noun* variety of flax *(Linum usitatissimum)*, with a short straw; it produces a good yield of seed used for producing oil

◊ **linoil** ['lɪnɔɪl] *noun* linseed oil

> COMMENT: after oil has been extracted, the residue (known as linseed cake) is used as a protein-rich feeding stuff. In 1993, 150,000 hectares were grown in the UK, with an average yield of 1.23 tonnes per hectare

Linum ['liːnəm] *Latin name for* flax

linuron ['lɪnjuːrɒn] *noun* residual herbicide which acts in the soil

lipase [['laɪpeɪz] *noun* enzyme which breaks down fats in the intestine

liquid ['lɪkwɪd] *adjective & noun* matter (like water) which is not solid and is not a gas; **liquid manure** = manure consisting of dung and urine in a liquid form (manure in semi-liquid form is slurry); **liquid fertilizers**

= non-pressurized solutions of normal solid fertilizers; liquid fertilizers are easier, quicker and cheaper to handle and apply than solid fertilizers

liquor ['lıkə] *noun* concentrated liquid substance; **silage liquor** = liquid which forms in silage and drains away from the silo

liquorice ['lıkərıs] *noun* root of a plant *(Glycyrrhiza glabra)* used in making sweets and soft drinks; it also has medicinal properties (formerly grown in Pontefract, Yorkshire, but now not common there)

listeria [lıs'tıərıə] *noun* genus of bacteria found on domestic animals and in prepared foods, which can cause meningitis

litchi *or* **lichee** *or* **lychee** [laı'tʃiː] *noun* subtropical fruit *(Litchi chinensis)* a native of China; it produces fruit with a hard red skin and a soft white juicy pulp surrounding a hard shiny brown seed

litre *or* *US* **liter** ['liːtə] *noun* metric measurement of liquids, equal to 0.22 gallon (NOTE: usually written **l** after figures: **25l**)

litter ['lıtə] **1** *noun* **(a)** bedding for livestock; straw is the best type of litter, although bracken, peat moss, sawdust and wood shavings can be used **(b)** group of young mammals born to one mother at the same time; *a pig litter of 34 has been recorded, although about 12 is average* **2** *verb* to have a litter of young

liver ['lıvə] *noun* large organ in the body, the main organ for removing harmful substances from the blood; **liver fluke** = parasitic trematode which lives in the liver and bile ducts of animals, such as *Fasciola hepatica* which infests sheep and cattle, causing loss of condition; **fatty liver (syndrome)** = condition in older cows, where the animal absorbs calcium too slowly and the liver is affected; goats are also affected

livery ['lıvəri] *noun* stable which keeps a horse for the owner and usually includes feeding, grooming, and exercising

livestock ['laıvstɒk] *noun* domesticated farm animals, which are reared to produce meat *or* milk *or* other products; *livestock production has increased by 5%; pastoralists move their livestock from place to place to find new grazing;* **livestock records** = each farm has to make simple records of all livestock, and these are then available for the MAFF returns which are compiled each year; **livestock unit** = part of a farm, where livestock are reared

◊ **liveweight** ['laıvweıt] *noun* the weight of a live animal (as opposed to deadweight, the weight of the carcass); **liveweight marketing** = the marketing of live animals

llama ['lɑːmə] *noun* pack animal in the Andes of South America; it is a ruminant, and belongs to the camel family

Llanwenog [læn'wenɒg] *noun* breed of sheep found in many parts of West Wales. The fleeces are considered to be the finest produced in the U.K. The wool has a very soft handle

Lleyn [leın] *noun* breed of sheep native to the Lleyn peninsula in North Wales; the animals are small, hornless and hardy; they are good milkers, and very productive, often producing triplets

loader ['ləudə] *noun* machine used to load crops, manure, etc., into trailers or spreaders; the front-end tractor-mounted loader is the most common

loam [ləum] *noun* **(a)** good dark soil, with medium-sized grains of sand, which crumbles easily and is very fertile and easily cultivated. Loam is intermediate in texture between clay and sandy soil. Loams warm up quite early in spring, and are fairly resistant to drought **(b)** *(in horticulture)* mixture of clay, sand and humus, used as a potting compost

◊ **loamy** ['ləumi] *adjective* crumbly fertile dark (soil)

locks [lɒks] *noun* small tufts of wool separated from the fleece during shearing

locust ['ləukəst] *noun* flying insect of the grasshopper family *(Locusta migratoria)* which occurs in subtropical areas, flies in swarms and eats large amounts of vegetation; **locust bean** = the broken-down pods of the carob tree, used as animal feed

lode [ləud] *noun* deposit of metallic ore

lodgepole pine ['lɒdʒpəʊl 'paɪn] *noun* slow growing tree which thrives on poor soil and is used as a pioneer crop *(Pinus contorta)*

lodging ['lɒdʒɪŋ] *noun* the tendency of cereal crops to bend over, so that they lie more or less flat on the ground. Lodging can be caused by wet weather or high winds, and also by disease such as eyespot; it can cause problems for harvesting the crop

loess ['ləuɪs] *noun* fine fertile material formed of tiny clay and silt particles deposited by the wind; soils developed in loess are prized for intensive crop production

loganberry ['ləugənberi] *noun* soft fruit, a cross between a raspberry and a blackberry

Loghtan ['lɒhtən] *see* MANX

Lomé Convention ['ləumeɪ kən'venʃn] *noun* an agreement reached in 1975 between the European Community and 66 developing nations (the ACP states); it gives entry into the EU for certain agricultural produce without duty, with sections on guaranteed prices

long-grain rice ['lɒŋgreɪn 'raɪs] *noun* varieties of rice with long grains, grown in tropical climates, such as India, (as opposed to short-grain rice grown in colder climates such as Japan)

Longhorn ['lɒŋhɔːn] *noun* dual-purpose hardy breed of cattle, with long down-curving horns; the animals are usually red or brown in colour, with white markings. The breed is now rare

long ton ['lɒŋ 'tʌn] *noun* British ton, equal to 1,016kg

longwool *or* **longwoolled sheep** ['lɒŋwul] *noun* name of several breeds of sheep with long wool; *see also* LEICESTER

Lonk [lɒŋk] *noun* breed of moorland sheep, found in the Pennines of Lancashire and Yorkshire; it is one of the Swaledale group, although larger than other varieties; it produces finer wool than most hill sheep; the face and legs are white with dark markings

loose-box ['luːsbɒks] *noun* stable for animals that are kept untied; a loose-box should have a hay rack, manger, water bowl and tying rings; it should also have a grooved floor to make cleaning and drainage easier. Loose-boxes are also useful for housing sick animals

loose-leaved ['luːs 'liːvd] *adjective* plant (such as a lettuce) with a loose collection of leaves and no heart

loose silky bent [luːs 'sɪlki 'bent] *noun* plant with thin green or purple stems which affects winter cereals *(Apera spica-venti)*

loose smut ['luːs 'smʌt] *noun* fungus *(Ustilago nuda)* affecting wheat and barley; masses of black spores collect on the diseased heads; the spores are dispersed in the wind, and only a bare stalk is left

lop [lɒp] **1** *noun* **lop ears** = long ears which hang down on either side of the animal's head **2** *verb* to cut the branches of a tree

louping-ill ['luːpɪŋ 'ɪl] *noun* infective parasitic disease of sheep *(Ixodes ricinus)*, carried by ticks in hill pastures. Animals suffer acute fever and nervous twitch and staggers; *also called* STAGGERS, TWITCH, TREMBLES

louse [laus] *noun* small insect of the *Pediculus* genus, which sucks blood and lives on the skin as a parasite on animals and humans (NOTE: the plural is **lice**)

◊ **louse disease** ['laus dɪ'ziːz] *noun* an external parasitic disease of cattle. Severe infection leads to loss of condition, wasting and anaemia

lovage ['lʌvɪdʒ] *noun Levisticum officinale,* herb used as a vegetable and for making herbal teas

low [ləu] **1** *noun* depression, area of low atmospheric pressure; *a series of lows are* *crossing the Atlantic towards Iceland* **2** *verb (of cow)* to make a noise; *see also* BLEAT, GRUNT, NEIGH

◊ **lowlands** ['ləʊlændz] *noun* area of low land where conditions are usually good for farming (as opposed to hills and mountains, or highlands); *vegetation in the lowlands or in the lowland areas is sparse*

◊ **low loader** ['ləʊ 'ləʊdə] *noun* farm trailer with its flat floor near the ground to make loading easier

LPR = LAND PRICE REVIEW

LSU = LIVESTOCK UNIT

LTA = LAND TRUSTS ASSOCIATION

lucerne [lu's3:n] *noun Medicago sativa*, plant of the Leguminosae family, grown to use as fodder

COMMENT: lucerne is the most important forage legume; it is called lucerne in Europe, Oceania and South Africa, and elsewhere it is called alfalfa. Lucerne is perennial, drought-resistant and rich in protein. It is mainly used for cutting, either for green feed or for hay or silage. It can also be grazed if carefully managed; lucerne is also dried artificially and made into pellets of feed

lugs [lʌgz] *noun* projections from the tyres of tractor wheels; they increase traction by digging into the soil and by keeping the tyre in contact with solid surfaces in muddy conditions

Luing [lɪŋ] *noun* hardy breed of beef cattle, found mainly in North-West Scotland

lump lime ['lʌmp 'laɪm] *noun* another name for burnt lime or quicklime

lumpy jaw ['lʌmpi 'dʒɔ:] *see* ACTINOMYCOSIS

lungworm ['lʌŋwɜ:m] *noun* parasitic worm which infests the lungs; infestation can be cured with anthelmintics (NOTE: also called **hoose)**

lupin ['lu:pɪn] *noun* leguminous plant (*Lupinus polyphillus*) grown as a crop for protein and seed oil

COMMENT: lupins were originally grown in the UK as green manure on acid sandy soils, and for some sheep folding; now grown for grain production. The seeds of lupin contain 30-40% protein and 10-12% edible oil. The white lupin is an early-ripening sweet type, but is difficult to harvest and must be combined carefully

lush [lʌʃ] *adjective* thick rich (vegetation); *the cattle were put to graze on the lush grass by the river; the lush tropical vegetation rapidly covered the clearing*

LY = LETHAL YELLOWING

lychee [laɪ'tʃi:] *see* LITCHI

lymph [lɪmf] *noun* colourless liquid containing white blood cells, which circulates in the lymph system from all body tissues, carrying waste matter away from tissues to the veins

lynchet ['lɪntʃɪt] *noun* **(a)** strip of land formed as the result of a movement of soil down a slope as a result of cultivation; negative lynchets form at the top of the slope and positive lynchets at the bottom **(b)** unploughed strip of land forming a temporary boundary between fields

COMMENT: lynchets on former prehistoric fields can still be seen in the form of steps on the sides of hills

lyophilize [laɪ'ɒfɪlaɪz] *verb* to freeze dry food, a method of preserving food by freezing it rapidly and drying in a vacuum

lysine ['laɪsi:n] *noun* essential amino acid in protein foodstuffs, essential for animal growth. It is the most important and often the first limiting amino acid in protein; of particular importance in pig and poultry diets, it is estimated to make up almost 7% of the amino acids in pig meat protein

Mm

macroclimate ['mækrəʊklaɪmət] *noun* climate over a large area; *compare* MICROCLIMATE

macronutrient [mækrəʊ'njuːtriənt] *noun* a nutrient which is needed by a plant in relatively large amounts, such as nitrogen, phosphorus, potassium, calcium, magnesium and iron

mad cow disease [mæd 'kaʊ dɪ'ziːz] *see* BOVINE SPONGIFORM ENCEPHALOPATHY

Maedi-Visna (MV) ['maɪdi 'vɪsnə] *noun* virus disease of sheep, which causes breathing difficulties

MAFF = MINISTRY OF AGRICULTURE, FISHERIES AND FOOD

maggot ['mægət] *noun* soft-bodied, legless larva of a fly, such as a bluebottle, warble fly or frit fly; maggots may attack crops and livestock

magnesium [mæg'niːziəm] *noun* chemical element, a white metal which is used in making alloys, and is also an essential element in biological life (in human diets, it is found in green vegetables). Magnesium is a necessary part of chlorophyll. If soil is deficient in magnesium, it can be added in the form of ground limestone, magnesium sulphate or farmyard manure (NOTE: chemical symbol is **Mg**; atomic number is **12**)

COMMENT: the addition of magnesium to soil may prevent deficiency diseases in crops or in livestock, such as interveinal yellowing of leaves in potatoes and sugar beet, and hypomagnesaemia or 'grass staggers' in grazing animals. Heavy spring applications of potash (potassium) fertilizers will increase the chance of grass staggers occurring

mahogany [mə'hɒgəni] *noun* dark tropical hardwood, now becoming rare

maiden (tree) ['meɪdən] *noun* tree in its first year after grafting or budding, when it is formed of a single stem

maincrop potatoes ['meɪnkrɒp] *noun* varieties of potato grown as a main crop, as opposed to 'earlies'; *see also* POTATO

Maine [meɪn] *see* BLEU

◊ **Maine-Anjou** *noun* breed of dual-purpose cattle developed in Brittany, now imported into the UK from France, and exported to many other countries; the animals are roan or red and white in colour

maintenance ration ['meɪntənəns 'ræʃən] *noun* the amount of food needed by an animal to keep in good condition

maize [meɪz] *noun* widely grown cereal crop *(Zea mays)*; **maize gluten** = animal feedingstuff obtained after maize has been milled; it is high in protein (NOTE: in US English called **corn**)

COMMENT: maize is a tall annual grass plant, with a strong solid stem. The male flowers form a tassel on the top of the plant and the females some distance away in the axils of some of the middle stem leaves. After wind pollination of the filament-like styles or silks, the grain develops into long 'cobs' of tightly packed seeds. In Great Britain the crop is grown for making silage, or for harvesting as ripened grain; some is grazed or cut as a forage crop, while a small proportion is sold for human consumption as 'corn on the cob'. Maize needs rich deep well-drained soils and ideally a frost-free growing season with a lot of sunshine before harvest. Maize is the only grain crop which was introduced from the New World into the Old World, and it owes its name of Indian corn to the fact that it was cultivated by American Indians before the arrival of European settlers. It is the principal crop grown in the United States, where it is used as feed for cattle and pigs. In Mexico it is the principal food of the people, being coarsely ground into flour from which tortillas are made. Maize is also a staple food grain in the wetter parts of Africa; in South Africa the cobs are known as 'mealies'

malathion [mælə'θaɪən] *noun* organophosphorous insecticide used to kill small aphids and mites; malathion is used especially to fumigate silos and grain stores before harvest, or is applied to the grain itself when it is being stored. It has a relatively low toxicity to mammals

male [meɪl] *noun* & *adjective* (animal) which produces sperm; (plant) that produces pollen

malodours [mæl'əʊdəz] *noun* unpleasant smells

QUOTE increasing concern with all types of pollution makes it important to have satisfactory methods for predicting the dispersion of malodours from agricultural sources, including livestock buildings and slurry stores
Farm Building Progress

malt [mɔːlt] **1** *noun* best-quality barley grains which have been through the malting process and are used in breweries to make beer and in distilleries to make whisky; **malt culms** = roots and shoots of partly germinated malting barley; a by-product of the malting process, the culms are used as a feedingstuff for livestock **2** *verb* to treat grain, such as barley, by allowing it to sprout and then drying it; the malted grain is used for making beer; **malted meal** = brown wheat flour mixed with flour made from barley

◊ **maltase** *noun* enzyme present in germinating grains, in saliva and in intestinal juices in the small intestine; it converts maltose into glucose

◊ **malting** *noun* the process by which barley grain is soaked in water, then sprouted on a floor to produce an enzyme; it is then dried in a kiln and the roots and shoots are removed to leave the malt grains; **malting barley** = best-quality barley used for malting

◊ **maltose** *noun* sugar formed by digesting starch *or* glucose

◊ **maltster** *noun* person who makes malt for sale to breweries

QUOTE malting barley prices in Scotland have achieved a premium as high as £70/t over feed, with less than one-third of the crop suitable for malting
Farmers Weekly

Malus ['meɪləs] *Latin name for* the apple tree

mammal ['mæməl] *noun* animal of the class Mammalia (such as the human being) which gives birth to live young, secretes milk to feed them, keeps a constant body temperature and is covered with hair. All farm livestock with the exception of poultry are mammals

mammary glands ['mæməri 'glændz] *noun* glands in females that produce milk; in cows, sheep and goats, the glands are located in the udder

manage ['mænɪdʒ] *verb* to control; to be in charge of; *the department is in charge of managing land resources;* **managed woodland** = woodland which is controlled by felling, coppicing, planting, etc.

◊ **management** ['mænɪdʒmənt] *noun* organized use of resources *or* materials; **pasture management** = the control of pasture by grazing, cutting, reseeding, etc.; **woodland management** = controlling an area of woodland so that it is productive (by regular felling, coppicing, planting, etc.)

◊ **manager** ['mænədʒə] *noun* person who manages an organization, etc.; **farm manager** = person who runs a farm on behalf of the owner

mandioca [mændi'ɒkə] *noun* name used in Brazil for cassava

mane [meɪn] *noun* long hair on the neck of a horse

manganese ['mæŋɡəniːz] *noun* metallic trace element which is essential for biological life as an enzyme activator (NOTE: chemical symbol is **Mn;** atomic number is **25**)

COMMENT: manganese deficiency is associated with high pH and soils which are rich in organic matter; it can cause grey leaf of cereals, marsh spot in peas and speckled yellowing of leaves of sugar beet. It is usually cured by applying manganese sulphate as foliar spray

mange [meɪndʒ] *noun* skin disease of hairy or woolly animals, caused by mites, including *Sarcoptes*, the itch mite

mangel *or* **mangold** *or* **mangel wurzel** ['mæŋɡəl *or* 'mæŋɡəʊld *or* 'mæŋɡəl 'wɜːzəl] *noun Beta vulgaris*, a plant similar to sugar beet, but with larger roots; mainly grown in Southern England as a fodder crop; **mangel fly** *or* **mangold fly** = fly whose yellow-white legless larvae cause blistering of the leaves of mangels and sugar beet; this holds back plant growth and in severe cases can kill the plant

COMMENT: varieties of mangels include Globes, Tankards (oblong-shaped), Longs and Intermediates. Mangels contain less than 15% dry matter and are normally harvested before maturity and dried off in a clamp

manger ['meɪndʒə] *noun* a trough in a stable, from which horses and cattle feed

mangetout [mɒnʒ'tuː] *noun* variety of pea, which is picked before the seeds are developed, and of which the whole pod is cooked and eaten

mango ['mæŋgəʊ] *noun* tropical tree (*Mangifera indica*) and the fruit it produces; the tree originated in India, but is grown widely in tropical countries; the fruit is large, yellow or yellowish-green, with a soft orange pulp surrounding the very large seed; the seeds and bark are also used medicinally

mangold ['mæŋgəʊld] *see* MANGEL

mangosteen ['mæŋgəʊstiːn] *noun* tree (*Garcinia mangostana*) which is native of Malaysia, but which is now cultivated in the West Indies; the fruit has a dark shiny rind and a soft sweet white flesh

manioc [mæniɒk] *noun* the French name for cassava, used as an animal feedingstuff; *see also* CASSAVA

manive ['mæniːv] *noun* cassava meal, used as an animal feedingstuff

manure [mə'njʊə] **1** *noun* animal dung used as fertilizer (in liquid form it is called 'slurry'); **artificial manure** = artificial fertilizer, a chemical substance manufactured for use to increase the nutrient level of the soil; **green manure** = rapid-growing green vegetation (such as rape or mustard plants) which is grown and ploughed into the soil to rot and act as manure; **manure cycle** = the process by which waste materials from plants, animals and humans are returned to the soil to restore nutrients **2** *verb* **(a)** to spread animal dung on land as fertilizer **(b)** green

manuring = growing green crops and ploughing them in to increase the organic content of the soil

◊ **manure spreader** *or* **muck spreader** *noun* trailer with a moving floor conveyor and a combined shredding and spreading mechanism, used to distribute manure over the soil; another type has an open, cylindrical-shaped trailer with a rotor shaft running from front to back; flails are attached to the rotor and throw the manure out from the side of the trailer. Slurry can be spread on the land with special pumping equipment, using a pipeline and slurry guns

COMMENT: all farm manures and slurries are valuable, and should not be regarded as a problem for disposal, but rather as assets to be used in place of expensive fertilizers which would otherwise need to be bought. Manure and slurry has to be spread in a controlled way, or dangerous pollution can result from runoffs into streams after rainfall

QUOTE Farm pollution of rivers rose last year (1988) to a record 4,141 reported cases in England and Wales. Animal slurry, the main form of farm pollution, is up to 100 times more damaging than untreated sewage, while the liquor from silage is 200 times more damaging. Mild and wet conditions last winter were partially blamed for forcing farmers to stockpile slurry until it could be spread

Guardian

Manx Loghtan ['mæŋks 'lɒhtæn] *noun* rare breed of sheep, which is native to the Isle of Man. The wool is mouse-brown and the animals are multi-horned

maple ['meɪpl] *noun* hardwood tree (*Acer* sp), of northern temperate regions, some varieties of which produce sweet sap which is used for making sugar and syrup

mapping ['mæpɪŋ] *noun* the process of collecting information and using it to produce maps; **soil mapping** = making maps showing different types of soil in an area

Maran ['mærɒn] *noun* heavy continental breed of fowl, which has a greyish-brown barred plumage; the eggs are dark brown

marbling ['mɑːblɪŋ] *noun* appearance of muscle with intramuscular fat, seen on the cut surface of meat

Marchigiana [mɑːkɪdʒi'ɑːnə] *noun* breed of white beef cattle from Italy, now imported into the UK and used for cross-breeding to improve beef-calf quality in dairy cows

mare [meə] *noun* female horse, five years old or more

Marek's (disease) ['mæreks dɪ'ziːz] *noun* virus disease of poultry, causing lameness and paralysis

> QUOTE an outbreak of Marek's in a vaccinated flock could have been triggered off by a worm infestation
> *Smallholder*

margin ['mɑːdʒɪn] *noun* difference between the amount of money received for a product and the money which it cost to produce; **margin over purchased feed (MOPF)** = amount of money received for produce (such as per litre of milk) shown as a percentage above the amount spent in purchasing feed for the animals

> QUOTE for the all-grass farmer with little choice of alternative enterprises, it is important to obtain the biggest possible margin a litre
> *Farmers Weekly*

> QUOTE if one takes a gross margin of £120 per animal as being acceptable, this means that a minimum trading margin (i.e. the difference between the buying and selling price per animal) of £250 would be needed
> *Practical Farmer*

marginal ['mɑːdʒɪnəl] *adjective* (land) at the edge of cultivated land (such as edges of fields, banks beside roads, etc.);(land) of poor quality which results from bad physical conditions, such as poor soil, high rainfall, steep slopes, and where farming is often hazardous; *cultivating marginal areas can lead to erosion*

mariculture ['mærɪkʌltʃə] *noun* farming in the sea or in a marine environment

> COMMENT: mariculture refers to aquaculture in seawater, such as raising oysters, lobsters and fish in special enclosures

market ['mɑːkɪt] *noun* place (often in the open air) where farm produce is sold

◊ **market garden** ['mɑːkɪt 'gɑːdən] *noun* place for the commercial growing of plants, usually vegetables, soft fruit, salad crops and flowers, found near a large urban centre which provides a steady outlet for the sale of its produce; **market gardening** = growing vegetables, salad crops, fruit for sale; **market gardener** = person who runs a market garden

◊ **marketing boards** *noun* in the UK certain commodities are marketed by boards set up by Acts of Parliament, such as the Potato Marketing Board

◊ **market town** *noun* town with a permanent or regular market, which serves as a trading centre for the surrounding area; some markets specialize in certain types of livestock or produce

marjoram ['mɑːdʒərəm] *noun* Mediterranean aromatic herb *(Origanum)*; the dried leaves are used as flavouring

markings ['mɑːkɪŋz] *noun* coloured patterns on the coat of an animal *or* in the feathers of a bird

marl [mɑːl] **1** *noun* fine soil formed of a mixture of clay and lime, used to improve the texture of light soils, such as sandy soils or black fen soils **2** *verb* to add marl to a light soil to improve its texture

Marota [mə'rəutə] *noun* breed of goat found in Brazil; the animals are white or light grey, and a reared both for milk and for their skins

marram grass ['mærəm 'grɑːs] *noun* type of grass *(Ammopila arenaria)* which is planted on sand dunes to stabilize them and prevent them being extended by the wind

marrow ['mærəu] *noun* large vegetable *(Cucurbita pepo)* of the pumpkin family, which may be grown as bush or trailing varieties

marrowstem kale ['mærəustem 'keɪl] *noun* variety of kale with a thick stem and large leaves, grown as feed for livestock in the autumn and winter months (though it is not winter hardy)

marsh [mɑːʃ] *noun* area of permanently wet land and the plants which grow on it; **marsh gas** = methane; **marsh spot** = disease affecting peas, caused by manganese deficiency

◊ **marshland** ['mɑːʃlænd] *noun* area of land covered with marsh

◊ **marshy soil** ['mɑːʃi 'sɔɪl] *noun* very wet soil

COMMENT: a marsh usually has a soil base, as opposed to a bog or fen which is made up of peat. Many former areas of marshland have been reclaimed and have been artificially drained by an system of ditches and sluices which allow surface water to escape to the sea, but prevent salt water entering the area. The drained soils are usually fertile and some of these areas are important for agriculture

martingale ['mɑːtɪŋɡeɪl] *noun* device used to regulate the way a horse's carries its head; it consists of a strap or straps, attached to the girth at one end, and at the other to the reins or to the noseband

mash [mæʃ] *noun* mixture of feeding meals mixed to provide all the necessary elements for a balanced diet; **wet mash** = mash fed mixed with water

Masham ['mæʃəm] *noun* crossbred type of sheep which results from a Wensleydale or Teeswater ram mated with a hill ewe of the Swaledale type; the breed is an economical ewe with a good lambing average and a useful fleece; there are black markings on the face and legs

mashlum ['mæʃlʌm] *noun* mixture of oats and barley (and sometimes wheat), sown to provide grain for feeding to livestock; *Also called* MASLIN, MESLEN, MESLIN

mast [mɑːst] *noun* (a) seeds of the beech tree, used as food by pigs and other animals (b) **mast cell** = large cell in connective tissue which carries histamine and reacts to allergens (c) **mast swine** = German term for fattening pig

mastitis [mæ'staɪtɪs] *noun* common bacterial disease affecting dairy animals in which the udders become inflamed and swollen, and the passage of the milk is blocked

COMMENT: common causes are staphylococci such as *Staphylococcus aureus* (staphylococcal mastitis), streptococci (*Streptococcus uberis*) or other bacteria (E. coli mastitis). The condition can be treated with antibiotics

mat [mæt] *noun* covering of undecayed grassland vegetation which forms on very acid soil, when the soil lacks the microorganisms necessary to break decaying matter down

◊ **matted** *adjective* with many fibres woven together

mate [meɪt] **1** *noun* animal which is paired with another of the opposite sex **2** *verb* (*of an animal*) to have sexual intercourse with another; **assortive mating** *or* **mating likes** = mating animals which have a similar appearance; **flock mating** = mating system which uses several males to mate with the females of a flock; **pen mating** = using one male animal to mate with a number of females

matorral [mətɔː'ræl] *noun* kind of shrubby vegetation which covers much of the arid northern regions of Mexico

Ma T'ou ['mɑː 'təʊ] *noun* breed of goat found in central China; the animals are white and very prolific breeders

mattock ['mætək] *noun* heavy hoe

mature [mə'tʃʊə] **1** *adjective* ripe **2** *verb* to become ripe

◊ **maturity** [mə'tʃʊrɪti] *noun* (a) time when a plant's seeds are ripe (b) time when an animal has become an adult; time when an animal is ready for slaughter

maw [mɔː] *noun* stomach, especially the last of a ruminant's four stomachs

may [meɪ] *noun* popular name for hawthorn, a common plant for making hedges

◊ **mayweeds** *noun* group of weeds which affect cereals, (*Chamomilla* spp, *Anthemis* spp, *Matricaria* spp); the weeds affect winter crops and vegetables, and are found on headlands. They can cause considerable problems to machinery; other names for them are 'wild chamomile' and 'dogdaisy'

MCPA 2-methyl-4chloro-phenoxy-acetic acid; a translocated herbicide, introduced in 1942

COMMENT: MCPA kills all the worst and strongest broad-leaved weeds. The following common weeds are easily controlled by MCPA: annual nettle, buttercups, charlock, dock seedlings, fat hen, orache, plantains, poppy thistles and wild radish

MCPP *see* MECOPROP

ME = METABOLIZED *or* METABOLIZABLE ENERGY *ME levels in concentrates*

meadow ['medəʊ] *noun* field of grass and other wild plants, grown for fodder; **water meadow** = meadow close to a river, which can be flooded; **meadow fescue** = perennial grass *(Festuca pratensis)* which has considerable importance for hay and grazing. It is a highly productive grass which flourishes when sown with timothy

◊ **meadowgrass** *noun* varieties of grass of the genus *Poa;* **annual meadowgrass** = widespread weed *(Poa annua)* found in all arable and grass crops; **smooth-stalked meadowgrass** = species of grass which can withstand quite dry conditions; a perennial grass with smooth greyish-green leaves and green purplish flowers

meal [miːl] *noun* finely ground compound feedingstuff for poultry and pigs, containing all the elements necessary for good health and steady growth

mealies ['miːlɪz] *noun* native South African term for maize

mealworm ['miːlwɜːm] *noun* larva of various beetles of the genus *Tenebrio* that infests and pollutes grain products

meat [miːt] *noun* animal flesh which is eaten as food; meat is formed of the animal's muscle; **meat-eating animals** = carnivores, animals (such as the cat family) which eat almost exclusively meat; **meat extender** = any foodstuff or mixture of foodstuffs added to meat preparations to increase their bulk; **meat and bone meal** = meal made from waste meat and bones, very rich in protein

◊ **-meat** *suffix* showing the flesh of an animal (used in particularly in the EU); *pigmeat, sheepmeat*

◊ **Meat and Livestock Commission** organization which provides services to livestock breeders, including the evaluation of breeding stock potential, carcass grading and classification; the Commission also carries out various research projects. Its staff also provide services for abattoirs and livestock auction markets. The Commission promotes the sale of British meat

Meatlinc ['miːtlɪnk] *noun* new breed of

sheep used as a terminal sire. Only the rams are sold

mechanization [mekənaɪ'zeɪʃn] *noun* introduction of machines for agricultural working purposes

COMMENT: mechanization has been an important factor in the contraction of the agricultural labour force. Mechanization has not only involved increases in the number and range of machines, but also dramatic increases in their size and power. This has enabled slopes previously regarded as too steep for ploughing to be cultivated. The increased size of tractors and combines has encouraged enlargement of fields and the removal of hedgerows. This has caused alarm amongst conservationists and led to increased erosion in wet weather in some areas

mecoprop (MCPP) ['mekəʊprɒp] *noun* herbicide used to control weeds such as chickweed and cleavers, as well as the weeds controlled by MCPA

meiosis *or US* **miosis** [maɪ'əʊsɪs] *noun* process of cell division which results in two pairs of cells, each with only one set of chromosomes; *compare* MITOSIS

Meishan [meɪ'ʃæn] *noun* breed of pig imported into the UK from China; it is a fast-maturing breed and is remarkably prolific and is one of the Taihu breeds; animals commonly have 18 to 20 teats

melon ['melən] *noun* plant of the cucumber family *(Cucumis melo)* with a sweet fruit; the flesh of the fruit varies from green to orange or white; **water melon** = plant of the genus *Citrullus vulgaris* with large green fruit with watery pink flesh

Mendel's laws ['mendəlz 'lɔːz] *noun* laws governing heredity

COMMENT: the two laws set out by Georg Mendel following his experiments growing peas, were (in modern terms): that genes for separate genetic characters assort independently of each other, and that the genes for a pair of genetic characters are carried by different gametes. For animal breeders, the main features of Mendelism is that it is based on simple and clearly-defined traits that are inherited as separate entities: these were traits such as colour, which are controlled by single genes

merchant ['mɜːtʃənt] *noun* person who sells a product; **seed merchant** *or* **corn merchant** = trader who sells seed or corn, usually wholesale

Merino [mə'riːnəʊ] *noun* breed of sheep which originated in North Africa and was then introduced into Spain. It is now bred in all parts of the world (especially in Australia, South Africa and New Zealand) for its dense soft fine fleece, with strong and curly fibres

Merinolandschaf [məriːnəʊ'læntʃæf] *noun* breed of sheep found in South Germany. Large travelling flocks are common

meslen *or* **meslin** ['meslɪn] = MASHLUM

meta- ['metə] *prefix meaning* which changes *or* which follows

◊ **metabolism** [me'tæbəlɪzm] *noun* chemical processes which are always taking place in organisms and which are essential to life; **basal metabolism** = energy used by a body at rest (i.e. energy needed to keep the body functioning and the temperature normal); this can be calculated while an animal is in a state of complete rest, by observing the amount of heat given out or the amount of oxygen taken in and retained

◊ **metabolic** [metə'bɒlɪk] *adjective* referring to metabolism; **basal metabolic rate (BMR)** = rate at which a body uses energy when at rest; **metabolic cycle** = cycle by which plants absorb sunlight, transform it into energy by photosynthesis and create carbon compounds; **metabolic size** = the size of an animal to which the metabolic rate is proportional

◊ **metabolize** [me'tæbəlaɪz] *verb* to change the nature of something by metabolism; *the liver metabolizes proteins and carbohydrates*

◊ **metabolized** *or* **metabolizable energy** *noun* the proportion of energy from feed which is used by an animal through its metabolism; it is the measure of energy following digestion, after the alimentary gases and urinary losses have been subtracted. Animals cannot be expected to transfer energy from feed with perfect efficiency as there will always be losses through undigested food and as alimentary gases. The energy needs of livestock can be calculated from their size

COMMENT: metabolism covers all changes which take place in an organism: the building of tissue (anabolism); the breaking down of tissue (catabolism); the conversion of nutrients into tissue; the elimination of waste matter; the action of hormones, etc.

metaldehyde [met'ældɪhaɪd] *noun* substance, sold in small blocks, used to light fires or used as a molluscicide to kill slugs and snails

metamorphosis [metə'mɔːfəsɪs] *noun* change, especially the change of a larva into an adult insect

meter ['miːtə] **1** *noun* device for counting *or* measuring; *farmers who are worried about the level of nitrates in crops or runoff water, can now test for it with a small hand-held meter, which gives a digital reading of nitrate concentration;* **electric meter** *or* **gas meter** *or* **water meter** = device which records how much electricity *or* gas *or* water has been used **2** *verb* to count *or* measure with a meter; **metered chop harvester** *see* PRECISION CHOP FORAGE HARVESTER

methane (CH₄) ['miːθeɪn] *noun* marsh gas, a colourless gas, produced naturally from rotting organic waste; **methane converter** = process which takes the gas produced by rotting waste in landfill sites and processes it into a form which can be used

COMMENT: methane is produced naturally from rotting vegetation in marshes, where it can sometimes catch fire, creating the phenomenon called will o' the wisp, a light flickering over a marsh. Large quantities may also be formed in the rumen of cattle. It occurs as the product of animal excretions in livestock farming. Excreta from livestock can be passed into tanks where methane is extracted leaving the slurry which is then used as fertilizer. The methane can be used for heating or as a power source. Methane is also a greenhouse gas, and it has been suggested that methane from rotting vegetation, from cattle excreta, from water in paddy fields, and even from termites' nests, all contribute to the greenhouse effect

QUOTE the principal sources of methane are - enteric fermentation in livestock and insects, ricefields and natural wetlands, biomass burning, landfills and gas and coal fields

Nature

QUOTE people are worried about the possible escapes of methane gas should the tip be established
Environment Now

methanol (CH₃OH) ['meθənɒl] *noun* alcohol, manufactured from coal, natural gas or waste wood, which can be used as fuel

COMMENT: methanol can be used as a fuel in any type of burner; its main disadvantage is that it is less efficient than petrol and can cause pollution if it escapes into the environment as it mixes easily with water. Production of methanol from coal or natural gas, does not help fuel conservation, since it reduces the earth's fossil fuel resources

methionine ['meθɪəniːn] *noun* one of the eight essential amino acids, essential for animal growth

metre *or US* **meter** ['miːtə] *noun* measurement of length, the basic unit of the metric system (one metre equals 39.37 inches)

metritis [mɪ'traɪtɪs] *noun* an infection of the lining of the womb in cattle; the symptoms are a white discharge and/or a high temperature; sows can also have mastitis (MMA) (NOTE: also called **whites)**

Meuse-Rhine-Ijssel ['mɜːz 'raɪn 'aɪsəl] *noun* dual-purpose breed of cattle, originating from the Netherlands. It is used by breeders in Britain to upgrade Dairy Shorthorn. The breed's dairy performance is similar to that of the British Friesian, and it has a fine beef conformation. Cattle are red and white in colour

mezzadria [me'tsædrɪə] *noun* system used in Southern Italy, where a vineyard is leased and the landlord is paid a half-share of the wine produced

MFA = MASTER OF FOXHOUNDS ASSOCIATION

Mg *chemical symbol for* magnesium

MGA = MAIZE GROWERS ASSOCIATION, MUSHROOM GROWERS ASSOCIATION

MHC = MOISTURE-HOLDING CAPACITY

microbe ['maɪkrəub] *noun* microorganism, very small organism which can only be seen with a microscope. (Viruses, bacteria, protozoa and fungi are all forms of microbe)

◊ **microbial** [maɪ'krəubɪəl] *adjective* referring to microbes; **microbial disease** = disease caused by a microbe; **microbial ecology** = study of the way in which microbes develop in nature; **microbial protein** = protein source in ruminants from dead rumen microbes, usually forming 70% to 100% of the ruminant's supply of protein

QUOTE nitrates and organic matter from dead and decaying vegetation and from animal droppings, stimulate microbial activity in the soil
New Scientist

microclimate [maɪkrəu'klaɪmət] *noun* climate found in a very small area, such as a pond *or* a tree *or* a field; *compare* MACROCLIMATE

◊ **microhabitat** [maɪkrəu'hæbɪtæt] *noun* single small habitat

◊ **micronutrient** [maɪkrəu'njuːtrɪənt] *noun* substance (such as iron, zinc, copper, etc.) which organisms only need in very small quantities, i.e. as trace elements

micron ['maɪkrɒn] *noun* measurement of thickness, one millionth of a metre, used in measuring the fineness of hair or wool (NOTE: usually written µ with figures: **25µ)**

QUOTE a new record price for wool was set at the Dunedin sale last week with a bale of 15.7 micron merino hogget wool sold for $210 a kilo
New Zealand Farmer

microorganism [maɪkrəu'ɔːgənɪzm] *noun* microbe, a very small organism which can only be seen with a microscope. (Viruses, bacteria, protozoa and fungi are all forms of microorganism)

midden ['mɪdən] *noun* heap of dung

Middle White ['mɪdl 'waɪt] *noun* breed of white pig which comes from a cross between the Large White and the Small White. A short compact breed with long upright ears and a turned-up snout; now a rare breed

mids ['mɪdz] *noun* middle-sized potatoes which are graded and sold for human consumption

Midwest [mɪd'west] *noun* region of the USA, between the Rocky Mountains on the west and the Alleghenies on the east, comprising the main farming area of the country

milch cow ['mɪltʃ 'kaʊ] *noun* cow which gives milk or is kept for milk production

mildew ['mɪldjuː] *noun* plant disease caused by a fungus which produces a fine powdery film on the surface of the infected parts; **downy mildew** = particularly strong form of mildew which affects lettuces and rots their leaves and stems

milk [mɪlk] **1** *noun* opaque white liquid secreted by female mammals during lactation; **milk composition** = the percentages of protein, lactose, fat, minerals and water which make up milk; the composition varies according to the breed of cow, but average percentages are: protein (3.4%), milk sugar (4.75%), fat (3.75%), minerals (0.75%), water (87.35%); **milk compositional quality scheme** = a scheme organized by the former Milk Marketing Board, which categorizes milk and pays farmers according to the quality of milk produced; **milk cooler** = stainless steel bulk storage tank, in which milk is cooled by running water passing over the outside of the tank; **milk producer** = farmer who is registered with the MAFF, and produces milk in compliance with the regulations concerning clean milk production. In 1994-5 the estimated number of registered producers was 27,800 in England and Wales; **milk products** = milk and other foodstuffs produced from it, which are sold for human consumption; the main milk products are liquid milk (homogenized. pasteurized, sterilized or UHT), butter, cheese, cream, condensed milk and milk powder; **milk recording** = keeping a record of the milk given by each cow at each milking, the quality of the milk is analysed each month; **milk ripe stage** *or* **milky stage** = stage in the development of grain, such as wheat, where the seed has formed but is still soft and white and full of white sap; **milk sinus** = the space in each teat into which the milk is secreted; **milk teeth** = first temporary teeth of young animals **2** *verb* to extract milk from a cow's udder; pressure on the teats makes the milk spurt out. Milking can be done by hand, but is usually done by machines in a milking parlour

◊ **milker** *noun* **(a)** cow which is giving milk; cow which is kept for milk **(b)** farmworker who supervises the milking of cows **(c)** part of the milking machine which is attached to the cow's teats with teat cups

COMMENT: in the UK, most milk comes from Friesian cows, and has been heat treated, pasteurized, sterilized or ultra-heat treated before it is sold to the public. It may also be calcium enriched or lactose reduced. Milk is sold in cartons or plastic bottles, either as homogenized, semi-skimmed or skimmed. In bottles it is sold with various coloured metal tops: 'silver top' is pasteurized with an average 3.9% fat: it has a noticeable cream line; 'red top' is similar to the silver, but the milk is homogenized to distribute the cream throughout; 'gold top' is pasteurized milk from Guernsey or Jersey breeds of cow, and has an average fat content of 4.9%; 'red and silver striped top': pasteurized semi-skimmed milk with average 1.6% fat content; 'blue and silver checked top': pasteurized skimmed milk, with an average 0.1% fat content; 'green top': untreated whole milk, with an average 3.9% fat (no longer sold in the UK). Sterilized whole milk with fat content of 3.9% is sold in bottles with crown caps or blue foil tops. UHT milk is also available as whole, semi-skimmed or skimmed: it is milk with a shelf life of 6 months, though when opened it should be kept cold and used as ordinary pasteurized milk

milk fever ['mɪlk 'fiːvə] *noun* disease of milk cows, milk goats and ewes. In spite of its name, the disease is not a fever, and may affect a dairy cow just before calving or during the seven days which follow calving. The first symptoms are restlessness, moving the hind feet up and down while standing; these symptoms are followed by loss of balance and later loss of consciousness. The disease is common at the third, fourth or fifth time of calving, and is caused by a metabolic disturbance or imbalance in the system, due to a low calcium content in the blood. The disease is treated by injections of calcium borogluconate

milk goat ['mɪlk 'gəʊt] *noun* goat which is reared for its milk

milking machine ['mɪlkɪŋ mə'ʃiːn] *noun* machine which imitates the sucking action of a calf, used to extract milk from the cow's udder. It uses a pulsator mechanism to apply pressure to the teats, causing the release of the milk. The milk is then passed into a collecting jar or may pass by pipeline to a large tank

milking parlour ['mɪlkɪŋ 'pɑːlə] *noun* building in which cows are milked, and often are also fed, washed and cleaned.

COMMENT: There a four basic designs of parlour: the herringbone parlour, where the cow stands at an angle of 45° to the milker, is commonest for large herds; the abreast parlour, where the cows stand side by side with their backs to the milker; the tandem parlour where they stand in line with their sides to the milker; the most expensive and complex of the four systems is the rotary parlour, where the cows stand on a rotating platform with the milker in the middle

Milk Marketing Board (MMB) until 1994, the board which organized the collection and buying of milk from farmers and its sale to customers. Replaced by Milk Marque

Milk Marque name of the new national cooperative which has replaced the Milk Marketing Board, with the aim of liberalizing the milk market. Farmers are to be allowed to trade with any buyer. Milk Marque is a farmer-owned milk producers' co-operative. It has around 20,000 dairy farmer members. England and Wales are divided into 89 districts and members in each district elect a representative on one of five area councils. Milk is collected each day and taken to dairies or creameries. Milk measurement and quality testing are carried out

milk quota (MQ) ['mɪlk 'kwəʊtə] *noun* system by which farmers are only allowed to produce certain amounts of milk, introduced to restrict the overproduction of milk in member states of the EU

COMMENT: quotas were introduced in 1984, and were based on each state's 1981 production, plus 1%. A further 1% was allowed in the first year. A supplementary levy or superlevy, was introduced to penalize milk production over the quota level. Reductions in the quota amount were made in 1986/7 and 1987/8. In the UK, milk quotas can be bought and sold, either together with or separate from farmland, and are a valuable asset. The government is responsible for the setting of quotas for milk production, according to the directives of the EU commission

milk sheep ['mɪlk 'ʃiːp] *noun* sheep which is reared for its milk

milk sugar ['mɪlk 'ʃʊgə] *noun* lactose, the sugar found in milk

milky stage ['mɪlki 'steɪdʒ] *see* MILK RIPE STAGE

milk yield ['mɪlk 'jiːld] *noun* the quantity of milk produced each year by a cow

COMMENT: In the UK, the average annual milk yield per dairy type cow increased from 3,989 litres per cow in 1974/5 to 5,141 litres per cow in 1991/2

mill [mɪl] **1** *noun* factory where a substance is crushed to make powder, especially one for making flour from the dried seeds of corn; **watermill** *or* **windmill** = mills driven by water *or* wind; **mill race** = channel of water which turns the wheel of a watermill; **mill wheel** = large wheel (with wooden bars) which is turned by the force of water **2** *verb* to process grain into a another form (wheat into flour, brown rice into white rice); **milling quality** = calculation of how easy it is to separate the white endosperm from the brown seed coat or bran, in the milling process; in general hard wheats are of higher milling quality than soft wheats; **milling wheat** = best-quality wheat used to make flour for making bread

◊ **millstone** *noun* heavy round slab of stone, used to grind corn

millet ['mɪlɪt] *noun* common cereal crop (*Panicum miliaceum*) grown in many of the hot, dry regions of Africa and Asia, where it is a staple food

COMMENT: the two most important species are finger millet and bulrush millet. Millet grains are used in various types of food. It can be boiled and eaten like rice; made into flour for porridge, pasta or chapatis; mixed with wheat flour to make bread. Millet can also be malted to make beer. Millets are also grown as forage crops, and the seed is used as a poultry feed. The main producers of millet are Nigeria, Mali and Niger

milo ['maɪləʊ] *noun US* sorghum

mineral ['mɪnərəl] *noun* inorganic solid substance which is found in nature; **mineral matter** = the solid part of the soil composed of stones, sand, silt and clay (as opposed to the vegetable matter, formed from dead or decaying plants); **mineral matter content** = the amount of minerals found in plants; **mineral nutrients** = nutrients (except carbon, hydrogen and oxygen) which are inorganic and are absorbed by plants from the soil

COMMENT: the most important minerals needed by the body are: calcium (found in milk and green plants) which helps the growth of bones and encourages blood clotting; iron (found in bread and liver) which helps produce red blood cells; phosphorus (found in bread and fish) which helps in the growth of bones and the metabolism of fats; and iodine (found in fish) which is essential to the functioning of the thyroid gland. In animals and plants, the various mineral elements play different roles, either forming bones and other structures, or preventing deficiency diseases. Minerals can be included in diets in the form of concentrates or as blocks to be licked by animals. Plants require nitrogen (N) for protein and other compounds; phosphorus (P) for the metabolic processes and the formation of cell membranes; potassium (K) for transport, and balancing cation; calcium (Ca) for cell division; sulphur (S) for protein; copper (Cu), chlorine (Cl), sodium (Na), manganese (Mn), zinc (Zn), boron (B), iron (Fe), and molybdenum (Mo) are also important in various ways

mineralization [mɪnərəlaɪˈzeɪʃn] *noun* breaking down of organic waste into its non-organic chemical components

minitandia [mɪnɪˈtændiə] *noun* small estate (in South America)

minimum [ˈmɪnɪməm] *noun* the smallest possible quantity; *the law provides only the minimum protection for workers*

◊ **minimal** [ˈmɪnɪməl] *adjective* very small; **minimal area** = smallest area for sampling in which specimens of all species can be found; **minimal disease herd** = herd of livestock with a very low level of infectious diseases

◊ **minimal cultivation** *noun* system of cultivation which subjects the land to shallow working and minimizes the number of passes of machinery. No ploughing is needed

COMMENT: although suitable for cereal production, minimal cultivation is not suitable for all crops or soil conditions. Crops like sugar beet and potatoes need a deeper tilth than that obtained by minimal cultivation

Ministry of Agriculture, Fisheries and Food (MAFF) British government department concerned with agriculture

Minorca [mɪˈnɔːkə] *noun* breed of poultry, originating in the Mediterranean; the birds are black or white in colour

miosis [maɪˈəʊsɪs] *noun US* = MEIOSIS

Miranda [mɪˈrændə] *noun* breed of cattle found in Portugal; the animals are dark brown in colour, with horns coloured white with black tips; they are bred for meat and for draught

mite [maɪt] *noun* tiny animal of the spider family *(Arachnidae)* which may be free-living in the soil or parasitic on animals or plants

mitosis [maɪˈtəʊsɪs] *noun* process of cell division, whereby a cell divides into two identical cells; *compare* MEIOSIS

mixed [mɪkst] *adjective* made up of different elements, types, sexes, etc.; **mixed cropping** = growing more than one species of plant on the same piece of land at the same time; **mixed culture** = growing several species of tree together on the same piece of land (as opposed to monoculture); **mixed farming** = farming involving arable and livestock farming and showing no particular specialization; **mixed grazing** = grazing system where more than one type of animal grazes the same pasture at the same time; **mixed woodland** = woodland containing conifers and deciduous trees

MMB = MILK MARKETING BOARD

Mn *chemical symbol for* manganese

Mo *chemical symbol for* molybdenum

mode of action [ˈməʊd ʌv ˈækʃən] *noun* the way in which a pesticide acts, as, for example, organophosphorous compounds disrupt the nerve impulses in insects

moder [ˈməʊdə] *noun* intermediate type of humus, which is partly acid mor and partly neutral mull

MOET = MULTIPLE OVULATION AND EMBRYO TRANSFER

mohair [ˈməʊheə] *noun* fine wool from a goat (over 30 microns); *compare* CASHMERE

Moiled [mɔild] *see* IRISH MOILED

moist [mɔist] *adjective* slightly damp; containing a small amount of water

◊ **moisture** ['mɔistʃə] *noun* slight amount of water as found in soil, grain, etc.

◊ **moisture content** *noun* the percentage moisture in harvested crops

COMMENT: the safe moisture content for storage of all grains is 14% or less. In the UK, only fully ripe grain in a very dry period is likely to be harvested at less than 14% moisture content. In a wet season, the moisture content may be as high as 30%, and these grains will have to be dried. The moisture content of hay is 80% when cut, and has to be reduced to below 20% for safe storage

moisture-holding capacity (MHC) ['mɔistʃə həuldɪŋ kə'pæsiti] *noun* the amount of water held by soil between field capacity and the permanent wilting point. The amount of moisture will vary according to the texture of the soil

molasses [mə'læsɪs] *noun* dark brown syrup, a by-product of sugar production left after sugar has been separated; it is used as a binding agent in compound animal feeds; molasses is also added to silage

mold [məuld] *US* = MOULD

mole [məul] *noun* (a) base SI unit of amount of a substance (b) small grey mammal which lives underground and which eats worms and insects (c) part of a mole plough, which cuts a round channel

◊ **mole drain** *noun* underground circular channel formed by a mole plough as it is pulled across a field; mole drains are normally drawn 3 to 4 metres apart, and are best used in fields with a clay subsoil

◊ **molehill** ['məulhɪl] *noun* small heap of earth pushed up to the surface by a mole as it makes its tunnel

◊ **mole plough** *noun* implement used to draw mole drains; the plough is pulled by a tractor, and consists of a vertical blade below which there is a torpedo or bullet-shaped mole. Behind this mole is an expander which widens the drain as it is cut

Molinia [mə'lɪniə] *noun* poor type of grass found in rough mountain or hill grazings; it is of little value as grazing

mollusc ['mɒlʌsk] *noun* any of many animals with soft bodies, usually living in shells; *slugs and snails are molluscs, as are oysters and other shellfish*

◊ **molluscicide** [mə'lʌskɪsaɪd] *noun* substance used to kill molluscs (such as slugs and snails)

molybdenum [mə'lɪbdənəm] *noun* metallic trace element, essential to biological life; excess molybdenum in grazing can, however, lead to insufficient copper uptake in sheep (NOTE: the chemical symbol is **Mo**; atomic number is **42**)

Mongolian ['mɒŋgəulɪən] *noun* breed of cattle, usually reddish-brown in colour; the animals are small and kept for milk and meat. Now not very common

mono- ['mɒnəu] *prefix* meaning single *or* one

◊ **monocotyledon** [mɒnəukɒtə'liːdən] *noun* one of the two classifications of plants, a plant such as grass which has a single cotyledon (or seed leaf); *compare* COTYLEDON, DICOTYLEDON

◊ **monocropping** *or* **monoculture** ['mɒnəkrɒpɪŋ *or* 'mɒnəkʌltʃə] *noun* system of cultivation where only one crop is grown on the same piece of land over a period of years; **conifer monoculture** = system of afforestation where only one type of conifer is grown (NOTE: opposite is **mixed culture, polyculture**)

COMMENT: monocropping involving cash crops, groundnuts, cotton, etc., exposes farmers in Africa to price fluctuations on the world market. Diversification is needed to stabilize farm incomes

◊ **monogastric** *adjective* animals with only one stomach, such as pigs, as opposed to ruminants

◊ **monopitch** *noun* type of piggery with artificially controlled natural ventilation

monsoon [mɒn'suːn] *noun* season of wind and rain in tropical countries; **monsoon forest** = tropical rainforest in an area where rain falls during the monsoon season

moor [muə] *noun* high land which is not cultivated, formed of acid soil covered with grass and low shrubs such as heather

◊ **moorland** ['muəlænd] *noun* area of land

covered with moor of low agricultural value; it is a habitat for game birds

MOPF = MARGIN OVER PURCHASED FEED

mor [mɔːr] *noun* type of humus which is acid and relatively undecomposed, found, for example, under coniferous forests

morbidity [mɔːˈbɪdɪti] *noun* being diseased; **morbidity rate** = number of cases of a disease per hundred thousand of population

Morrey system [ˈmɒreɪ ˈsɪstəm] *noun* paddock system of rotational grazing, used in the management of dairy herds

mortality [mɔːˈtælɪti] *noun* death; **high mortality rate** = high percentage of animals which die (as a percentage of a group)

mosaic virus [məˈzeɪɪk ˈvaɪrəs] *noun* virus disease of plants which makes spots on the leaves and can seriously affect some crops

mosquito [mɒsˈkiːtəʊ] *noun* insect which sucks blood and passes viruses or parasites into the bloodstream

COMMENT: in tropical countries diseases such as malaria and yellow fever are transmitted in this way. Mosquitoes breed in water, and they spread rapidly in lakes or canals made by dams and other irrigation schemes. Because irrigation is more widely practised in tropical countries, mosquitoes are increasing and diseases such as malaria are spreading

moss [mɒs] *noun* very small plant without roots, which grows in damp places and forms mats of vegetation; **moss peat** = dried and sterilized peat formed from the remains of mosses, sold in bags for horticultural purposes

mould *or US* **mold** [məʊld] *noun* **(a)** any of various plants of the Kingdom Fungi; especially mildew, fungus which produces a fine powdery film on the surface of an organism **(b)** soft soil rich in humus; **leaf mould** = soft fibrous material formed from rotten and broken-down dead organic matter such as leaves

◊ **mouldboard** *noun* part of a plough, which lifts and turns the slice of earth, so

making the furrow. Mouldboards are made of steel, and are made in many different styles, each producing a different surface finish; the main ones are general-purpose, digger and semi-digger

moult [məʊlt] **1** *noun* losing feathers *or* hair at a certain period of the year; **moult plumage** = small feathers which remain on a bird when it is moulting **2** *verb* to lose feathers *or* hair at a certain period of the year; *most animals moult at the beginning of summer;* **moulting season** = time of year when an animal *or* bird moults

mount [maʊnt] *verb* to attach an implement to a tractor so that it is held by the tractor and has no other support

COMMENT: if the implement is simply pulled by the tractor, and is supported by its own wheels, then it is said to be 'trailed' (a rotary cultivator can be either mounted or trailed). If the implement is supported both by the tractor and by its own wheels it is said to be 'semi-mounted'

mountain [ˈmaʊntən] *noun* **(a)** high land **(b) commodity mountain** = surplus of a certain agricultural product produced in the EU, for example the 'butter mountain'

movement licence [ˈmuːvmənt ˈlaɪsəns] *noun* licence which is needed in order to move animals from areas of infectious disease; restriction on the movement of animals is a measure used to prevent the spread of disease

◊ **movement record** *noun* record kept by a farmer of all movements of animals on and off the farm premises (compulsory in the UK since 1925)

mow [məʊ] *verb* to cut grass or a forage crop

◊ **mower** *noun* machine used to cut grass and other upright crops

COMMENT: the cut crop is left in a swath behind the cutting mechanism for further treatment in the process of making hay or silage. The main types of mowers are: *cutter bar mowers*, which are rear-mounted and consist of a framework with a hinged cutter bar; the cutter bar has a number of fingers with a cutting edge on each side; these support the knife; *rotary mowers* are made with two or four rotors, each having two, three or four turning blades; *flail mowers* have a high-speed rotor fitted with swinging flails which cut the grass and leave it bruised in a fluffy swath; *cylinder mowers*, used for lawns, but rarely used on farms

Moxoto [mɒk'səʊtəʊ] *noun* Brazilian breed of goat, which is light brown with a black stripe along the back and belly; they are reared both for meat and for their skins

MPA = MEAT PROMOTION EXECUTIVE

MQ = MILK QUOTA

MRC = MEDICAL RESEARCH COUNCIL

MSC = MODIFIED STARCH CELLULOSE

muck [mʌk] *noun* farmyard manure; **muck spreader** *or* **manure spreader** = trailer with a moving floor conveyor and a combined shredding and spreading mechanism, used to distribute manure over the soil; another type has an open, cylindrical-shaped trailer with a rotor shaft running from front to back; flails are attached to the rotor and throw the manure out from the side of the trailer. Slurry can be spread on the land with special pumping equipment, using a pipeline and slurry guns

mucosal disease [mju:'kəʊzəl dɪ'zi:z] *noun* livestock disease caused by a virus, often fatal

mud [mʌd] *noun* very wet clay

◊ **muddy** ['mʌdi] *adjective* covered with mud

◊ **mudflats** ['mʌd 'flæts] *noun* areas of flat mud, usually near the mouths of rivers

mulberry ['mʌlbəri] *noun* tree *(Morus nigra)* with dark fruit, similar to blackberries; **white mulberry** = *Morus alba*, a tree grown for its leaves, on which silkworms feed

mulch [mʌltʃ] **1** *noun* organic material (such as dead leaves *or* straw) used to spread over the surface of the soil to prevent evaporation or erosion **2** *verb* to spread material (such as straw *or* dead leaves) over the surface of the soil to prevent evaporation

COMMENT: black plastic sheeting is often used by commercial horticulturalists, but the commonest mulches are organic. Apart

from preventing evaporation, mulches reduce weed growth and encourage worms

mule [mju:l] *noun* **(a)** hybrid (and usually sterile) offspring of a male ass and a mare **(b)** crossbred sheep from a Blue-faced Leicester ram and a Swaledale ewe; mules have speckled faces and a high lambing rate

mulesing ['mju:lzɪŋ] *noun* operation to cut away loose skin on sheep to prevent blowfly attacks

mull [mʌl] *noun* well-decomposed humus mixed with mineral matter; it occurs under neutral or alkaline conditions, such as under deciduous forests

multi-cropping ['mʌlti 'krɒpɪŋ] *noun* multiple cropping, growing more than one crop on the same piece of land in one year; *wet rice is often multi-cropped*

multigerm seed ['mʌltɪdʒɜ:m 'si:d] *noun* seed which occurs as a cluster of seeds fused together and which produces more than one plant when it germinates; a common example is the sugar beet seed; the multiple plants must be reduced to one by a process called 'singling'

multi-horned ['mʌlti 'hɔ:nd] *adjective* animal, such as a Jacob's sheep, with more than two horns

multiple ['mʌltɪpl] *adjective* happening many times; **multiple cropping** = growing more than one crop on the same piece of land in one year; **multiple ovulation and embryo transfer (MOET)** = method of insemination where embryos are transferred to cows (as opposed to artificial insemination); **multiple suckling** = system in dairy breeding where nurse cows suckle several beef calves at the same time

mung *or* **mungbean** [mʌŋ] *noun* *Phaseolus aureus*, green gram, an Indian pulse used as a vegetable or also to produce bean sprouts

Murray Grey ['mʌreɪ 'greɪ] *noun* breed of beef cattle, silver grey in colour; a polled early-maturing hardy breed; the carcass has a high proportion of lean meat

Musa ['mju:zə] *Latin name for* banana and plantain

muscle [mʌsəl] *noun* organ in the body

which contracts to make a part of the body move; **muscle fibre** = long fibre cells which make up muscle

COMMENT: it is the muscle on an animal which forms the meat which is eaten

Muscovy ['mʌskəvi] *noun* utility breed of duck; large in size, and coloured either black and white, or black, blue and white

mushroom ['mʌʃrʊm] *noun* common edible fungus, often grown commercially; **mushroom compost** = compost used for the commercial growing of mushrooms, sold when it has been used, for further use as a horticultural mulch

musk melon ['mʌsk 'melən] *noun* variety *(Cucumis melo)* of melon, with large scented fruit

must [mʌst] *noun* grape juice which has been extracted for wine, but which has not started to ferment

mustard ['mʌstəd] *noun* species of brassica *(Sinapsis)*, whose seeds are among the most important spices; also used as green manure

COMMENT: much of the mustard grown commercially is rape *(Brassica napus)*; the seeds of black mustard *(Brassica nigra)* are ground to produce the yellow spice; while white mustard *(Brassica alba)* is grown as a salad crop (used in mustard and cress)

mutant ['mjuːtənt] *noun & adjective* (i) gene in which mutation has taken place; (ii) organism carrying a mutant gene

◊ **mutagen** ['mjuːtədʒen] *noun* agent which can make genes mutate

◊ **mutate** [mjuːˈteɪt] *verb* to undergo a genetic change; *bacteria can mutate*

suddenly, and become increasingly able to infect

◊ **mutation** [mjuːˈteɪʃn] *noun* change in the DNA which changes the gene

mutton ['mʌtən] *noun* meat of a mature sheep, produced from older sheep such as ewes which are finished for breeding

muzzle ['mʌzl] *noun* projecting part of an animal's head, especially the mouth, jaws and nose

MV = MAEDI-VISNA

myc- *or* **myco-** ['maɪkəʊ] *prefix* referring to fungus

◊ **mycelium** [maɪˈsiːliəm] *noun* mass of threads which forms the main part of a fungus

◊ **mycology** [maɪˈkɒlədʒi] *noun* study of fungi

◊ **mycorrhiza** *noun* types of fungi which are associated with plant roots and help in nutrient uptake

mycotic dermatitis [maɪˈkɒtɪk dɜːməˈtaɪtɪs] *noun* disease affecting sheep, caused by fungal-like bacteria which multiply on the skin and cause inflammation. Severe infections cause fleece loss

myiasis ['maɪəsɪs] *noun* infestation of animals by the larvae of flies, such as wool maggots, bot flies or warble flies

Myrtaceae [mɜːˈteɪʃii] *noun* family of Australian plants, including the eucalyptus

myxomatosis [mɪksəməˈtəʊsɪs] *noun* usually fatal virus disease affecting rabbits, transmitted by fleas

Nn

N *chemical symbol for* nitrogen

QUOTE in an attempt to reduce nitrate
leaching and minimize nitrogen
starvation, beet growers might be able to
apply N as a foliar sprays
Farmers Weekly

QUOTE for potatoes, sugar beet and
spring cereals, the first dressing of N is
applied at a critical time, often in the
seedbed or immediately post drilling
Arable Farming

Na *chemical symbol for* sodium

NABIM = NATIONAL ASSOCIATION OF
BRITISH AND IRISH MILLERS

NAC = NATIONAL AGRICULTURAL
CENTRE

NACNE = NATIONAL ADVISORY
COUNCIL ON NUTRITION AND
EDUCATION

naked grain ['neɪkɪd 'greɪn] *noun* grain
(such as wheat) that is easily separated or
threshed out from its husk (i.e., in its
caryopsis state)

Nandi ['nændi] *noun* East African breed of
cattle, short horned and mainly red in
colour, a dairy breed

nanny goat ['næni 'gəʊt] *noun* female goat

**National Agricultural Centre
(NAC)** the site of the annual Royal Show
(at Stoneleigh, in Warwickshire), owned by
the RASE

National Farmers' Union (NFU)
organization representing the interests of
British farmers in negotiations with the
government and other agencies

**National Institute of Agricultural
Botany (NIAB)** organization which tests
all new varieties of crops; after successful
testing, the varieties are made available to
farmers

NASPM = NATIONAL ASSOCIATION
OF SEED POTATO MERCHANTS

national list ['næʃnl 'lɪst] *noun* list of
agricultural crop varieties tested by the
NIAB and available for sale; under EU
regulations, all seeds sold to farmers or
horticulturists must be tested and certified

national park ['næʃnl 'pɑːk] *noun* large
area of unspoilt land, owned and managed
by a special planning authority which has
control over agriculture as well as the
normal planning powers. Parks exist for
people to enjoy and for the protection and
conservation of scenery, flora and fauna,
and where established farming is effectively
maintained

**National Union of Agricultural and
Allied Workers (NUAAW)**
independent section of the Transport and
General Workers Union representing the
interests of farmworkers in negotiating
terms and conditions of their employment

native ['neɪtɪv] *adjective* which belongs to
an area; **native breeds** = breeds which have
been developed in a country, and not
brought in from other countries

nature ['neɪtʃə] *noun* (i) essential quality of
something; (ii) kind *or* sort; (iii) plants,
animals and their environment in general;
nature conservation = active management
of natural resources to ensure their quality
is maintained and that they are wisely used;
Nature Conservancy Council (NCC) =
official body in the UK which is responsible
for the conservation of nature; **nature
reserve** = special area where the wildlife is
protected (National Nature Reserves are
designated by the NCC); **nature trail** =
footpath with signposts and explanatory
notices, designed to explain the fauna and
flora of a piece of countryside to the general
public; some farms have designed trails as a
means of supplementing farm income

◊ **natural** ['nætʃərəl] *adjective* **natural
immunity** = immunity from disease
inherited by newborn offspring from birth,

acquired in the womb or from the mother's milk; **natural resources** = part of the environment which can be used commercially (such as coal); **natural selection** = evolution of a species, whereby characteristics which help individual organisms to survive and reproduce are inherited and those characteristics which do not help are not passed on; **natural vegetation** = vegetation which exists in a natural state (such as a rainforest) and has not been planted or managed by people

◊ **naturalize** ['nætʃrəlaɪz] *verb* to introduce a species into an area where it has not lived before so that it becomes established as part of the ecosystem

navel-ill ['neɪvəlɪl] *noun* disease of young livestock, especially newborn calves, kids, and lambs; it causes abscesses at the navel and swellings in some joints; *also known as* JOINT-ILL

navy bean ['neɪvɪbiːn] *noun* dried seed of the common bean *(Phaseolus vulgaris);* also called haricot bean; used in particular for canning as baked beans

NCC = NATURE CONSERVANCY COUNCIL

NCDL = NATIONAL CANINE DEFENCE LEAGUE

N'Dama [n'dɑːmə] *noun* West African breed of cattle, bred for meat; the animals are small and fawn coloured, with no hump

neat [niːt] *noun* old term referring to any kind of cattle

neck collar ['nek 'kɒlə] *noun* leather band put round the neck of a horse or cow, to hold the animal in a stall; **neck rot** = disease affecting bulb onions during storage; the onions become soft and begin to rot from the stem downwards

necrosis [nɪ'krəʊsɪs] *noun* death of tissue *or* cells in an organism

nectar ['nektə] *noun* sweet liquid produced by flowers to attract bees

nectarine ['nektəriːn] *noun* smooth-skinned variety of peach *(Prunus persica nectarina)*

NEDO = NATIONAL ECONOMIC DEVELOPMENT OFFICE

neem tree ['niːm 'triː] *noun* evergreen tree *(Azadirachta indica),* up to 20 metres in height found in the more arid areas of India. Its products are used in medicine, toiletries and for timber and fuel

neigh [neɪ] **1** *noun* sound made by a horse **2** *verb (of horse)* to make a noise; *see also* BLEAT, GRUNT, LOW

nematode ['nemətəʊd] *noun* type of roundworm, some of which are parasites of animals, such as hookworms, while others live in the roots of plants; **nematode disease** = disease of the alimentary tract and lungs, caused by nematodes; infection is transmitted from one group of animals to another by means of infective larvae in herbage

◊ **nematicide** [ne'mætɪsaɪd] *noun* substance which kills nematode worms

◊ **Nematodirus disease** [nemətəʊ'daɪrəs dɪ'ziːz] *noun* disease of lambs caused by parasitic roundworms; the animals suffer diarrhoea and loss of condition

QUOTE root-knot nematodes are among the most widespread pests that limit agricultural productivity. Almost all plants that account for the majority of the world's food crop - including the potato - are susceptible to these pests
Appropriate Technology

nemoriculture [nə'mɒrɪkʌltʃə] *noun* primitive stage of human culture, concerning mainly food-gathering in forests

nest [nest] **1** *noun* construction built by birds and some fish for their eggs; construction made by some social insects, such as ants and bees, for the colony to live in; **nest box** = open-fronted box in which a hen lays eggs; the box may be a single unit or part of a series of boxes **2** *verb* to build a nest

net blotch ['net 'blɒtʃ] *noun Pyrenophora teres,* a fungal disease of barley, with dark brown blotches affecting the leaves

nettle ['netl] *noun* plant, especially one of the genus *Urtica* which possesses stinging hairs; *see also* HEMP NETTLE, RED DEADNETTLE

neutral ['njuːtrəl] *adjective* neither acid

nor alkali; *a pH value of 7 is neutral;* **neutral soil** = soil which is neither acid nor alkaline, that is, where the pH value is neutral

◊ **neutralize** ['nju:trəlaız] *verb* **(a)** to make an acid neutral; *acid in drainage water can be neutralized by limestone;* **neutralizing value** = measurement of the capability of a lime material to neutralize soil acidity; it is the same as the calcium oxide equivalent **(b)** to counteract the effect of something; *(in bacteriology)* to make a toxin harmless by combining it with the correct amount of antitoxin; *alkali poisoning can be neutralized by applying acid solution*

> QUOTE alkaline soils, such as soils rich in limestone, can neutralize acid directly
> *Scientific American*

new blood ['nju: 'blʌd] *noun* new strain of a breed introduced by outcrossing

Newcastle disease ['nju:kɑ:səl dɪ'zi:z] *noun* fowl pest, an acute febrile contagious disease of fowls; affected birds suffer loss of appetite, diarrhoea and respiratory problems. Mortality rates are high

New Hampshire Red [nju: 'hæmpʃə 'red] *noun* breed of poultry with red plumage, lighter in weight than Rhode Island Red; mainly a layer, producing brownish-tinted eggs

NFE = NITROGEN-FREE EXTRACT used in the chemical analysis of animal feeding stuffs; the nitrogen-free extract consists mainly of soluble carbohydrates (sugars) and starch

NFFO NON FOSSIL FUEL OBLIGATION (technologies designed to ensure diversity of power supply, such as hydro power, energy crops and wind power)

NFU = NATIONAL FARMERS' UNION

Nguni [ŋ'gu:ni] *noun* breed of small dairy cattle found in southern Africa; the animals are various colours

NIAB = NATIONAL INSTITUTE OF AGRICULTURAL BOTANY

nicotine ['nɪkəti:n] *noun* toxic substance in tobacco, also used as an insecticide

◊ **Nicotiana** [nɪkəti'ɑ:nə] *Latin name for the* tobacco plant

NIEMP = NATIONAL INVESTIGATION INTO THE ECONOMICS OF MILK PRODUCTION

Nigerian [naɪ'dʒi:əriən] *noun* West African breed of goat, mainly reared for meat; varieties within the breed include the Kano Brown, the Red Sokoto and the Bornu White

nightshade ['naɪtʃeɪd] *noun* plants of the family *Solanaceae* which, if eaten by stock, are likely to cause sickness or death; **deadly nightshade** = poisonous plant *(Atropa belladonna)* which is sometimes eaten by animals

night soil ['naɪt 'sɔɪl] *noun* human excreta, collected and used for fertilizer in some parts of the world, particularly in China and the Far East

Nilgiri nettle [nɪl'gi:ri 'netl] *noun* perenial herb *(Girardinia diversifolia)* found in Nepal; a source of fibre used in weaving; *also known as* HIMALAYAN NETTLE

NIMBY ['nɪmbi] = NOT IN MY BACKYARD *phrase* used to describe people who encourage the development of agricultural land for building houses or factories, provided it is not near where they themselves are living

nip bar ['nɪp 'bɑ:] *noun* bar fitted to moving mechanisms to prevent parts of the body being drawn into the machine

nipplewort ['nɪplwɜ:t] *noun* an annual weed, *Lapsana communis*

nitrate ['naɪtreɪt] *noun* (i) anion with the formula NO_3; (ii) natural constituent of plants; beets, cabbage, cauliflower and broccoli can contain up to 1mg/kg; (iii) chemical compound containing the nitrate ion, such as sodium nitrate; **nitrate-sensitive area (NSA)** = region of the country where nitrate pollution is likely, and where the use of nitrate fertilizers is strictly controlled

> COMMENT: nitrates are a source of nitrogen for plants; they are used as fertilizers but can run off into rivers and as they are highly soluble in water easily enter the water supply

◊ **nitrification** [naɪtrɪfɪ'keɪʃn] *noun* process by which bacteria found in the soil break down complex nitrogen compounds and form nitrates which plants can absorb;

nitrification inhibitor = chemical product used to slow down the release of nitrate in organic manure

◊ **nitrify** ['nɪtrɪfaɪ] *verb* to convert nitrogen to a nitrate; **nitrifying bacteria** = bacteria which convert nitrogen into nitrate

◊ **nitrite** ['naɪtraɪt] *noun* (i) anion with the formula NO_2; (ii) chemical compound containing the nitrite ion, such as sodium nitrite, found in many plant foods

COMMENT: nitrites are formed by bacteria from nitrogen as an intermediate stage in the formation of nitrates; they are formed during the process of nitrification

nitrogen ['naɪtrədʒən] *noun* chemical element which is essential to biological life (a gas which is the main component of air and is an essential part of protein); **nitrogen cycle** = process by which nitrogen enters living organisms (the nitrogen is absorbed into green plants in the form of nitrates, the plants are then eaten by animals, and the nitrogen is returned to the ecosystem through the animal's excreta or when an animal or a plant dies); **nitrogen deficiency** = lack of nitrogen in soil, found where organic matter is low; it results in thin, weak growth of plants, especially grasses, cereals, kales and cabbages; **nitrogen fertilizers** = many straight and compound fertilizers contain nitrogen, but it is quickly changed by bacteria in the soil to the nitrate form; **nitrogen fixation** = process by which nitrogen in air is converted by bacteria in certain plants into nitrogen compounds, and when the plants die the nitrogen is released into the soil and acts as a fertilizer; **nitrogen-fixing plants** = plants, such as lucerne or beans, which form an association with bacteria which convert nitrogen from the air into nitrogen compounds which pass into the soil (NOTE: chemical symbol is **N**; atomic number is **7**)

◊ **nitrogenous   fertilizers** [naɪ'trɒdʒənəs 'fɜːtɪlaɪzəz] *noun* fertilizers (such as sulphate of ammonia) which are based on nitrogen

◊ **nitrogen-free   extract   (NFE)** ['naɪtrədʒənfriː 'ekstrækt] *noun* used in the chemical analysis of animal feeding stuffs, the nitrogen-free extract consists mainly of soluble carbohydrates (sugars) and starch

COMMENT: nitrogen is taken into an animal's body by digesting protein-rich foods; excess nitrogen is excreted in urine.

When the intake of nitrogen and the excretion rate are equal, the body is in nitrogen balance or protein balance. Nitrogen is supplied to the soil by fertilizers, organic matter, nodule bacteria on legumes, and by nitrogen-fixing microorganisms in the soil. The nitrogen cycle is the series of processes by which nitrogen is converted into nitrates which are absorbed into green plants. The plants are then eaten by animals, which themselves are eaten by other animals. The nitrates are returned to the ecosystem in excreta or when animals or plants die

nitrogen oxide (NO_x) ['naɪtrədʒən 'ɒksaɪd] *noun* oxide (such as nitric oxide (NO) *or* nitrogen dioxide (NO_2) formed when nitrogen is oxidized

COMMENT: in general nitrogen oxides form the major part of air pollution, though nitric oxide (produced by burning fossil fuel) is not directly dangerous to humans. Nitrogen dioxide is produced by car engines and is toxic. Nitrogen oxides are also produced when farmland is sprayed with fertilizers; the bacteria in the soil feed on the fertilizer and produce the gas

QUOTE NO_x and hydrocarbons can react in sunlight to form ozone
Environment Now

NMRS = NATIONAL MILK RECORDS SCHEME

node [nəʊd] *noun* joint on the stem of a plant where a leaf is attached

◊ **nodule** ['nɒdjuːl] *noun* small node found on the roots of leguminous crops; the nodules contain types of bacteria which can convert nitrogen from the air into nitrogen compounds

nomad ['nəʊmæd] *noun* person whose home is a large area of land, who moves from place to place in it without settling in any one spot (e.g. a person who hunts game *or* a herdsman who drives his animals from place to place to find grazing)

◊ **nomadic** [nəʊ'mædɪk] *adjective* referring to nomads; *the herdsmen in the area lead a nomadic existence; nomadic pastoralists move their livestock around to feed on available grazing*

◊ **nomadism** [nəʊmæ'dɪzm] *noun* habit of certain animals which move around from place to place without having a fixed habitat

nominated service [ˈnɒmɪneɪtɪd ˈsɜːvɪs] *noun* artificial insemination with semen from a named and tested male animal

non- [nɒn] *prefix meaning* not; **non-EU** = not in the EU; **non-selective weedkiller** = weedkiller which kills all plants

non-centrifugal   **sugar** [nɒnsentrɪˈfjuːgəl ˈʃʊgə] *noun* dark semi-solid sugar made by boiling the juices obtained from crushed sugar cane; India is the principal producer

non-persistent   **pesticide** [nɒnpɜːˈsɪstənt ˈpestɪsaɪd] *noun* pesticide which does not remain toxic for long so does not enter the food chain

non-selective herbicide [nɒnsɪˈlektɪv ˈhɜːbɪsaɪd] *noun* chemical herbicide which kills all vegetation

noose [nuːs] *noun* loop in a rope, with a loose knot which allows it to tighten (as in a halter or a lasso)

Norfolk horn [ˈnɔːfək ˈhɔːn] *noun* rare breed of sheep adapted to dry heathland. Black-faced and horned

Norfolk rotation [ˈnɔːfək rəʊˈteɪʃn] *noun* system for farming, using arable farming for fodder crops, and involving the temporary sowing of grass and clover; the system was introduced into England in the early 18th century; the rotation involved root crops (turnips or swedes), then cereal (barley), followed by ley (usually red clover), and ended with cereal (usually wheat). The Norfolk rotation provided a well-balanced system for building up and maintaining soil fertility, controlling weeds and pests, providing continuous employment and profitability

Normandy [ˈnɔːməndi] *noun* breed of cattle from north-west France; the coat is white with red-brown patches. The animals are reared for meat and milk, from which the Camembert cheese is made

North Country Cheviot [ˈnɔːθ kʌntri ˈtʃiːviət] *noun* large-sized breed of sheep with fine good-quality wool; this variety of the Cheviot is found in Caithness and Sutherland

◊ **Northern Dairy Shorthorn** [ˈnɔːðən deəri ˈʃɔːthɔːn] *noun* dairy breed of cattle, which comes from the old Teeswater cattle, with perhaps a little Ayrshire blood; now established as a pure breed. The most popular colour is light roan, but red, white and mixtures of shades are found. The animals are thrifty, hardy and suitable for harsh upland conditions

◊ **North Devon** [ˈnɔːθ ˈdevən] *see* DEVON

◊ **North Ronaldsay** [ˈnɔːθ ˈrɒnəldseɪ] *noun* rare breed of small sheep, which varies in colour from white through grey, brown and black, and also combinations of these colours; the tail is short, and most of the animals have horns

noseband [ˈnəʊzbænd] *noun* broad leather band worn around the horse's nose and above the bit; used to prevent a horse from opening its mouth too wide

notify [ˈnəʊtɪfaɪ] *verb* to report something officially; *the MAFF must be notified of any outbreak of foot and mouth disease*

◊ **notifiable disease** [nəʊtɪˈfaɪəbl dɪˈziːz] *noun* **(a)** serious infectious disease of animals and poultry which in Great Britain has to be reported to the police when an outbreak is confirmed on a farm **(b)** diseases and pests of plants which must be notified to the MAFF

> COMMENT: the following are some of the notifiable diseases: anthrax; foot and mouth disease; Newcastle disease; rabies; sheep pox; sheep scab; swine fever

nozzle [ˈnɒzl] *noun* metal end of a pipe or sprayer, usually adaptable to allow different flows of liquid

NOAH = NATIONAL OFFICE OF ANIMAL HEALTH

NPBA = NATIONAL PIG BREEDERS' ASSOCIATION

NPK *abbreviation* for a compound or mixed fertilizer containing nitrogen, phosphorus and potassium

NRA = NATIONAL RIVERS AUTHORITY

NSA = NITRATE-SENSITIVE AREA Thirty new areas are proposed by a government scheme which will restrict nitrogen use to 150 kg/ha for five years. EU directive in 1994 aimed at reducing nitrate pollution on up to 2 million hectares of farmland in the UK

NSDO = NATIONAL SEED DEVELOPMENT ORGANISATION

NUAAW = NATIONAL UNION OF AGRICULTURAL AND ALLIED WORKERS

Nubian goat ['nju:biən 'gəʊt] *noun* breed of goat of mixed Egyptian and Indian origin, now crossed with British goats to produce the Anglo-Nubian breed

nucleus ['nju:kliəs] *noun* central body in a cell, containing DNA and RNA, and controlling the function and characteristics of the cell (NOTE: the plural is **nuclei**)

nurse cow ['nɜːs 'kaʊ] *noun* cow used to suckle the calves of others

◊ **nurse crop** ['nɜːs 'krɒp] *noun* crop grown to give protection to young plants of a perennial crop which is being established; nurse crops provide shade and act as windbreaks; in the tropics, bananas can be used as a nurse crop for cocoa

nursery ['nɜːsəri] *noun* place where plants are grown until they are large enough to be planted in their final positions; **tree nursery** = place where trees are grown from seed until they are large enough to be planted out; **nursery bed** = bed in which seedlings are planted out from the seedbed until they are large enough to be put in permanent positions

nut [nʌt] *noun* **(a)** the seed of certain trees (a fruit with a single seed in a hard woody shell) **(b)** any hard edible seed contained in a fibrous or woody shell (such as groundnuts) **(c)** small cube of compressed meal; a convenient form of animal feed

nutmeg ['nʌtmeg] *noun* tropical spice obtained from the tree *(Myristica fragrans)* which is native to Indonesia; the hard seeds are ground and used for flavouring

nutrient ['nju:triənt] *noun* (i) substance (such as protein *or* fat *or* vitamin) in food which is necessary to provide energy or to help the body grow, repair and maintain itself; (ii) substance which a plant needs to allow it to grow and produce seed; the major nutrients needed by plants are carbon, hydrogen, oxygen, nitrogen, phosphorus, potassium, calcium, magnesium and sulphur; minor nutrients are trace elements such as boron, copper or manganese; **nutrient stripping** = removing nutrients from sewage to prevent eutrophication of water in reservoirs

QUOTE in the rainforest, despite the profusion of plants and trees, the underlying soils are poor, almost all the nutrients being bound up in the vegetation. Once the forests have been cut down, those few nutrients that remain in the soil are quickly washed away, transforming the land into a barren wasteland
Ecologist

QUOTE poor farmers burn patches of rainforest to release some of the nutrients locked in the biomass back into the soil, but those nutrients that are released are often exhausted after only one year of intensive agriculture
New Scientist

nutrition [nju:'trɪʃn] *noun* **(a)** nourishment *or* food which an animal eats **(b)** study of diet

◊ **nutritional** [nju:'trɪʃənəl] *adjective* referring to nutrition; *the nutritional quality of meat;* **nutritional disorder** = disorder related to food and nutrients

◊ **nutritious** *or* **nutritive** [nju:'trɪʃəs *or* 'nju:trətɪv] *adjective* (feed) which provides nutrients

nymph [nɪmf] *noun* stage following the larval stage in certain flies, such as the dragonfly

Oo

O 1 *symbol for* oxygen **2** Below Average (in the EUROP carcass classification system)

oak [əʊk] *noun* common hardwood tree of the genus *Quercus* found in temperate regions; it provides valuable timber

oarweed ['ɔːwiːd] *noun* common seaweed (*Laminaria digitata*) used as food

OAS = ORGANIC AID SCHEME

oasis [əʊ'eɪsɪs] *noun* place in an arid desert where the water table is near the surface and where vegetation can grow; in the oases of the hot desert regions, the date palm forms an important food supply; **oasis effect** = loss of water from an irrigated area due to hot dry air coming from an unirrigated area nearby

oasthouse ['əʊsthaʊs] *noun* building containing a kiln for drying hops; it is a circular or square building with a characteristic conical roof

oats [əʊts] *noun* hardy cereal crop (*Avena sativa*) grown in most types of soil in cool wet northern temperate regions; best quality oats can be used for making biscuits, oatcakes and porridge. Mainly used for animal feed

◊ **oatmeal** ['əʊtmiːl] *noun* feeding stuff produced when the husk is removed from the oats kernel by a rolling process; oatmeal is particularly good for horses and valuable for cattle and sheep, but not as suitable for pigs because of its high fibre content

OBF = OFFICIALLY BRUCELLOSIS FREE

oca ['əʊkə] *noun Oxalis tuberosa*, a tuber used as a vegetable in South America, especially in Peru and Ecuador

oe- [iː] *prefix* (NOTE: words beginning with **oe-** are written **e-** in American English)

OECD = ORGANISATION FOR ECONOMIC COOPERATION AND DEVELOPMENT

oestrogen ['iːstrədʒən] *noun* hormone produced in the ovaries of animals which stimulates the reproductive system; also produced in small amounts by the testis; also called 'bulling hormone'

oestrus ['iːstrəs] *noun* the time of heat in a female, when she will accept mating by the male; *see also* ANOESTRUS, HEAT, SUBOESTRUS

Oestrus ['iːstrəs] *noun* family of flies, including the bot fly

offal ['ɒfəl] *noun* **(a)** inside parts of an animal, such as liver, kidney or intestines, when used as food (as opposed to meat, which is muscle) **(b) wheat offals** = the embryo and seed coat of the wheat grain, used as animal feed

officinalis *or* **officinale** [ɒfɪsɪ'nɑːlɪs *or* ɒfɪsɪ'nɑːli] *Latin word meaning* 'used in medicine', often part of the generic name of plants

off-going crop ['ɒfgəʊɪŋ 'krɒp] *noun* crop sown by a tenant farmer before leaving the farm at the end of his tenancy; he is permitted to return and harvest the crop and remove it (NOTE: also known as **following crop** or **away-going crop**)

oil [ɔɪl] *noun* liquid which cannot be mixed with water; **corn oil** = vegetable oil obtained from maize grains, used for cooking and as a salad oil; **essential oils** = oils from scented plants used in cosmetics and as antiseptics; **palm kernel oil** = vegetable oil produced from the kernels of the oil palm nut, used for making margarine, and cooking fat; the residue after extraction of the oil is used as a livestock feed; **palm oil** = edible oil produced from the seed or fruit of an oil palm; red in colour, with only 5-12% polyunsaturated fatty acids, it is widely used in cooking fats and margarines; the residue after extraction is used as livestock feed; **sunflower oil** = oil extracted from sunflower seeds: seeds contain between 25 and 45% of semi-drying oil which is high in unsaturated fats

◊ **oil palm** ['ɔɪl 'pɑːm] *noun Elaeis guineensis,* a tree indigenous to West Africa, which yields large supplies of oil; the fruit produces two distinct oils, palm oil from the pericarp and palm kernel oil from the hard kernel

◊ **oilseed cake** *or* **oilcake** ['ɔɪlsiːd 'keɪk *or* 'ɔɪlkeɪk] *noun* a feedingstuff concentrate, high in protein, made from the residue of seeds which have been crushed to produce oil; **oilseeds** *or* **oilseed crops** = crops from which vegetable oils are extracted from the fruit or seed; **oilseed rape** = important oil-yielding crop; the seed contains about 42% of oil, which is extracted by crushing and is used in making margarine, cooking fats and salad oils. The residual cake is used for stock feeding. Oilseed rape will grow in a wide range of soil and climatic conditions; there are both spring and winter varieties; it is also useful as a break crop between cereals; *see also* DOUBLE LOWS

COMMENT: there are three types of oil: fixed vegetable and animal oils; essential volatile oils; and mineral oils. The most important oil-producing crops are the coconut palm, the oil palm, groundnuts, linseed, soya beans, maize and cotton seed. Other sources are olives, rape seed, lupin, sesame and sunflowers. After the nuts or seed have been crushed to extract the oil, the residue may be used as a livestock feed or as a fertilizer

okra ['ɒkræ] *noun* immature fruit pods of an annual herb plant *(Hibiscus esculentus)* with green or red stems. The fruits are ridged oblong capsules

Old English game ['əʊld ɪŋglɪʃ 'geɪm] *noun* breed of poultry, now mainly a fancy breed; the birds are coloured black and white with blue wing tips

oligotrophic [ɒlɪgəʊ'trɒfɪk] *adjective* (water) which contains few nutrients; *compare* EUTROPHIC

olive ['ɒlɪv] *noun* Mediterranean tree *(Olea europaea),* with fruit from which an edible oil can be produced; a considerable quantity of fruit is grown for direct consumption

OM = ORGANIC MATTER

omasum [əʊ'meɪsəm] *noun* the third stomach of a ruminant, which acts as a filter, and where much of the water in food is taken out before the food passes onto the abomasum; *see also* ABOMASUM, RETICULUM, RUMEN (NOTE: also called **the Bible** or **the Book**)

omnivore *or* **omnivorous animal** ['ɒmnɪvɔː *or* ɒm'nɪvərəs 'ænɪməl] *noun* animal (such as the pig) which eats anything, both vegetation and meat; *compare* CARNIVORE, HERBIVORE

once grown seed ['wɒns grəʊn 'siːd] *noun* seed obtained from plants grown from a certified seed and intended for use by the farmer on his own farm, and not for resale

Ongole ['ɒŋgəʊli] *noun* breed of Indian cattle from Northern Madras. White in colour with dark neck and hump. Used for beef production and also as a draught animal

onion ['ʌnjən] *noun* vegetable crop *(Allium cepa),* grown either for cooking or for eating in salads; the ripe onion consists of the edible swollen leaf bases surrounded by scale leaves. It is harvested when the growing tops have fallen over; it is the dormant bulbs which are harvested and eaten; **onion couch** = grass weed *(Arrhenatherum elatius)* which grows to 24-48ins and develops long oat-like hairs like flower heads. Affects cereals; **onion fly** = an insect pest *(Hylemyia antiqua)* which causes damage to onions by the maggots which eat into the developing bulb; **onion set** = a seed onion, a small onion grown from seed, which has been dried, and which is planted the following year so that it will root and grow on to maturity

on-off grazing ['ɒnɒf 'greɪzɪŋ] *see* ROTATIONAL GRAZING

on-the-hoof [ɒnðə'huːf] *adjective* (animals) sold live for slaughter

open ['əʊpən] *adjective* not closed; **open country** *or* **open land** = (i) any large area of land which is not built over; (ii) land which has not been enclosed; **open countryside** = area of country without many trees or high mountains; **open furrow** = furrow shaped like a V, with the furrow slices laid in opposite directions to each other

◊ **open fields** ['əʊpən 'fiːldz] *noun* fields which are not separated by hedges or walls, but by banks of earth; formerly fields were divided into strips, each worked by a farmer; the system was used originally by the Saxons

COMMENT: In recent years the removal of many field boundaries in the interest of farm consolidation has led to an increase in the size of the average British field, and created large open fields again. Hedges have been removed to allow large farm machinery to be used more economically, and the loss of hedgerows has had a marked effect on the wildlife in the countryside

optimum ['ɒptɪməm] *adjective* best; *the optimum temperature for germination*

orache ['ɒrɪtʃ] *noun* common weed (*Atriplex patula*) which affects sugar beet and maize crops, and makes harvesting the crop difficult (NOTE: also known as **dungleweed)**

orange ['ɒrɪndʒ] *noun* fruit of the *Citrus aurantium*, a native tree of China; its nutritional value is due mainly to its high vitamin C content; grown in semi-tropical and Mediterranean regions, it is eaten as fresh fruit or used for juice and for making preserves. The USA, Brazil, Spain, Morocco and Israel are large exporters of oranges

COMMENT: blood oranges are coloured by the presence of anthocyanins; mandarin oranges such as satsumas and tangerines have loose peel. The Seville orange is a bitter orange, grown in Spain and used by marmalade manufacturers

orchard ['ɔːtʃəd] *noun* area of land used for growing fruit trees

COMMENT: orchards were once a common feature of most farms, but now fruit is commercially produced by specialized commercial growers. The modern orchard consists of trees grafted onto dwarfing rootstock, shaped by pruning and closely planted in rows which are separated to allow room for tractors and sprayers to pass. Apples, plums, pears and cherries are the most important tree fruits in Britain, with Kent, Worcestershire and parts of East Anglia being the most important growing

regions. In the USA, oranges and other citrus fruits are grown in orchards in the Southern States, in particular in Florida and California

orf [ɔːf] *noun* virus disease affecting sheep, cattle and goats and easily passed on to humans; the disease causes scabs and ulcers which affect the mouth, nose and eyes; in its later stages, legs, genitals and udders may be affected

organ ['ɔːgən] *noun* part of the body which is distinct from other parts and has a particular function (such as the liver *or* an eye *or* the ovaries, etc.)

◊ **organic** [ɔː'gænɪk] *adjective* (i) (substance) which comes from an animal *or* plant; (ii) chemical compound which contains carbon; (iii) (food) which has been cultivated naturally, without any chemical fertilizers *or* pesticides; **organic fertilizers =** plant nutrients which are returned to the soil from dead or decaying plant matter and animal wastes, such as compost farmyard manure and bone meal

◊ **organically** [ɔː'gænɪkli] *adverb* (food) grown using natural fertilizers and not chemicals

◊ **organic farming** [ɔː'gænɪk 'fɑːmɪŋ] *noun* method of farming which does not involve using chemical fertilizers *or* pesticides

COMMENT: organic farming uses natural fertilizers and rotates stock farming (i.e. raising of animals) with crop farming; soil nutrients are maintained by the addition of plant and animal manures, and pest control is achieved by the use of naturally derived pesticides, and by crop rotation, which allows natural predation to take place. Organic farming may produce lower yields than traditional or intensive farming, but the lower yields may be offset by the high cost of the chemical fertilizers used in intensive farming; it may become more economic than conventional farming due to premium prices which are paid for organic products. In areas of overproduction, organic farming has the advantage of reducing crop production without loss of quality and without taking land out of agricultural use. At the present time, less than 10% of Britain's farmland is organically cultivated. The main factor in controlling conversion to organic farming is the capital cost. A government scheme to encourage farmers to convert to organic agriculture has begun. Payments will be made to farmers in England over a 5 year period to assist with the costs of converting land to organic production. The scheme is also designed to stimulate a form of production which emphasises soil improvement and the control of pests and diseases. In 1994 there were 18,590 hectares registered as organic land in England

◊ **organic matter (OM)** [ɔː'gænɪk 'mætə] *noun* **(a)** part of the dry matter content of a feedingstuff **(b)** the remains of plants and animals in the soil

COMMENT: Organic matter is acted on by bacteria, fungi, earthworms, and it decomposes to form humus. Humus is finally broken down by an oxidation process. The organic matter content of soil varies according to soil type, and usually increases with clay content. Peaty soils have a high organic matter content and some are totally organic matter

QUOTE last year organic cereal production was estimated to be 10,000t, of which half was oats and half wheat. About 1000t of hard organic milling wheat was imported from North America to supplement UK production
Farmers Weekly

organochlorine [ɔːgænəʊ'klɔːriːn] *noun* chlorinated hydrocarbon, any of several chemical compounds containing chlorine, used as an insecticide (such as aldrin, dieldrin, etc.)

COMMENT: organochlorine insecticides are very persistent, with a long half-life of up to 15 years, while organophosphorous insecticides have a much shorter life. Chlorinated hydrocarbon insecticides not only kill insects, but also enter the food chain and kill small animals and birds which feed on the insects

organophosphorous compound [ɔːgænəʊ'fɒsfərəs 'kɒmpaʊnd] *noun* an organic compound containing phosphorus

organophosphate *or* **organo-phosphorous insecticide** [ɔːgænəʊ'fɒsfeɪt *or* ɔːgænəʊ'fɒsfərəs ɪn'sektɪsaɪd] *noun* any of several synthetic chemical insecticides (such as malathion and parathion), based on chemical compounds including phosphate, which attack the nervous system but are not as persistent as the organochlorines

COMMENT: organophosphates are not as persistent as organochlorines and so do not enter the food chain. They are, however, very toxic and need to be handled carefully, as breathing in their vapour may be fatal. This group of insecticides has a wide range of animal and plant applications; they are especially effective in protecting growing plants against attack by aphids, and for use against both internal and external parasites. Some of the organophosphorous compounds are purely contact poisons, while others are taken up by the treated plants and so act as systemic pesticides

orphaned animal ['ɔːfənd 'ænɪməl] *noun* young animal whose mother has died, and is therefore either fostered onto another animal or has to be hand-reared

Orpington ['ɔːpɪŋtən] *see* BUFF

Oryza ['ɒrɪzæ] *Latin name for* rice

OS = ORDNANCE SURVEY

osier ['əʊzɪə] *noun* a species of willow, the shoots of which are used in making baskets

osmosis [ɒz'məʊsɪs] *noun* movement of a solution from one part through a semipermeable membrane to another part, where enough of the molecules in solution pass through the membrane to make the two solutions balance; much of the water movement into and from plant cells is due to osmosis

◊ **osmotic pressure** [ɒz'mɒtɪk 'preʃə] *noun* the force exerted to enable the flow of water through a cell wall in the process of osmosis

OSR = OILSEED RAPE

ost- *or* **osteo-** ['ɒstɪəʊ] *prefix* referring to bone

◊ **osteomalacia** [ɒstɪəʊmə'leɪʃə] *noun* condition where the bones become soft because of lack of calcium or phosphate

ostrich ['ɒstrɪtʃ] *noun* large flightless bird *(Struthio camelus)* raised in farms for its meat

out [aʊt] *noun* place where a plough is lifted out of the soils on reaching the headland; *see also* INS

outback ['aʊtbæk] *noun* *(in Australia)*

large area of wild or semi-wild land in the centre of the continent; *many wild animals which used to live in the outback are becoming rare as the outback is reclaimed for farming*

◊ **outbreak** ['aʊtbreɪk] *noun* series of cases of a disease which start suddenly; *there is an outbreak of foot and mouth disease in the area*

outbreeding ['aʊtbriːdɪŋ] *noun* breeding of animals that are less closely related than the average of the population from which they come (NOTE: opposite is **inbreeding)**

◊ **outcrossing** ['aʊtkrɒsɪŋ] *noun* bringing some new genetic variation ('new blood') into a flock or herd; usually done by introducing a new male

outfall (sewer) ['aʊtfɔːl] *noun* pipe which takes sewage (either raw *or* treated) and discharges it into a river *or* lake *or* the sea; *27% of outfalls discharge totally untreated sewage; sewage should be discharged through outfall pipes which are placed far enough from the shore*

outfields ['aʊtfiːldz] *noun* in hill farms, the fields furthest from the homestead, cropped only from time to time and allowed to lie fallow for long periods

outhouse ['aʊthaʊs] *noun* farm building which is not attached to the main farmhouse, and may be used for storage or for keeping poultry

out-of-season ['aʊtʌv'siːzən] *adjective & adverb* (plant) grown or sold at a time when it is not naturally available from outdoor cultivation; *out-of-season strawberries are imported from Spain; glasshouses provide out-of-season tomatoes*

outstation ['aʊtsteɪʃən] *noun* (*New Zealand & Australia*) sheep station separate from the main station

outwintering ['aʊtwɪntərɪŋ] *noun* the practice of keeping cattle and sheep outdoors in fields during the winter months; *a herd of outwintered heifers*

ova ['əʊvə] *see* OVUM

ovary ['əʊvəri] *noun* (a) organ in the female animal which produces ova (b) the core of the carpel of a flower, in which seeds are produced

oven-ready poultry ['ʌvənredi 'pəʊltri] *noun* poultry which has been slaughtered and dressed so that it can be cooked without any further preparation

over- ['əʊvə] *prefix* too much

◊ **overcropping** [əʊvə'krɒpɪŋ] *noun* growing too many crops on poor soil, which has the effect of weakening the soil still further

◊ **overcultivated** [əʊvə'kʌltɪveɪtɪd] *adjective* (land) which has been too intensively cultivated and so is exhausted

◊ **overexploit** [əʊvəɪk'splɔɪt] *verb* to cultivate soil so much that it becomes exhausted; *we overexploit land in the same way as we overexploit the sea; irrigated land given over to cash crops can be overexploited*

◊ **overexploitation** [əʊvəeksplɔɪ'teɪʃn] *noun* action of exploiting land too much; *overexploitation has reduced yields by half*

◊ **overfeed** [əʊvə'fiːd] *verb* to give animals too much feed

◊ **overgraze** [əʊvə'greɪz] *verb* to graze a pasture so much that it is no longer rich enough to provide grazing for cattle

◊ **overgrazing** [əʊvə'greɪzɪŋ] *noun* action of grazing a pasture so much that it loses nutrients; *overgrazing has led to soil erosion and desertification; overgrazing exhausts the productive species and allows unpalatable plants to become established*

◊ **overgrown** [əʊvə'grəʊn] *adjective* (seedbed *or* field) which is covered with weeds or other unwanted vegetation

◊ **overlying** [əʊvə'laɪɪŋ] *noun* crushing of piglets by the sow which lies on top of them

◊ **overproduction** [əʊvəprə'dʌkʃn] *noun* producing too much; *beef mountains and wine lakes are caused by overproduction in the EU*

◊ **overtopping** [əʊvə'tɒpɪŋ] *noun* cutting too much off the top of a plant when preparing it (as in the case of sugar beet)

oviduct ['əʊvɪdʌkt] *noun* duct through which ova pass after leaving the ovary; in birds, the duct through which the egg passes from the ovary to the outside

oviparous [əʊ'vɪpərəs] *adjective* (animal) which carries and lays eggs; *compare* VIVIPAROUS

Ovis ['əʊvɪs] *Latin name for* the sheep genus

ovulate ['ɒvjuːleɪt] *verb* to release a mature ovum from an ovary

◊ **ovulation** [ɒvjuːˈleɪʃn] *noun* release of an ovum from the ovary

◊ **ovule** ['ɒvjuːl] *noun* female cell within the ovary which, when fertilized, becomes the egg or seed

◊ **ovum** ['əʊvəm] *noun* an egg (NOTE: the plural is **ova)**

COMMENT: at regular intervals in mammals (in the human female, once a month) ova, or unfertilized eggs, leave the ovaries and move down to the uterus. Ovulation is regular in the mare, sow, ewe and cow

ox ɒks] *noun* male or female domestic cattle, and also the castrated male, especially when used as a draught animal (NOTE: plural is **oxen)**

ox-eye ['ɒksaɪ] *noun* any flower with a round yellow centre, such as the ox-eye daisy

Oxford Down ['ɒksfəd 'daʊn] *noun* the largest of the down breeds of sheep, produced by crossing Southdown improved stock with the longwoolled Cotswold; it has a dark-brown face and legs and a conspicuous topknot

oxidase ['ɒksɪdeɪz] *noun* enzyme which encourages oxidation by removing hydrogen and adding it to oxygen to form water

oxidize ['ɒksɪdaɪz] *verb* to react with oxygen, and convert a substance into an oxide

oxygen ['ɒksɪdʒən] *noun* chemical element, a common colourless gas which is present in the air, essential to biological life and is widespread in rocks (NOTE: chemical symbol for oxygen is **O**; atomic number is **8)**

◊ **oxygenate** ['ɒksɪdʒəneɪt] *verb* to treat (blood) with oxygen

◊ **oxygenation** [ɒksɪdʒəˈneɪʃn] *noun* becoming filled with oxygen

COMMENT: oxygen is an important constituent of living matter, as well as water and air. Oxygen is absorbed from the air into the bloodstream through the lungs and is carried to the tissues along the arteries; it is essential to normal metabolism. Oxygen is formed by plants from carbon dioxide in the atmosphere during photosynthesis and released back into the air

oxytoxin [ɒksɪˈtɒksɪn] *noun* hormone which activates the release of milk in the udder and the contractions in the uterus during birth; also possibly important in contracting the uterus during mating. Its action is blocked by the release of adrenalin

Pp

P 1 *chemical symbol for* phosphorus **2** Poor (in the EUROP carcass classification system)

Pabna ['pæbnə] *noun* dairy cow of Bangladesh, derived from crosses of local cows with Sahiwal, Haryana or Red Sindhi bulls

packhouse ['pækhaus] *noun* building used for grading, cleaning and packing produce on a farm, before it is sent to the customer

paddock ['pædək] *noun* small enclosed field, usually near farm buildings; **paddock grazing** = rotational grazing system which uses paddocks of equal area for grazing, followed by a rest period

paddy *or* **padi** ['pædi] *noun* (i) growing rice crop, a cultivated grass, the main grain crop of the monsoon area of south-east Asia; (ii) rice grains which have been harvested, but not processed in any way; **paddy (field)** = field filled with water, in which rice is grown; *paddies are breeding grounds for mosquitoes; scientists believe that paddy fields contribute to the methane in the atmosphere see also* RICE

pak-choy ['pæk'tʃɔɪ] *noun* Chinese cabbage *(Brassica chinensis)* of which the leaves can be cooked or eaten raw in salads

palatable ['pælətəbl] *adjective* which is good to eat; *some types of grass are less palatable than others; big bales preserve the grass in an almost cut state which is very palatable*

◊ **palatability** [pælətə'bɪlɪti] *noun* being good to eat

pale [peɪl] *noun* (i) fence stake; (ii) pointed piece of wood used for fencing; (iii) husk on grass or cereal seeds

pale leaf spot ['peɪl 'liːf spɒt] *noun* white spots which form on leaves of clover plants due to potash deficiency

pale persicaria ['peɪl pɜːsɪ'keərɪə] *noun* weed found in spring-sown crops (NOTE: also known as **pale willow weed)**

palm [pɑːm] *noun* large tropical plant growing as tall as a tree, with branching fern-like leaves, producing fruits which give oil and other foodstuffs; **coconut palm** *or* **date palm** = palms which produce coconuts *or* dates; **palm kernel oil** = vegetable oil produced from the kernels of the oil palm nut, used in the manufacture of margarine, and cooking fat; the residue after extraction of the oil is used as a livestock feed; **palm oil** = edible oil produced from the seed or fruit of an oil palm; red in colour, with only 5-12% polyunsaturated fatty acids, it is widely used in cooking fats and margarines; the residue after extraction is used as livestock feed

palynology [pælɪ'nɒlədʒi] *noun* study of pollen

pampas ['pæmpəs] *noun* temperate grasslands of South America, found in Argentina and Uruguay

COMMENT: the eastern pampas, which has a higher rainfall, has a natural covering of tall coarse grass, known as pampas grass. Vast areas of the pampas are now cultivated, and cattle and sheep are reared on them

pan [pæn] *noun* **(a)** wide shallow pot for growing seeds **(b)** hard cemented layer of soil, impervious to drainage, lying below the surface; it is formed by the deposition of iron compounds or by ploughing at the same depth every year; pan may be broken up by using a subsoiler

pan- [pæn] *prefix* meaning generalized *or* affecting everything

pandemic [pæn'demɪk] *noun* disease which affects many parts of the world; *compare* ENDEMIC, EPIDEMIC

panicle ['pænɪkl] *noun* inflorescence where

several small flowers branch from the same stem

Panicum ['pænɪkəm] *Latin for* millet

pannage ['pænɪdʒ] *noun* (i) pasturage for pigs in a wood or forest; (ii) the corn and beech mast on which pigs feed

papain [pə'peɪɪn] *noun* substance derived from the pawpaw, applied to meat to make it tender

Papaver [pə'pɑːvə] *Latin for* poppy

papaya *or* **pawpaw** [pə'paɪə *or* 'pɔːpɔː] *noun* tree *(Carica papaya)*, native of tropical South America, but now found in all tropical regions. The greenish-yellow melon-shaped fruits have a soft pulp which is eaten raw; the fruit is also a valuable source of papain, which is used in brewing and in tenderizing meat. The leaves are also used in cooking to make meat tender

para rubber tree ['pærə 'rʌbə triː] *noun* tree producing high yields of rubber *(Hevea brasiliensis)*

paraquat ['pærəkwɒt] *noun* non-selective contact herbicide; it becomes quite inert on contact with the soil, hence its agricultural use as a pre-emergent herbicide and for clearing rough land before ploughing

parasite ['pærəsaɪt] *noun* plant *or* animal which lives on or inside another organism (the host) and draws nourishment from it; *see also* ECTOPARASITE, ENDOPARASITE

◊ **parasitic** [pærə'sɪtɪk] *adjective* referring to parasites; *a parasitic plant;* **parasitic gastro-enteritis (PGE)** = infection of the stomach caused by roundworms, especially *Osteragia;* it can be cured by anthelmintics

◊ **parasiticide** [pærə'sɪtɪsaɪd] *noun & adjective* (substance) which kills parasites

◊ **parasitism** ['pærəsaɪtɪzm] *noun* state in which one organism (the parasite) lives on another organism (the host) and derives its nourishment and other needs from it

◊ **parasitize** ['pærəsɪtaɪz] *verb* to live as a parasite on (another organism); *sheep are parasitized by flukes*

◊ **parasitoid** ['pærəsaɪtɔɪd] *noun* animal which is a parasite only at one stage in its development

◊ **parasitology** [pærəsaɪ'tɒlədʒi] *noun* scientific study of parasites

COMMENT: the commonest parasites affecting animals are external parasites such as lice and ticks, and various types of internal parasite such as roundworms and flukes. Many diseases of humans (such as malaria and amoebic dysentery) are caused by infestation with parasites. Viruses are parasites on animals, plants and even on bacteria. Fungal diseases in plants, such as mildews and rusts, are caused by the action of parasitic fungi on their hosts

parathion [pærə'θaɪən] *noun* organophosphorous compound used to control insect pests

paratyphoid [pærə'taɪfɔɪd] *noun* disease of pigs caused by infection with bacteria salmonella; young pigs run a high fever and may die within 24 hours

parent ['peɪərənt] *noun* **parent cell** *or* **mother cell** = original cell which splits into daughter cells by mitosis; **parent material rock** = the unweathered base rock which breaks down to form a constituent part of the surface soil

park [pɑːk] *noun* area of open land used as a place of recreation; **country park** = area in the countryside set aside for the public to visit and enjoy; **national park** = large area of unspoilt land, owned and managed by a special planning authority which has control over agriculture as well as the normal planning powers. Parks exist for people to enjoy and for the protection and conservation of scenery, flora and fauna, and where established farming is effectively maintained

◊ **parkland** ['pɑːklænd] *noun* grassland with scattered trees, typical of large landed estates

parlour ['pɑːlə] *noun* **milking parlour** = building in which cows are milked, and often are also fed, washed and cleaned; **parlour systems** = the four basic designs of parlour: the herringbone parlour, the abreast parlour, the tandem parlour and the rotary parlour; *see note at* MILKING PARLOUR

parrot mouth ['pærət 'maʊθ] *noun* malformation of the upper jaw of horses, preventing proper mastication. The condition prevents the horse from grazing

parsley ['pɑːsli] *noun* common herb
(Petroselinum crispum) used for
garnishing and flavouring; **parsley piert** =
common weed *(Aphanes arvensis)*
affecting winter cereals; also known as
lamb's foot

parsnip ['pɑːsnɪp] *noun* plant *(Pastinaca
sativa)* whose long white root is eaten as a
vegetable

Parthenais ['pɑːtəneɪ] *noun* breed of
cattle originating in France; it produces
calves for a suckler herd, and is known for
easy calving and high growth rate

parthenocarpy [pɑːθenəʊˈkɑːpi] *noun* the
production of fruit without seeds

parthenogenesis [pɑːθenəʊˈdʒenəsɪs]
noun reproduction by unfertilized ova (as in
aphids)

partial drought ['pɑːʃəl 'draʊt] *noun* in
the UK, period of at least 29 consecutive
days when the mean rainfall does not
exceed 2.54mm

particle ['pɑːtɪkl] *noun* very small piece of
material; *soil with very fine particles*

partition [pɑːˈtɪʃn] *noun* moveable wall
which divides a room, such as a partition in
a stable

part-time farming ['pɑːttaɪm 'fɑːmɪŋ]
noun type of farming, where the farmer has
a regular occupation other than farming;
common throughout much of central and
Eastern Europe. In the UK, part-time
farmers are mainly wealthy people who
farm as a hobby or as a second form of
business

parturition [pɑːtjuˈrɪʃn] *noun* giving birth
to offspring, when the foetus leaves the
uterus, called by different names according
to the animal; *see also* CALVING,
FARROWING, FOALING, LAMBING

passion fruit ['pæʃn 'fruːt] *noun*
Passiflora edulis, granadilla, a climbing
plant with purple juicy fruit. It is native to
Brazil

pasta ['pæstə] *noun* food made from flour
from durum wheat, such as spaghetti,
lasagne, etc.

pastern ['pæstɜːn] *noun* thin part of a
horse's leg, between the fetlock and the
hoof

pasteurellosis [pɑːstʃərəˈləʊsɪs] *noun* a
clostridial disease mainly affecting young
lambs, adult sheep and store lambs; it may
be caused by contaminated food or water;
symptoms are high temperature, difficult
breathing and death may follow a few days
after the symptoms become apparent

pasteurization [pɑːstʃəraɪˈzeɪʃn] *noun*
heating of food *or* food products to destroy
bacteria, but without affecting palatability

◊ **pasteurize** ['pɑːstʃəraɪz] *verb* to kill
bacteria in food by heating it; *the
government is telling people to drink only
pasteurized milk*

COMMENT: Pasteurization is carried out by
heating food for a short time at a lower
temperature than that used for sterilization:
the two methods used are heating to 72°C
for fifteen seconds (the high temperature
short time method) or to 65° for half an hour,
and then cooling rapidly. This has the effect
of killing tuberculosis bacteria.
Pasteurization is used principally in the
preservation of milk for human
consumption

pastoral ['pɑːstərəl] *adjective (of land)*
available for pasture

◊ **pastoralist** ['pɑːstərəlɪst] *noun* farmer
who keeps grazing animals on pasture; *the
people most affected by the drought in the
Sahara are nomadic pastoralists*

COMMENT: pastoral farming ranges from
large-scale highly scientific systems such
as cattle- or sheep-ranching to primitive
systems of nomadism. In relatively dry or
inhospitable regions, the need to find
pastures for grazing animals causes the
farmers to lead a nomadic existence

pasture ['pɑːstʃə] **1** *noun* grassland grown
for or used for the grazing of livestock, as
opposed to meadows which are grown for
hay or silage; *the cows are in the pasture or
the cows are put to pasture;* **alpine pastures** =
grass fields in high mountains which are
used by cattle farmers in the summer;
pasture management = the control of
pasture by grazing, cutting, reseeding, etc.
2 *verb* to put animals in a pasture; *their
cows are pastured in fields high in the
mountains*

◊ **pasturage** ['pɑːstʃərɪdʒ] *noun* place
where animals are pastured

◊ **pastureland** ['pɑːstʃələnd] *noun* land which is given over to pasture

QUOTE scarcity of pastureland which cattle need in abundant quantities often leads to deforestation and hill erosion
Environmental Action

patch [pætʃ] *noun* (a) small area; *patches of bare soil in grass; animal with a white coat and red patches* (b) small cultivated area, with one type of plant growing in it, such as a pumpkin patch or onion patch

patent flour ['peɪtənt 'flaʊə] *noun* very fine good-quality wheat flour

path- *or* **patho-** [pæθ] *prefix* referring to disease

◊ **pathogen** ['pæθədʒən] *noun* germ, virus, etc.; a microorganism which causes a disease

◊ **pathogenesis** [pæθə'dʒenəsɪs] *noun* origin *or* production *or* development of a disease

◊ **pathogenetic** [pæθədʒə'netɪk] *adjective* referring to pathogenesis

◊ **pathogenic** [pæθə'dʒiːnɪk] *adjective* which can cause *or* produce a disease; **pathogenic bacteria** *or* **organisms** = bacteria *or* organisms responsible for transmitting a disease

◊ **pathogenicity** [pæθədʒə'nɪsəti] *noun* ability of a pathogen to cause a disease

Patna ['pætnə] *noun* common variety of long-grain Indian rice

pause [pɔːz] *noun* rest period in a bird's laying cycle

Pb *chemical symbol for* lead

PDA = POTASH DEVELOPMENT ASSOCIATION

pea [piː] *noun Pisum sativum,* an important grain legume; **pea and bean weevil** = pest *(Sitona* sp) affecting peas, beans and other legumes. The eggs are laid in soil near the plants, allowing the larvae to feed on the roots. The adult weevils feed on the leaves, making U-shaped notches in the edges of the leaves

COMMENT: Peas are grown for pulses and for their immature seeds which are eaten

fresh as a green vegetable and are also often frozen; occasionally the young pods are eaten as mangetouts. Peas are also grown for forage and may be used for hay and silage; they are often grown following a cereal crop in rotation, and enrich the soil with nitrogen. Most peas are harvested, transported and processed on the same day and the majority are taken for freezing and canning. Vining peas are Britain's most important contract vegetable crop

peach [piːtʃ] *noun* small deciduous tree *(Prunus persica)* found particularly in Mediterranean areas, though it will grow as far north as southern England. The fruit are large and juicy, with a downy skin; they cannot be kept for any length of time. Peaches are divided into two groups: the freestone (where the flesh is not attached to the stone), and the clingstone. The nectarine is a form of peach with a smooth skin; **peach-leaf curl** = fungal disease which affects peaches, where the leaves swell and become red

pear [peə] *noun* a pome fruit of the genus *Pyrus* used for dessert fruit, cooking or for fermenting to make perry; in the UK, William's Bon Chretien, Conference and Doyenné du Comice are popular dessert varieties, while William's is also commonly used for canning

peat [piːt] *noun* wet, partly-decayed mosses and other plants which form the soil of a bog

◊ **peatland** ['piːtlænd] *noun* area of land covered with peat bog

◊ **peaty** ['piːti] *adjective* containing peat

COMMENT: acid peats are formed in waterlogged areas where marsh plants grow, and where decay of dead material is slow. Black fen soils found in East Anglia are very fertile. These soils contain silts and calcium carbonate in addition to the remains of vegetation. Peat can be cut and dried in blocks, which can then be used as fuel. It is also widely used in horticulture, after drying and sterilizing

pecan ['piːkæn] *noun* North American tree *(Carya illinoensis)* which produces sweet nuts which are eaten as dessert nuts and used in many forms of confectionery

peck [pek] **1** *noun* measure of capacity of dry goods, equal to a quarter of a bushel or two gallons (used as measure of grain) **2**

verb (of birds) to pick up food with the beak; **pecking order** = the order of social dominance in a group of birds, and also animals (the equivalent in cattle is the 'bunt order')

pectin ['pektɪn] *noun* cellulose substance which is found in cell walls of certain fruit, and which forms a jelly when heated with sugar; this is important in making jam

ped [ped] *noun* an aggregate of soil particles

pedigree ['pedɪgriː] *noun* the ancestral line of animals bred by breeders, or of cultivated plants; **pedigree market** = the market for animals bred for breeding, not for slaughter (as opposed to the commercial market); **pedigree records** = records of pedigree stock kept by the breeder and by breed societies; pedigree animals are registered at birth and given official numbers; **pedigree selection** = the selection of animals for breeding based on the records of their ancestors

pedology [pə'dɒlədʒi] *noun* the study of the soil

◊ **pedologist** [pə'dɒlədʒɪst] *noun* scientist who has specialized in the study of the soil

peel [piːl] **1** *noun* outer layer of a fruit; also the skin of a potato; *oranges have a thick peel; lemon peel is used as flavouring* **2** *verb* to remove the peel from a fruit or potato

PEG = PRODUCERS ENTITLEMENT GUARANTEE

Pekin [piː'kɪn] *noun* breed of table duck; it has buff coloured feathers and bright orange feet, legs and bill

pellet ['pelɪt] *noun* form of feedingstuff (usually mash) which has been moistened and pressed to form small grains

◊ **pelleted seed** ['pelɪtɪd 'siːd] *noun* seed coated with clay to produce pellets of uniform size and density; pelleting is done to make the sowing of very fine seed easier

pen [pen] **1** *noun* small enclosure for animals or poultry; **pen mating** = using one male animal to mate with a number of females; *compare* FLOCK MATING **2** *verb* to enclose animals, such as sheep, in a pen

Penicillium [penɪ'sɪliəm] *noun* fungus from which penicillin is derived

◊ **penicillin** [penɪ'sɪlɪn] *noun* common antibiotic produced from a fungus

COMMENT: penicillin is effective against many microbial diseases, such as mastitis in cattle

pepper ['pepə] *noun* **(a)** spice (either black or white) from the fruit of the pepper vine **(b)** fruit of the *Capsicum,* either red, yellow or green **(c) chilli peppers** = small red fruit of the *Capsicum frutescens* which are very pungent; the dried fruit are used to make Cayenne pepper

peppermint ['pepəmɪnt] *noun* aromatic herb *(Mentha piperata)* which is cultivated to produce an oil used in confectionery, drinks and toothpaste

pepsin ['pepsɪn] *noun* enzyme in the stomach which breaks down the proteins in food into peptones

◊ **peptic** ['peptɪk] *adjective* referring to digestion *or* to the digestive system

◊ **peptone** ['peptəʊn] *noun* substance produced by the action of pepsins on proteins in food

perch [pɜːtʃ] **1** *noun* wooden rail on which a bird sits **2** *verb (of a bird)* to sit on a branch or wooden rail

Percheron ['pɜːʃərɒn] *noun* heavy breed of horse, developed in Normandy; it is grey in colour

Perendale ['penəndeɪl] *noun* New Zealand breed of sheep

perennial [pə'renɪəl] **1** *adjective* term applied to plants which persist for more than two years; **perennial agriculture** = system of agriculture (as in the tropics) where there is no winter and several crops can be grown on the same land each year; **perennial irrigation** = system which allows the land to be irrigated at any time. This may be by primitive means such as shadufs, or by distributing water from barrages by canal and ditches; **perennial ryegrass** = a grass *(Lolium perenne)* which forms the basis of the majority of long leys in the UK; the most important grass in good permanent pasture and often sown mixed with other grasses and clover; perennial ryegrass has a long growing season, is quick to become established and responds well to fertilizers. It is best suited to grazing and is highly palatable for animals **2** *adjective &*

noun (plant) which lives for a long time, flowering each year without dying; *compare* ANNUAL, BIENNIAL

performance test [pə'fɔːməns 'test] *noun* record of growth rate in an individual animal over a given period of time, when fed on a standard ration. Performance testing gives the breeder a better chance of identifying genetically superior animals

pericarp ['perikɑːp] *noun* fibrous layers next to the outer husk of a cereal grain, removed during milling. It is the major constituent of bran

permafrost ['pɜːməfrɒst] *noun* ground which is permanently frozen as in the Arctic regions, where the top layer of soil melts and softens in the summer while the soil beneath remains frozen

permanent ['pɜːmənənt] *adjective* which exists always; **permanent hardness** = hardness of water, caused by calcium and magnesium, which remains even after the water has been boiled; **permanent pasture** = land which remains solely as grassland over a long period of time, and is not ploughed

PPS = PERMANENT PRODUCER SAMPLE sample survey carried out every 5 years on the structure of dairy farms in England and Wales. It covers such items as size and breed of herds, land use and tenure, milking labour, techniques of milking and capital investment

permeability [pɜːmiə'bɪləti] *noun (of a rock or soil)* ability to allow water to pass through; *(of a membrane)* ability to allow fluid containing chemical substances to pass through

◊ **permeable** ['pɜːmiəbl] *adjective* (rock or soil) which allows a liquid to pass through; (membrane) which can allow substances dissolved in fluids to pass through (NOTE: US English is **pervious**)

perpetual-flowering [pə'petjuəl 'flauəriŋ] *adjective* variety of plant which bears flowers more or less all year round

perry ['peri] *noun* fermented pear juice

persimmon [pə'simən] *noun* native tree *(Diospyros kaki)* of Japan and China, which produces reddish-orange fruit, similar in appearance to tomatoes; the fruit are eaten either as dessert or may be cooked; they are very rich in vitamins

persist [pə'sist] *verb* to continue for some time; to remain active for some time; *some substances persist in toxic forms in the air for weeks; plants which persist for more than two years are called perennials*

◊ **persistence** [pə'sistəns] *noun* ability of a pollutant to remain active for a period of time

◊ **persistency** [pə'sistənsi] *noun* ability of a plant to survive for a long time, even when the soil is cultivated; *ryegrasses are used for leys where persistency is not important*

◊ **persistent** [pə'sistənt] *adjective* which continues for some time; **persistent insecticide** = insecticide that remains toxic (either in the soil or in the body of an animal) and is passed from animal to animal through the food chain (NOTE: the opposite is **non-persistent)**

QUOTE the persistence of very low ozone levels into the southern hemisphere's summer could have serious ecological effects
New Scientist

pervious ['pɜːviəs] *adjective US* (membrane) which can allow substances dissolved in fluids to pass through (NOTE: GB English is **permeable)**

pest [pest] *noun* **(a)** living organism which is in any way disruptive to the growth of crops or animals, or which may cause trouble to stored produce; *a spray to remove insect pests;* **pest control** = the control of pests (by killing them) **(b)** name given to some diseases, such as fowl pest

COMMENT: the word is a relative term: a pest to one person may not be a pest to another, so foxes are pests to chicken farmers, but not to naturalists. It has been estimated that around 30% of the total potential world harvest is lost through the activities of pests, and losses are significantly higher than this in poorer countries where pesticides are less readily available

pesticide ['pestisaid] *noun* toxic agent used to control a wide variety of plant, animal and microbial organisms which might have an adverse effect on growing or harvested crops, or on livestock

◊ **Pesticides Safety Directorate** department of MAFF, which evaluates and approves pesticides

◊ **Pesticide Safety Precaution Scheme (PSPS)** agreement between the agrochemical industry and the government, supported by Health and Safety regulations, which designates products as safe to use, provided recommended precautions are taken during their use; *see also* FEPA

COMMENT: there are three basic types of pesticide. *1.* organochlorine insecticides, which have a high persistence in the environment of up to about 15 years (DDT, dieldrin and aldrin); *2.* organophosphates, which have an intermediate persistence of several months (parathion, carbaryl and malathion); *3.* carbamates, which have a low persistence of around two weeks (Temik, Zectran and Zineb). Most pesticides are broad-spectrum, that is they kill all insects in a certain area and may kill other animals like birds and small mammals. Pesticide residue levels in food in the UK are generally low. Pesticide residues have been found in bran products, bread and baby foods, as well as in milk and meat. Where pesticides are found, the levels are low and rarely exceed international maximum residue levels

petal ['petl] *noun* outer coloured part of the corolla of a flower

Pe-tsai ['peɪ'tsaɪ] *noun* Chinese cabbage (*Brassica pekinensis*) whose leaves are used as vegetable

PGE = PARASITIC GASTRO-ENTERITIS

PGR = PLANT GROWTH REGULATOR

PGRO = PROCESSORS AND GROWERS RESEARCH ORGANISATION

pH ['piː'eitʃ] *noun* the negative log of the hydrogen ion concentration, a measure of the concentration of hydrogen ions in a solution, which shows how acid or alkaline it is; *there was a sudden increase in pH; polluted water with a low pH value is more acid and corrosive;* **pH meter** = meter which measures the pH value of a solution; **pH test** = test to see how acid or alkaline a solution is

COMMENT: the pH is shown as a number. A value of 7 is neutral. Lower values indicate increasing acidity and higher values indicate increasing alkalinity. So 0 is most acid and 14 is most alkaline. Acid rain has been known to have a pH of 2 or less, making it as acid as lemon juice. Plants vary in their tolerance of soil pH; some grow well

on alkaline soils, some on acid soils only, and some can tolerate a wide range of pH values. The soil pH value for rye and lupins is approximately 4.5, for oats and potatoes 5.0, for wheat, beans, peas, turnips and swedes 5.5, for clover, maize and oilseed rape 6.0, and for barley, sugar beet and lucerne 6.5

QUOTE the sulphuric and nitric acids in cloud droplets can give them an extremely low pH. Water collected near the base of clouds in the eastern US during the summer typically has a pH of about 3.6 but values as low as 2.6 have been recorded. In the greater Los Angeles area the pH of fog has fallen as low as 2
Scientific American

QUOTE lack of fermentation means that the pH will often remain high - about 5, rather than dropping to 4 as in clamp silage. This can be dangerous when there is soil contamination because listeria bacteria in the soil will thrive at this pH level
Farmers Weekly

QUOTE though soybean gives best yield at 6 to 6.5 pH, yet compared to most other pulses, it can be grown without much loss in soils from 5.5 to 7.5 pH. It needs liming if the soil is below 5.5 pH. In other words, the tolerance of soybeans to broader pH range from moderate acidity to moderate alkalinity enables it to be grown from medium to high rainfall regions of the country
Indian Farming

phacelia [fə'siːliə] *noun* plant used as a ground cover crop (introduced into the UK from the USA)

phag- *or* **phago-** ['fægəʊ] *prefix* referring to eating

◊ **-phage** [feɪdʒ] *suffix* which eats

◊ **phagocyte** ['fægəʊsaɪt] *noun* cell (such as a white blood cell) which can surround and destroy other cells, such as bacteria cells

◊ **phagocytic** [fægə'sɪtɪk] *adjective* (i) referring to phagocytes; (ii) (cell) which destroys other cells

Phaseolus [fæzɪ'əʊləs] *Latin name for* beans, such as the French bean and butter bean

pheasant ['fezənt] *noun* game bird *(Phasianus colchicus)* with long tail feathers

phenotype ['fiːnətaɪp] *noun* physical characteristics of an organism which its genes produce, such as brown eyes, height, etc.; *compare* GENOTYPE

phenoxyherbicide [fiːnɒksɪ'hɜːbɪsaɪd] *noun* herbicide, based on 2,4,5-T

pheromone ['ferəməʊn] *noun* sex attractant produced by females to attract males. Artificial pheromones are used to disorientate male insects and prevent mating. Synthetic boar pheromones can be used both to stimulate and detect oestrus in sows and gilts

phloem ['fləʊəm] *noun* living tissue in a plant which carries organic substances from leaves to other parts of the plant (as opposed to xylem, which is dead tissue)

phosphate ['fɒsfeɪt] *noun* salt of phosphoric acid, which is an essential plant nutrient and which is formed naturally by weathering of rocks

COMMENT: phosphates escape into water from sewage, especially in waste water containing detergents, and encourage the growth of algae by eutrophication. Natural organic phosphates are provided by guano and fishmeal; otherwise, phosphates are mined (especially in Russia, the USA and Mexico). Artificially produced phosphates are used in agriculture: the main types of phosphate fertilizer are ground rock phosphate, hyperphosphate, superphosphate, triple superphosphate and basic slags. Phosphate deficiency is one of the commonest deficiencies in livestock, and gives rise to osteomalacia (also known as creeping sickness)

phosphoric acid [fɒs'fɒrɪk 'æsɪd] *noun* acid which forms phosphates

◊ **phosphorus** ['fɒsfərəs] *noun* toxic chemical element which is essential to biological life, being present in bones and nerve tissue (pure phosphorus will cause burns if it touches the skin and can poison if swallowed); **phosphorus cycle** = cycle by which phosphorus atoms are circulated through living organisms (NOTE: the chemical symbol is **P**; atomic number is **15)**

COMMENT: phosphorus is an essential part of DNA and RNA, and when an organism dies, the phosphorus contained in its tissues returns to the soil and is taken up by plants in the phosphorus cycle. Phosphorus deficiency in plants causes stunted growth, discoloration of leaves and small or misshapen fruit

phot- *or* **photo-** [fəʊt] *prefix* referring to light

◊ **photoperiodicity** [fəʊtəʊpiːriə'dɪsɪti] *noun* way in which plants and animals react to changes in the length of the period of daylight from summer to winter

◊ **photorespiration** [fəʊtəʊrespɪ'reɪʃn] *noun* respiration that occurs in plants in the light; in C_4 plants photorespiration is very low

photosensitization [fəʊtəʊsensɪtaɪ'zeɪʃn] *noun* disease of livestock caused by activation of photodynamic agents in the skin by light. Skin becomes pink and inflamed and may develop deep cracks

photosynthesis [fəʊtəʊ'sɪnθəsɪs] *noun* process by which green plants convert carbon dioxide and water into sugar and oxygen using sunlight as energy

◊ **photosynthesize** [fəʊtəʊ'sɪnθəsaɪz] *verb* to carry out photosynthesis; *acid rain falling on trees reduces their ability to photosynthesize*

QUOTE light powers photosynthesis, the process by which a plant converts carbon dioxide and water into sugars, starch and oxygen
Scientific American

phylloxera [fɪ'lɒksərə] *noun Viti folii*, an aphid which attacks vines; it threatened to destroy the vineyards of Europe in the 19th century, but the vines were saved by grafting susceptible varieties onto resistant American rootstock

phyt- *or* **phyto-** ['faɪtəʊ] *prefix* referring to plants *or* coming from plants

◊ **phytome** ['faɪtəʊm] *noun* a plant community

◊ **phytophagous** [faɪ'tɒfəgəs] *adjective* herbivorous, (animal) which eats plants

phytotoxic [faɪtəʊ'tɒksɪk] *adjective* (substance) which is poisonous to plants

◊ **phytotoxicant** [faɪtəʊ'tɒksɪkənt] *noun* substance which is poisonous to plants

◊ **phytotoxin** [faɪtəʊ'tɒksɪn] *noun* plant which is poisonous to animals

pick [pɪk] *verb* to take ripe fruit or vegetables from plants

◊ **picker** ['pɪkə] *noun* person who picks fruit or vegetables; *a strawberry picker*

◊ **picking** ['pɪkɪŋ] *noun* work of taking fruit or vegetables from plants; *teams of workers are employed on raspberry picking in July*

pickle ['pɪkl] *verb* to preserve food by keeping it in vinegar; **pickling cabbage** *or* **pickling onions** = cabbage *or* onions specially grown for preserving

pick-up attachment ['pɪkʌp ə'tætʃmənt] *noun* attachment used on a combine to lift grass swath and feed it into the main elevator; it is fitted over the combine cutter bar

◊ **pick-up baler** ['pɪkʌp 'beɪlə] *noun* machine which picks up cut grass and makes small bales; the machine which makes big bales is called a 'big baler'

◊ **pick-up reel** ['pɪkʌp 'riːl] *noun* part of a combine harvester with spring tines, used to improve cutting efficiency in tangled crops

pick-your-own (PYO) ['pɪkjə'əʊn] *noun* system where customers are allowed onto the farm, to pick what they need from the field; it is used with most types of soft-fruit and many vegetables

COMMENT: the existence of a large car-owning population has helped in the growth of PYO farms, especially close to large urban populations

piebald ['paɪbɔːld] *adjective & noun* (animal) whose coat has two colours, especially black and white, in irregular shapes

piecework ['piːswɜːk] *noun* work for which workers are paid for the products produced *or* for the piece of work done, and not at an hourly rate; potato pickers and strawberry pickers are paid in this way

Piedmontese [piːdmən'tiːz] *noun* breed of beef cattle from north-west Italy; the animals are light or dark grey, with black horns, ears and tail

piert ['pɪɜːt] *see* PARSLEY PIERT

Pietrain ['piːtreɪn] *noun* Belgian breed of pig imported into the UK for cross-breeding; it has prick ears and irregular dark spots over the whole body. It is very lean but carries the stress or hacothane gene. A very muscular breed

pig [pɪg] *noun* animal of the *Suidae* family kept exclusively for meat production; **pig futures** = sales on the commodity market of pigmeat for future delivery; **pig improvement scheme** = scheme introduced by the Meat and Livestock Commission to improve pig stock; the scheme involves extensive performance testing (NOTE: the males are called **boars,** the females are **sows,** the young are **piglets.** Pigs reared for pork meat are called **porkers** and those reared for bacon are **baconers)**

COMMENT: there have been changes in pig types and breeds because of the needs of the meat industry. The main breeds used in the UK include the Large White, Landrace, Welsh, Hampshire, Duroc and Pietrain. In pig farming, the common practice is to use a white quality-type pig which may be crossed with other breeds to produce good quality commercial pigs. A large number of crossbred sows are being used. Good quality pigs are needed for fresh meat and for manufacturing, and they are reared to provide pork, bacon, hams, pies and sausages. Basically, there are three types of pig farming: some pig production is concerned with the specialized business of pig breeding and some farmers only breed and sell the young pigs as weaners or stores; but most pig farming consists of rearing pigs for pork and fresh meat, or for bacon, ham and other processed foods. Pigs mature quickly and are marketed at between four and six months of age. Bacon pigs are marketed at about 90kg liveweight and porkers from around 60kg. Most pigs are intensively housed in prefabricated buildings with mechanical ventilation and often automated feeding, although there is a movement towards 'freerange' pig farming. Pig rearing is carried on in most parts of the UK, but is being increasingly concentrated in the Eastern part of the country. With over 14 million pigs slaughtered each year, home production is worth nearly £1 billion or 7% of the total value of UK agricultural output. Rare breeds (non-commercial breeds) include the British Saddleback,

Large Black, Gloucester Old Spot, Berkshire, Tamworth, Middle White and Chester White

QUOTE A new breed of pig has been created for the growing number of farmers who are abandoning factory farming for the free range production of pork and bacon. About 70,000 sows (10% of the national breeding herd) are now reared outdoors. Farmers have found that pigs become more profitable when reared outdoors in small straw-like shelters, than in vast environmentally controlled sheds
Sunday Telegraph

◊ **piggery** ['pɪgəri] *noun* place where pigs are housed

COMMENT: pigs can be kept successfully under all sorts of conditions, but housing systems must provide efficient feeding facilities. Individual feeding of sows is essential, and confining dry sows in stalls is common. Farrowing sows and their litters need careful housing; farrowing should take place in a special crate which prevents overlying of piglets (where the sow crushes the piglets by lying on them). Breeding pigs may be reared outdoors and housed in moveable arks. Fattening pigs can be housed in yards, usually covered with straw bedding and provided with a shelter, but more commonly they are kept in special buildings with controlled environments and ventilation, feeding, watering and efficient disposal of dung through special channels. There are many systems of pig housing, ranging from the traditional cottage piggery, with its low-roofed pens and open yard; the Danish type, a totally enclosed building with a controlled environment; the flat deck type, used for rearing young pigs, with special heating and ventilation controls; the Solari piggery, with fattening pens on each side of a central feeding passage, housed in an open-sided Dutch barn; and the monopitch type, incorporating artificially controlled natural ventilation

◊ **piglet** ['pɪglət] *noun* young pig; **piglet anaemia** = metabolic disease caused by milk deficiency; it can cause death of piglets. Iron compounds are administered as treatment

◊ **pigman** ['pɪgmæn] *noun* term formerly used for a male farm worker who looks after pigs (now replaced, in advertisements, by 'pigperson')

◊ **pigmeat** ['pɪgmiːt] *noun* term used in the EU for meat from pigs

◊ **pigperson** ['pɪgpɜːsən] *noun* farm worker who looks after pigs (term used in job advertisements)

◊ **pig pox** ['pɪg 'pɒks] *noun* disease of pigs caused by infection with one of two different viruses, most commonly the swinepox virus. Young pigs have red spots on belly, face and head which turn into blisters. Fever may follow

◊ **pig production** ['pɪg prə'dʌkʃən] *noun* commercial farming of pigs

COMMENT: sows are mature enough to breed at around six months of age. They are usually mated with the boar every 140-150 days. After a gestation period of 115 days a litter is born and is usually allowed to suckle for 21-28 days. Four to seven days after separating the sow from her piglets, she comes into oestrus and allows the boar to mate. Each of the resulting litters will consist of about ten piglets, and over her lifetime the sow will mother six to eight litters. Sows may be kept in stalls, sometimes completely closed, and sometimes open-ended. Sows are tethered in open-ended stalls. Sows are also kept in covered yards on straw. New systems are being introduced, such as the use of a collar for each sow, with its computer code number; the number is relayed to a computer which allocates a measured amount of feed to the sow. Herds of sows are also kept in open fields, using arks for shelter

◊ **pigstock** ['pɪgstɒk] *noun* herds of pigs; **pigstock person** = farm worker who looks after pigs

◊ **pigsty** *or US* **pigpen** ['pɪgstaɪ *or* 'pɪgpen] *noun* small building for keeping a single pig

◊ **pigweed** ['pɪgwiːd] *noun* common name for 'knotgrass'

pigeon ['pɪdʒən] *noun* bird which is regarded as a pest, and can cause severe damage to small plants; the wood pigeon eats cereal seed and causes damage to young and mature crops of peas and brassicas. Destruction of nests and pigeon shoots are the only effective way of keeping these birds under control

◊ **pigeon pea** ['pɪdʒən 'piː] *noun* tropical legume *(Cajanus cajan)* native to Africa; a drought-resistant plant producing seeds which are eaten cooked. In India it is known as 'red gram' and used to make dhal

pinch out ['pɪntʃ 'aʊt] *verb* to remove small shoots by pinching between finger and thumb (done when training young plants)

pine [paɪn] *noun* **(a) pine tree** = evergreen tree of the genus *Pinus* growing in temperate latitudes; *the north of the country is covered with pine forests;* **pine kernels** = seeds of the Mediterranean pine *(Pinus pinea)* eaten raw or roasted and salted, and also used in cooking and confectionery **(b)** = PINING

pineapple ['paɪnæpl] *noun* fruit of a tropical plant *Ananas comosus,* native to South America and now grown in many tropical areas; the fruit are eaten both raw and canned, or in the form of juice extracted by crushing; Mexico, the Philippines, Cote d'Ivoire, Malaysia and Hawaii are major producers

pinewood ['paɪnwʊd] *noun* forest of pine trees

pining ['paɪnɪŋ] *noun* disease of sheep, caused by shortage of cobalt; the animals lose condition, and have poor fleece, dull eyes and general weakness

pint [paɪnt] *noun* non-metric measure of liquids and dry goods, such as seeds; equal to 0.56 litres (NOTE: usually written **pt** after figures: **4pts of water)**

pinworm ['pɪnwɜːm] *noun* thin parasitic worm *Enterobius,* which infests the large intestine; also called 'threadworm'

Pinzgauer ['pɪntsgaʊə] *noun* breed of cattle from Austria; the animals are coloured reddish or dark brown with a distinctive white strip along the back; both a meat and draught breed

pioneer crop ['paɪənɪːə 'krɒp] *noun* crop grown to improve general soil fertility, prior to the sowing of another more valuable crop; pioneer crops are grazed by livestock with the result that dunging improves the soil's fertility

pipe [paɪp] *verb* to bring water to an area in pipes (as opposed to irrigation canals); *water is piped to fields to provide irrigation*

piperazine [pɪ'perəziːn] *noun* drug used to treat worm infestation

pippin ['pɪpɪn] *noun* name given to several dessert apples

pistachio [pɪ'stæʃɪəʊ] *noun* small tree *(Pistacia vera)* native of central Asia and now cultivated in Mediterranean regions; the nuts are eaten salted or in confectionery

pistil ['pɪstɪl] *noun* the female part of a flower

Pisum ['paɪsəm] *Latin name for* pea

pit [pɪt] *noun* the stone in certain fruit (cherries, plums, peaches, or dried fruit such as raisins and dates)

◊ **pitted** ['pɪtɪd] *adjective* fruit with the stones removed

pitch [pɪtʃ] *noun* dark sticky substance obtained from tar, used to make objects watertight; **pitch pole** = harrow with double-ended tines

placement drill ['pleɪsmənt 'drɪl] *noun* machine which drills seeds and fertilizer at the same time, and places the fertilizer close to the side of and below the rows of seeds

placenta [plə'sentə] *noun* tissue which grows inside the uterus in mammals during pregnancy and links the embryo to the mother

◊ **placental** [plə'sentəl] *adjective* referring to the placenta

COMMENT: the placenta allows an exchange of oxygen and nutrients to be passed from the mother to the embryo. It does not exist in marsupials

plague [pleɪg] *noun* serious contagious disease; **cattle plague** = disease of cattle, eradicated from the UK in 1877, but still found in parts of Asia and Africa

plain [pleɪn] *noun* wide area of level or gently rolling country, usually at a low altitude

plane tree ['pleɪn triː] *noun Platanus acerifolia,* a common temperate deciduous hardwood tree, frequently grown in towns because of its resistance to air pollution

plant [plɑːnt] **1** *noun* organism containing chlorophyll with which it carries out photosynthesis; *botanists discovered several new species of plant in the jungle;* **plant breeding** = development of new and improved varieties and strains of plants; **plant food ratio** = ratio of nitrogen to phosphate and potash in a fertilizer; **plant nutrients** = minerals whose presence in the soil is essential for the healthy growth of plants; **plant population** = the number of plants in an area, such as the number of plants per hectare **2** *verb* to put a plant into the earth; *the ground needs to be dug over carefully before the young trees are planted see also* TRANSPLANT

◊ **planter** ['plɑːntə] *noun* **(a)** person who plants; especially a person who plants and looks after a plantation **(b)** device for planting; *see* POTATO PLANTER

plantain ['plæntein] *noun* **(a)** name given to various types of banana, used for cooking and brewing; it has a lower sugar content than dessert bananas **(b)** common weed *(Plantago major)*, also called 'ribwort'

> COMMENT: plantains are an important crop in Nigeria, Zaire and the Cameroons

plantation [plɑːn'teiʃn] *noun* **(a)** estate, especially one in the tropics on which large-scale production of cash crops takes place; plantations specialize in the production of a single crop, and crops such as cocoa, coffee, cotton, tea and rubber are typical plantation crops; **coffee plantation** = plantation of coffee bushes **(b)** area of land planted with trees for commercial purposes (NOTE: plantations of conifers are often by mistake called **forests**)

> QUOTE extensive cash crop plantation schemes such as rubber and oil palm estates form a vital part of the economy
> *Ecologist*

plastic ['plæstik] **1** *adjective* describing the state of a soil when it is too wet (the soil deforms and does not recover) **2** *noun* man-made material, used as a cover to protect young crops; thin films of polythene may be used to cover and warm soil; black plastic sheeting is used as a form of mulch, and also to cover clamps and bales

plate and flicker ['pleit n 'flikə] *noun* type of machine used for distributing fertilizer

◊ **plate mill** ['pleit 'mil] *noun* type of mill used for grinding grain; the machine is made of two circular plates, one of which is fixed, while the other rotates against it

plateau ['plætəu] *noun* extensive almost

level area of high land (NOTE: plural is **plateaux**)

plot [plɒt] *noun* small area of cultivated land, which has been clearly defined; *the allotments are divided into plots of 20m x 5m;* **nursery plot** *or* **seed plot** = area of cultivated soil used for growing plants on before being planted out, or for sowing seed

plough *or* US **plow** [plau] **1** *noun* agricultural implement used to turn over the surface of soil into ridges and furrows in order to cultivate crops; **plough body** = the main part of the plough, consisting of the frog, mouldboard, share and landside **2** *verb* to turn over the soil with a plough; **deep ploughing** = ploughing very deep into the soil (used when reclaiming previously virgin land for agricultural purposes)

◊ **plough in** [plau 'in] *verb* to cover a crop, stubble or weeds with soil, by turning over the surface with a plough

◊ **ploughland** ['plaulænd] *noun* arable or cultivated land

◊ **ploughman** ['plaumən] *noun* man who ploughs

◊ **plough pan** ['plau 'pæn] *noun* hard layer in the soil caused by ploughing at the same depth every year

◊ **ploughshare** ['plauʃeə] *noun* heavy metal blade of a plough, which cuts the bottom of the furrow

◊ **plough up** ['plau 'ʌp] *verb* to plough a pasture, usually in order to use it for growing crops

> COMMENT: the modern plough is usually fully mounted on a tractor's hydraulic system; some are semi-mounted, with the rear supported by one or more wheels and some may be trailed. The principal parts of a plough are the beam or frame, made of steel, to which are attached a number of parts which engage the soil, such as the disc coulter, the share and the mouldboard which turns the furrow slice. There are three main types of plough: conventional, with right-handed mouldboards, reversible, with left- and right-handed mouldboards, and disc ploughs, which are used for deep rapid cultivation of hard dry soils, and are not common in the UK. There are many designs of plough body, the main ones being the general purpose plough, the semi-digger, the digger and the Bar Point.

The three main methods of ploughing are: *systematic*, where the field is divided into lands by shallow furrows; *round and round ploughing*, in which fields are ploughed from the centre to the outside or from the edge to the centre (see also SQUARE PLOUGHING); *reversible ploughing*, where the field is ploughed up and down the same furrow, giving a very level surface

plow [plaʊ] *US* = PLOUGH

pluck [plʌk] *verb* **(a)** to remove the feathers from a bird's carcass (usually done by machine, but still also done by hand); **dry pluck** = removing the feathers when the bird is dry, so avoiding harming the skin; **wet pluck** = removing the feathers when the carcass is wet (easier than dry plucking, but may harm the skin) **(b)** to remove the internal organs from an animal carcass after slaughter **(c)** to remove the leaves from a plant (such as the tea plant)

plum [plʌm] *noun* stone fruit (*Prunus domestica*)

◊ **plum pox** ['plʌm 'pɒks] *noun* viral disease affecting plums, damsons and peaches. The fruit has dark blotches and ripens prematurely. It is often sour

COMMENT: there are many varieties of cultivated plums, and they vary in colour, shape, size and flavour. Pond's Seedling, Monarch and Pershore are cooking varieties, while the rich-flavoured dessert varieties include Victoria, Laxton and Kirke's Blue. American varieties include Chickasaw and the Oregon Plum. In the UK, plums are grown mainly in the south and in the west Midlands; in the USA, they are common in the Pacific states

plumage ['pluːmɪdʒ] *noun* the feathers of a bird

plumule ['pluːmjuːl] *noun* young shoot of a seed plant from which the stem and leaves develop

Plymouth rock ['plɪməθ 'rɒk] *noun* large heavy hardy dual-purpose breed of poultry, originally coming from the USA. The feathers are rich lemon-buff

PMB = POTATO MARKETING BOARD

pneumatic distributor [njuː'mætɪk dɪ'strɪbjʊtə] *noun* machine which conveys fertilizer from a hopper to nozzles for spreading by a stream of air; both trailed and mounted models are made; **pneumatic (grain) drill** = machine which sows grain, the seed being moved from a hopper down the drill pipe by compressed air

pneumonia [njuː'məʊnɪə] *noun* inflammation of a lung, where the tiny alveoli of the lung become filled with fluid; **bacterial pneumonia** = form of pneumonia caused by pneumococcus; **calf pneumonia** = disease caused by a virus, and affecting dairy-bred and suckled calves; **viral** *or* **virus pneumonia** = type of inflammation of the lungs caused by a virus, in particular a disease affecting pigs and calves; the animals suffer from a hard cough and are unthrifty. In calves, shivering, loss of appetite and noisy breathing are symptoms

poach [pəʊtʃ] *verb* **(a)** *(of cattle)* to trample the ground in wet weather; heavy soils such as clays are particularly susceptible to poaching **(b)** to steal game or fish which is privately owned

◊ **poacher** ['pəʊtʃə] *noun* person who takes game or fish illegally

pocket ['pɒkɪt] *noun* large sack of dry hops

pod [pɒd] **1** *noun* legume, the long casing for several seeds (as for peas or beans) **2** *verb* **to pod up** = to begin to develop pods

podsol *or* **podzol** ['pɒdsɒl *or* 'pɒdzɒl] *noun* type of acid soil where iron and aluminium oxides and/or humus have been leached from the bleached top layer into a lower layer which is characterized by bright colours due to iron; *compare* CHERNOZEM

◊ **podzolic soil** *or* **podzolized soil** [pɒd'zɒlɪk *or* 'pɒdzəlaɪzd 'sɔɪl] *noun* soil from which iron and aluminium oxides have been leached from the topsoil in moist cool climates

◊ **podzolization** [pɒdzəlaɪ'zeɪʃn] *noun* process by which iron and aluminium oxides leach from the top layers of acid soil

COMMENT: on the whole podzols make poor agricultural soils, due to their low nutrient status and the frequent presence of an iron pan; large areas of the coniferous forest regions of Canada and Russia are covered with podzols

poi [pɔɪ] *noun* (*in Polynesia*) meal made from taro

point [pɔɪnt] *noun* the forward end of a ploughshare

◊ **point of lay** ['pɔɪnt ʌv 'leɪ] *noun* term referring to pullets that are approaching the time when they will lay their first eggs

polder ['pəʊldə] *noun* piece of low-lying land which has been reclaimed from the sea, and is surrounded by earth banks to protect it from encroachment. In the Netherlands polders are fertile and form valuable agricultural land or pasture for cattle

poll [pəʊl] **1** *noun* the top of an animal's head; **Poll Dorset** = Australian breed of sheep similar to the Dorset Horn, but with no horns **2** *verb* to dehorn an animal; **polled stock** = (i) animals which are naturally hornless; (ii) animals which have had their horns removed

pollard ['pɒlɑːd] **1** *noun* **(a)** tree of which the branches have been cut back to a height of about two metres above the ground **(b)** waste bran left after flour has been ground from wheat **2** *verb* to cut back the branches on a tree every year, or every few years, at a height of about two metres from the ground; *compare* COPPICE

COMMENT: pollarding allows new shoots to grow as in coppicing, but high enough above the ground to prevent them being eaten by animals. Willow trees are often pollarded

pollen ['pɒlən] *noun* cells in flowers which convey the male gametes; **pollen count** = measurement of the amount of pollen in a sample of air; **pollen analysis** = scientific study of pollen, especially of pollen found in peat *or* coal deposits; *see also* PALYNOLOGY; **pollen beetle** = pest of Brassica, which makes buds wither; the beetle feeds on buds and flower parts

◊ **pollinate** ['pɒlɪneɪt] *verb* to transfer pollen from male to female reproductive organs in a flower

◊ **pollination** [pɒlɪ'neɪʃn] *noun* action of pollinating a flower; **self-pollination** = pollination where the pollen from an anther goes to a stigma on the same plant; *see also* CROSS-POLLINATION

◊ **pollinator** ['pɒlɪneɪtə] *noun* organism such as a bee *or* bird *or* other plant, which helps pollinate a plant; *some apple trees need to be planted with pollinators as they are not self-fertile; birds are pollinators for many types of tropical plant*

◊ **pollinosis** [pɒlɪ'nəʊsɪs] *noun* hay fever, inflammation of the nose and eyes caused by an allergic reaction to pollen *or* fungus spores *or* dust in the atmosphere

COMMENT: grains of pollen are released by trees in spring and by flowers and grasses during the summer. The pollen is released by the stamens of a flower and floats in the air until it finds a female flower. Pollen in the air is a major cause of hay fever. It enters the eyes and nose and releases chemicals which force histamines to be released by the sufferer, causing the symptoms of hay fever to appear

pollute [pə'luːt] *verb* to discharge harmful matter in abnormally high concentrations into the environment, often done by people but can occur naturally; *polluting gases react with the sun's rays; polluted soil must be removed and buried*

◊ **pollutant** [pə'luːtənt] *noun* substance or agent which pollutes

◊ **polluter** [pə'luːtə] *noun* person *or* company which pollutes; *certain industries are major polluters of the environment;* **polluter pays principle** = principle that if pollution occurs, the person *or* company responsible should be made to pay for consequences of the pollution

◊ **pollution** [pə'luːʃn] *noun* presence of abnormally high concentrations of harmful substances in the environment, often put there by people; *pollution of the atmosphere has increased over the last 50 years; soil pollution round mines poses a problem for land reclamation;* **atmospheric pollution** = pollution of the air

QUOTE farm pollution of rivers rose in 1988 to a record 4,141 cases reported in England and Wales according to a report by the Water Authorities Association and the MAFF. Animal slurry, the main form of farm pollution, is more than 100 times more damaging than untreated sewage, while the liquid from silage is 200 times more damaging

Guardian

QUOTE in November 1988, a £50 million package of grants was announced by the Agriculture Minister for stopping farm pollution. Farmers can claim up to half the cost of building or improving facilities for the storage, treatment and disposal of slurry and silage effluent

Guardian

Polwarth ['pɒlwɔːθ] *noun* Australian breed of sheep (from Lincoln and Merino) which gives a fine wool

poly- ['pɒli] *prefix* **(a)** meaning many *or* much *or* touching many organs **(b)** meaning made of polythene

polyculture ['pɒlɪkʌltʃə] *noun* rearing or growing of more than one species of plant or animal on the same area of land at the same time; *compare* MONOCULTURE

polyphagous [pə'lɪfəgəs] *adjective* (animal) which eats more than one type of food

◊ **polysaprobic** [pɒlɪsæ'prəubɪk] *adjective* (organism) which can survive in heavily polluted water

polytunnel ['pɒlɪtʌnəl] *noun* cover for growing plants, like a large greenhouse, made of a rounded plastic roof attached to semi-circular supports

pome [pəum] *noun* type of fruit which develops from the axis of the flower

COMMENT: common fruit like apples and pears are pomes, in which the fleshy part develops from the receptacle of a flower and not from the ovary as in other fruit

◊ **pomology** [pə'mɒlədʒi] *noun* study of growing pome fruit

pomegranate ['pɒmɪgrænɪt] *noun* a semi-tropical tree *(Punica granatum)* native to Asia, but now cultivated widely; the fruit are round and yellow, with masses of seeds surrounded by sweet red flesh

pond [pɒnd] *noun* small area of fresh water; *see also* DEW POND

pony ['pəuni] *noun* small breed of horse, often ridden by children; also a wild breed in some parts of the world; **pony-trekking** = recreational activity where people hire ponies to ride along country paths, now sometimes organized from farms as a form of diversification

porcine ['pɔːsaɪn] *adjective* referring to pigs; **porcine somatotropin (PST)** = hormone administered to feeder pigs, which has been shown to increase feed efficiency, the ration of lean meat to carcass weight, and market weight; **porcine spongiform encephalopathy (PSE)** = brain disease which has been induced in pigs experimentally

pore [pɔː] *noun* (i) tiny hole in the skin through which sweat passes; (ii) tiny space in a rock formation; **pore space** = the space in the soil not filled by soil particles, but which may be filled with water or air

pork [pɔːk] *noun* fresh meat from pigs (as opposed to cured meat, which is bacon or ham); *see also* PIGMEAT

◊ **porker** ['pɔːkə] *noun* pig specially reared for fresh meat (as opposed to bacon or other processed meats)

Portland ['pɔːtlənd] *noun* rare breed of sheep; both sexes are horned, with brown or tan faces and legs

post [pəust] *noun* solid wooden or concrete pole, placed in a hole in the ground, and used to support a fence or a gate; **post hole digger** = implement driven by a tractor, shaped like a very large screw which bores holes in the ground in which posts are placed

pot [pɒt] **1** *noun* **flower pot** = container in which a plant is grown; **pot-bound plant** = plant which is in a pot which is too small and which its roots fill **2** *verb* to put a plant into a pot; **potting compost** = special prepared compost for plants in pots

◊ **pot on** ['pɒt 'ɒn] *verb* to take a plant from one pot and repot it in a larger one so that it can develop

potash ['pɒtæʃ] *noun* general term used to describe potassium salts; potash salts are crude minerals and contain much sodium chloride; **potash fertilizers** = fertilizers based on potassium

COMMENT: the main types of potash used are muriate of potash (potassium chloride), sulphate of potash (potassium sulphate), kainit, nitrate of potash (potassium nitrate) or saltpetre, and potash salts; the quality of a potash fertilizer is shown as a percentage of potassium oxide K_2O equivalent. Potash deposits are mined in Russia, the USA, Germany and France

potassium [pə'tæsiəm] *noun* metallic element, essential to biological life; **potassium nitrate** = fertilizer recommended for horticultural crops; **potassium permanganate** = purple-coloured salt, used as a disinfectant (NOTE: chemical symbol is **K**; atomic number is **19**)

COMMENT: potassium is one of the three major soil nutrients needed by growing plants (the others are nitrogen and phosphorus); potassium also plays an important part in animal physiology and occurs in a variety of plant and animal foodstuffs. Potassium deficiency causes spotted leaves, weak growth and small fruit

potato [pə'teɪtəʊ] *noun* tuber of *Solanum tuberosum*, a most important starchy root crop; apart from being grown for human consumption, potatoes also are used to produce alcohol and starch, and are used as stock feed; **potato blight** = fungus disease (*Phytophthora infestans*) which kills potato foliage and rots the tubers; **potato clamp** = heap of stored potatoes covered with straw and earth; **potato cyst nematode** = *Globodera rostochiensis*, a pest found in most soils that have grown potatoes; the eggs hatch in the spring, the larvae invade the roots. The leaves of plant eventually yellow and are stunted; **potato harvester** = machine which lifts the crop onto a sorting platform, where up to six pickers sort the potatoes from soil and stones; the potatoes are then raised onto a trailer

COMMENT: the potato originated in South America; its introduction into almost every country of the world has proved of great importance on account of its extreme productiveness, its easy cultivation and its remarkable powers of acclimatization. Varieties of potato can be cultivated from the tropics to the furthest limits of agriculture, even further than the polar limit of barley. Huge quantities are grown in Russia, Poland, Germany, and the USA. Potatoes vary in shape; they may be round, oval, kidney-shaped or just irregular. They also vary in colour, from cream to brown or red. In the UK, varieties are classified as the first and second earlies (harvested from May to August), and early and late maincrop (harvested in September/October). About 20% of the potatoes grown in the UK are earlies. Seed potatoes are grown under a certification scheme. In the UK, potatoes are grown mainly for human consumption, either in tubers, or for processing into chips, crisps, for canning or dehydration

potato elevator digger [pə'teɪtəʊ 'elɪveɪtə 'dɪgə] *noun* machine which lifts potatoes

COMMENT: a wide flat share runs under the potato ridge and lifts the soil onto a rod-link conveyor. Most of the soil is returned to the ground and the potatoes move on to a second elevator, which returns the potatoes to the ground for hand picking

Potato Marketing Board (PMB) government organization in the UK, which regulates the area of land to be planted with potatoes; all potato growers must register with the board and the board implements guaranteed price arrangements and support buying when necessary

potato planter [pə'teɪtəʊ 'plɑːntə] *noun* machine for planting potatoes

COMMENT: a coulter makes a furrow and the seed potatoes are carried forward by a conveyor mechanism from the hopper and placed at regular intervals in the row. Potatoes are carried in self-filling cups from the hopper to the ground. Many potato planters have a fertilizer attachment. Automatic planters are best for handling unchitted seed

potato spinner [pə'teɪtəʊ 'spɪnə] *noun* machine for lifting the potato crop

COMMENT: a spider wheel rotates and the loosened potato ridge is pushed sideways. The potatoes hit a net at the side of the spinner and fall to the ground, where they are left for the hand pickers

potential transpiration [pə'tenʃəl trænspɪ'reɪʃn] *noun* the calculated amount of water taken up from the soil and transpired through the leaves of plants; the amount varies according to the climate and weather conditions

poult [pəʊlt] *noun* young turkey, less than 8 weeks old

poultry ['pəʊltri] *noun* general term for domestic birds kept for meat and egg production; chickens are the most common; turkey and geese are mainly kept as meat birds; ducks and guinea fowl also produce significant quantities of meat and eggs (NOTE: no plural)

◊ **poultryman** ['pəʊltrɪmən] *noun* farm worker who looks after and raises poultry

pound [paʊnd] *noun* unit of currency used in the UK and several other countries; **green pound** = value of the British pound as used for calculating agricultural subsidies and prices in the EU

poussin ['puːsæn] *noun* young chicken killed for the table

powder ['paʊdə] *noun* material which has been reduced to very fine grains

◊ **powdered sulphur** ['paʊdəd 'sʌlfə] *noun* sulphur which is used to dust on plants to prevent mildew

◊ **powdery mildew** ['paʊdəri 'mɪldjuː] *noun* fungal disease *(Erysiphe graminis)* affecting cereals and grasses; another form also affects sugar beet and brassicas

power ['paʊə] *noun* energy which makes substances move; **power harrow** = harrow with two tine bars driven from the power take-off; the tines move alternately in a rotary manner. Power harrows are used to prepare seedbeds of fine tilth; **power take-off (p.t.o.)** = mechanism providing power to drive field machines from a tractor

pox [pɒks] *see* FOWL POX, SHEEP POX

prairie ['preəri] *noun* area of grass-covered plain in North America, mainly treeless, with many different species of grasses and herbs; the prairie lands of the United States and Canada are responsible for most of the North America's wheat production (NOTE: in Europe, the equivalent is the **steppe**)

QUOTE within a lifetime, most of Canada's wild prairies have been lost or damaged by extensive ranching of cattle and urban or industrial development
Guardian

PRB = PLANT ROYALTY BUREAU

Préalpes-du-sud ['preɪælp dʊ 'sʊd] *noun* a hardy breed of sheep found in Alpes-Provençales region of France

precision chop forage harvester [prɪ'sɪʒən tʃɒp 'fɒrɪdʒ 'hɑːvəstə] *noun* type of harvester which cuts the crop with flails, chops it into precise lengths and blows it into a trailer; it may be self-propelled, off-set trailed or mounted. It is used for harvesting green material for making silage

◊ **precision drill** [prɪ'sɪʒən 'drɪl] *noun* seed drill which sows the seed separately at set intervals in the soil

pre-emergent [priːɪ'mɜːdʒənt] *adjective* before a plant's leave appear from the seed in the soil; **pre-emergent herbicide** = herbicide, such as paraquat, used to clear weeds before the crop leaves have emerged

pregnancy toxaemia ['pregnənsi tɒk'siːmiə] *noun* metabolic disorder affecting ewes and does during late pregnancy. Animals wobble and fall, breathing is difficult, and death may follow. Also known as 'twin lamb disease', it is associated with lack of feed in late pregnancy

premium ['priːmiəm] *noun* special extra payment; **variable premium** = extra payment which varies according to production quality; *see also* BEEF PREMIUM SCHEME

preserve [prɪ'zɜːv] *verb* to keep something in the same state; to stop food from changing *or* rotting

◊ **preservation** [prezə'veɪʃn] *noun* keeping of something in the same condition; *food preservation allows some types of perishable food to be eaten during the winter months when fresh food is not available;* **tree preservation order (TPO)** = order from a local government department which prevents a tree from being felled

◊ **preservative** [prɪ'zɜːvətɪv] *noun* substance added to food to preserve it by slowing natural decay caused by bacteria, fungi, etc. (in the EU, preservatives are given E numbers E200 - 297); salt and sugar are two additives commonly used in the preservation of foods; both have the effect of eliminating or reducing microbial activity

press [pres] **1** *noun* **(a)** machine for crushing fruit or seeds to extract juice or oil, such as an olive press; **cider press** = device for crushing apples to extract the juice to make cider **(b) furrow press** = special type of very heavy ring roller attached to the plough, used to press the furrow slices **2** *verb* to crush fruit or seeds to extract juice or oil; *residue from pressing is used for animal feed*

prick ears ['prɪk 'ɪəz] *noun* ears of an animal which stand up straight (as opposed to lop ears)

◊ **prick out** ['prɪk 'aʊt] *verb* to transplant seedlings from trays or pans into pots or flowerbeds

prill *or* **prilled fertilizer** [prɪl] *noun* form of fertilizer sold in small round granules

primary ['praɪməri] **1** *adjective* (something) which is first, and leads to another (secondary); **primary commodities** = basic raw materials *or* food; **primary forest** = forest which originally covered a region, before changes in the environment brought about by people; **primary host** = some parasites need two hosts, the primary host being the one where the parasite reaches sexual maturity; *see also* SECONDARY HOST **primary products** = products (such as wood, milk, fish) which are basic raw materials; **primary production** = amount of organic matter which is formed by photosynthesis **2** *noun* the **primaries** = main feathers on a bird's wing; also called 'flight feathers'

primed seed ['paɪmd 'siːd] *noun* seed which has been moistened to start the germination process before sowing

primitive ['prɪmətɪv] *adjective* ancient, not developed or modernized; **primitive breeds** = old breeds of livestock which have not been bred commercially, but which are the descendants of wild livestock

prion ['priːən] *noun* infective agent, which may be a form of protein, responsible for scrapie in sheep and BSE in cattle

process ['prəʊses] *verb* to treat produce in a way so that it will keep longer or become more palatable; **processed meats** = meat products such as bacon, sausages, etc.

COMMENT: food can be processed in many different ways: some of the commonest are drying, freezing, canning, bottling and chilling

prod [prɒd] *noun* spiked metal rod, used to make cattle move forward

produce ['prɒdjuːs] *noun* foodstuffs such as vegetables, fruit, eggs, etc., which are produced commercially; **farm produce** = foodstuffs grown on farms

◊ **producer** [prə'djuːsə] *noun* **(a)** **producer-retailer** = person who produces a commodity for sale directly to the public, as through a farm shop or by milk delivery **(b)** organism (such as a green plant) which takes energy from outside an ecosystem and channels it into the system (the first level in the food chain)

product ['prɒdʌkt] *noun* (i) thing which is produced; (ii) result *or* effect of a process

production [prə'dʌkʃn] *noun* **(a)** **production diseases** = metabolic disorders of animals which are caused by high levels of production **(b)** act of producing something; **primary production** = amount

of organic matter formed by photosynthesis; **production ration** = quantity of fodder needed to make a farm animal produce a product such as milk or eggs (always more than the basic maintenance ration)

productive [prə'dʌktɪv] *adjective* which produces a lot; **productive soils** = soils which are very fertile and produce large crops

◊ **productivity** [prɒdʌk'tɪvɪti] *noun* rate at which something is produced; *with new strains of rice, productivity per hectare can be increased;* **primary productivity** = (i) rate at which plants produce organic matter through photosynthesis; (ii) amount of organic matter produced in a certain area over a certain period of time (such as a crop during a growing season); **net productivity** = difference between the amount of organic matter produced by photosynthesis and the amount of organic matter used by plants in their growth

QUOTE rising global temperatures could completely alter the face of the earth. Many of the world's most fertile regions are likely to become drier and less productive, while regions like India and the Middle East are expected to become wetter and more fertile

Ecologist

profusion [prə'fjuːʒn] *noun* mass; *a profusion of small white flowers*

progeny ['prɒdʒəni] *noun* offspring of an animal or plant; **progeny test** = the evaluation of the breeding value of an animal or plant variety by examining the performance of its progeny

progesterone [prə'dʒestərəʊn] *noun* female sex hormone which stimulates the uterus prior to fertilization and maintains the animal in a state of pregnancy (gestation)

prolactin [prəʊ'læktɪn] *noun* hormone produced by the pituitary gland which stimulates lactation

prolific [prə'lɪfɪk] *adjective* (animal or plant) which produces a large number of offspring or fruit

promote [prə'məʊt] *verb* to encourage a reaction to take place; *growth-promoting hormones are used to increase the weight of beef cattle*

◊ **promoter** [prə'məutə] *noun* **growth promoter** = substance which is used to make animals *or* plants grow

◊ **promotion** [prə'məuʃn] *noun* encouraging something to take place; *growth promotion may change the nutritional quality of wheat*

prong [prɒŋ] *noun* one of the long pieces of metal which make up a fork

propagate ['prɒpəgeɪt] *verb* to produce new plants

◊ **propagation** [prɒpə'geɪʃn] *noun* making new plants grow

◊ **propagator** ['prɒpəgeɪtə] *noun* device in which seed can be sown or cuttings taken to produce a number of new plants

COMMENT: the commonest of the various forms of plant propagation are by *seminal propagation*: growing from seed or from tubers (such as potatoes) and by *vegetative propagation* (as by taking cuttings, grafting, dividing, budding roses, layering, etc.)

prostaglandin [prɒstə'glændɪn] *noun* hormone that is used to synchronize oestrus and trigger the birth process or abortion

protease ['prəutieɪz] *noun* digestive enzyme that breaks down proteins in food

protect [prə'tekt] *verb* to keep something safe from harm; **protected cropping** *or* **protected cultivation** = crops grown under some form of protection, such as in greenhouses, or under polythene sheeting; in the UK, the main areas are in Guernsey and along the south coast; commonest protected crops are tomatoes, cucumbers, lettuces and mushrooms

◊ **protectant fungicide** [prə'tektənt 'fʌndʒɪsaɪd] *noun* fungicide applied to the leaves of plants (it can be washed off by rain, so removing the protection)

protein ['prəuti:n] *noun* nitrogen compound which is present in and is an essential part of all living cells in the body, formed by the condensation of amino acids; **protein balance** = situation when the nitrogen intake in protein is equal to the excretion rate (in the urine); **protein deficiency** = lack of enough proteins in the diet; **protein efficiency ration** = measure of

the nutritional value of proteins carried out on young growing animals; defined as the gain in weight per gram of protein eaten; **protein equivalent** = measure of the digestible nitrogen of an animal feedingstuff in terms of protein; **protein quality** = measure of the usefulness of a protein food for various purposes, including growth, maintenance, repair of tissue, formation of new tissue and production of eggs, wool and milk

◊ **proteolysis** [prəutɪ'ɒlɪsɪs] *noun* breaking down of proteins in food by proteolytic enzymes

◊ **proteolytic** [prəutiəu'lɪtɪk] *adjective* referring to proteolysis; **proteolytic enzyme** *or* **protease** = digestive enzyme which breaks down proteins in food

COMMENT: proteins are necessary for growth and repair of the tissue of animals' bodies; they are mainly formed of carbon, nitrogen and oxygen in various combinations as amino acids. Certain foods (such as beans, meat, eggs, fish and milk) are rich in protein. Although ruminant animals can make use of cellulose and non-protein nitrogen (which human beings cannot metabolize), animal feeds are often supplemented with grains, pulses and other foods that contain protein

protoplasm ['prəutəuplæzm] *noun* living material which makes up the largest part of the cells of plants and animals

◊ **protoplasmic** [prəutəu'plæzmɪk] *adjective* referring to protoplasm

◊ **protoplast** ['prəutəuplɑːst] *noun* basic cell unit in a plant formed of a nucleus and protoplasm

proud [praud] *adjective* referring to excessive growth or development in crops or livestock

provender ['prɒvəndə] *noun* general terms used to describe dry feeds for livestock

proven sire ['pruːvn 'saɪə] *noun* a bull, boar or ram which has been shown to sire progeny which produces milk, meat or wool of high quality

proximate analysis ['prɒksɪmət ə'næləsɪs] *noun* method of chemical analysis used on animal feedingstuffs, which measures the amounts of ash, crude fibre, crude protein, ether extract, moisture content and nitrogen-free extract

prune [pru:n] **1** *noun* black-skinned dried plum **2** *verb* to remove pieces of a plant, in order to keep it in shape, or to reduce vigour; **pruning knife** = special knife with a curved blade, used for pruning fruit trees

◊ **pruners** ['pru:nəz] *noun* type of secateurs, used for pruning fruit trees

Prunus ['pru:nəs] *Latin name* family of trees including the plum, peach, almond, cherry, damson, apricot

PSE = PORCINE SPONGIFORM ENCEPHALOPATHY

PSPS = PESTICIDE SAFETY PRECAUTION SCHEME

PST = PORCINE SOMATOTROPIN

p.t.o. = POWER TAKE-OFF *the p.t.o. shaft from a tractor drives a baler or rotavator*

puddling ['pʌdlɪŋ] *noun* preparing the land in rice fields to prevent the water from percolating through the soil

pullet ['pulɪt] *noun* young female fowl, from hatching until a year old

pulling peas ['pulɪŋ 'pi:z] *noun* peas harvested by removing the pods when fresh and sold as young peas in pods

pulp [pʌlp] *noun* the soft flesh of a fruit

◊ **pulpy kidney disease** [pʌlpi 'kɪdni di'zi:z] *noun* disease caused by a strain of the same bacteria which cause lamb dysentery; it occurs in older lambs and can be fatal

pulsator [pʌl'seɪtə] *noun* part of a milking machine which causes the suction action and release of the milk from the udder

pulses ['pʌlsɪz] *noun* general term for certain seeds that grow in pods; the species of vegetable which have this type of fruit are very numerous and include trees as well as tender plants; the term is often applied to edible seeds of leguminous plants (lentils, beans and peas) used as food for human or animal consumption

QUOTE pulses are generally susceptible to soils which are acidic in nature and need liming if the soil pH is below 6; though soybean gives best yield at 6 to 6.5 pH, yet compared to most other pulses it can be grown without much loss in soils from 5.5 to 7.5 pH

Indian Farming

pumpkin ['pʌmpkɪn] *noun* large round yellow vegetable, eaten both as a vegetable and in pies as a dessert

punch [pʌntʃ] *see* SUFFOLK

pungent ['pʌndʒənt] *adjective* with a sharp taste or smell (as, for example, mustard)

pupa ['pju:pə] *noun* chrysalis, a stage in the development of certain insects (such as butterflies) where the larva becomes encased in a hard shell (NOTE: the plural is **pupae** ['pju:pi:])

◊ **pupal** ['pju:pəl] *adjective* referring to a pupa

◊ **pupate** [pju:'peɪt] *verb (of insect)* to move from the larval to the pupal stage

pure breeding ['pjuə 'bri:dɪŋ] *noun* mating together of purebred animals of the same breed

◊ **purebred** ['pjuəbred] *adjective* (animal) which is the offspring of parents which are themselves the offspring of parents of the same breed

put to ['pʌt 'tu:] *verb (in breeding)* to bring a female animal to be impregnated by a male; *ewes are put to the ram*

pygmy beetle ['pɪgmi 'bi:tl] *noun* beetle pest affecting sugar beet *(Atomaria linearis)*

PYO = PICK-YOUR-OWN

pyrethrum [paɪ'ri:θrəm] *noun* **(a)** annual herb, grown for its flowers which are used in the preparation of an insecticide **(b)** organic pesticide, developed from a form of chrysanthemum, which is not very toxic and non-persistent

pyrexia [paɪˈreksiə] *noun* livestock disease the cause of which is unknown but may be a virus infection. Spots appear on head, neck, back, and udder. There is severe itching and animals lose body condition (NOTE: also known as **haemorrhagic syndrome)**

Pyrus [ˈpaɪrəs] *Latin name for* the pear

Qq

quadrat ['kwɒdrət] *noun* area of land measuring one square metre, chosen as a sample for research into plant populations; *the vegetation of the area was sampled using the quadrat method*

quail [kweɪl] *noun* small game bird *(Coturnix coturnix)*, now reared to produce oven-ready birds and also for their eggs

quality grain ['kwɒlɪti 'greɪn] *noun* application of quality standards when selling grain; good quality is indicated by a high specific weight

quarantine ['kwɒrəntiːn] **1** *noun* period (originally forty days) when an animal *or* person *or* ship just arrived in a country has to be kept separate in case it carries a serious disease, to allow the disease time to develop and so be detected; *the animals were put in quarantine on arrival at the port; a ship in quarantine shows a yellow flag called the quarantine flag* **2** *verb* to put a person *or* animal in quarantine

COMMENT: animals coming into Britain are quarantined for six months because of the danger of rabies. People who are suspected of having an infectious disease can be kept in quarantine for a period which varies according to the incubation period of the disease. The main diseases concerned are cholera, yellow fever and typhus

quart [kwɔːt] *noun* measure of liquids and dry goods, such as grain, equal to two pints, or 1.14 litres

Quercus ['kwɜːkəs] *Latin name for* the oak tree

quick-freeze [kwɪk'friːz] *verb* to preserve food products by freezing them rapidly

quicklime ['kwɪklaɪm] *noun* calcium oxide, chemical used in many industrial processes, and also spread on soil to reduce acidity

quince [kwɪns] *noun* small tree *(Cydonia vulgaris)* native of western Asia; its hard pear-shaped sour fruit are rich in pectin and used to make jellies and other preserves

quinine ['kwɪniːn] *noun* alkaloid found in the bark of a South American tree *(Cinchona);* **quinine poisoning** = illness caused by taking too much quinine

◊ **quininism** *or* **quinism** ['kwɪniːnɪzm *or* 'kwɪnɪzm] *noun* quinine poisoning

COMMENT: quinine was formerly used to treat the fever symptoms of malaria, but is not often used now because of its side-effects. Symptoms of quinine poisoning are dizziness and noises in the head. Small amounts of quinine have a tonic effect and are used in tonic water

quintal ['kwɪntəl] *noun* unit of weight sometimes used to measure bulk agricultural commodities, equal to 100kg

QUOTE the average productivity of wheat and paddy in Uttar Pradesh is around 14 and 12 quintals per hectare respectively. However, high-yielding varieties combined with improved technologies push up the yields to around 50 and 60 quintals per hectare respectively
Indian Farming

quota ['kwəʊtə] *noun* fixed amount of something which is allowed; *the government has imposed a quota on the fishing of herring;* **quota system** = system where imports *or* supplies are regulated by fixing maximum amounts; **milk quotas** = system by which farmers are only allowed to produce certain amounts of milk, introduced to restrict the overproduction of milk in member states of the EU

COMMENT: quotas were introduced in 1984, and were based on each state's 1981 production, plus 1%. A further 1% was allowed in the first year. A supplementary levy or superlevy, was introduced to penalize milk production over the quota level. Reductions in the quota amount were made in 1986/7 and 1987/8. In the UK, milk quotas can be bought and sold, either together with or separate from farmland, and are a valuable asset

Rr

R Average (in the EUROP carcass classification system)

rabbit ['ræbɪt] *noun* common furry herbivorous rodent *(Oryctolagus cuniculus)*

> COMMENT: rabbits are raised for meat and for their fur; wild rabbits are a major pest in some parts of the world, and in Australia, myxomatosis was introduced to attempt to eradicate the wild rabbit population

rabi ['rɑːbi] *see* KOHLRABI

rabies ['reɪbiːz] *noun* frequently fatal notifiable viral disease transmitted to humans by infected animals

> COMMENT: rabies affects the mental balance, and the symptoms include difficulty in breathing or swallowing and an intense fear of water (hydrophobia) to the point of causing convulsions at the sight of water. Rabies is not present in Britain

race [reɪs] *noun* **(a)** general term for a breed or variety of plant or animal; *see also* LANDRACE **(b)** improvised wooden way along which animals are made to walk, such as when being loaded into a van (also called a 'raceway')

raceme ['ræsiːm] *noun* inflorescence on which the flowers are carried on short stalks (as opposed to a panicle where several small flowers branch from the same stem)

raceway ['reɪsweɪ] *see* RACE

rack [ræk] *noun* frame of wooden or metal bars which holds fodder, and from which animals can eat

raddle ['rædl] *noun* flexible length of wood used for making hurdles or fences

radicle ['rædɪkl] *noun* the portion of a plant embryo which develops from the primary root

radish ['rædɪʃ] *noun Raphanus sativus,* small plant with red or white roots used mainly in salads

Radnor ['rædnə] *noun* breed of small hill sheep similar to the Welsh Mountain

rafter ['rɑːftə] *verb* to plough land in such a way that the furrow is turned over onto the unploughed land next to it

ragwort ['rægwɜːt] *noun* a weed *(Senecio jacobea)* found in grassland; it can cause poisoning of cattle and sheep, and must therefore be controlled

rain [reɪn] *noun* water which falls from clouds as small drops; **acid rain** = rain which contains a higher level of acid than normal; *see also* ACID; **artificial rain** = rain which is made by 'seeding' clouds with crystals of salt and other substances

◊ **rainfall** ['reɪnfɔːl] *noun* amount of water which falls as rain on a certain area over a certain period

◊ **rainforest** ['reɪnfɒrɪst] *noun* thick tropical forest which grows in regions where the rainfall is very high; *poor farmers have cleared hectares of rainforest to grow cash crops*

◊ **rain gun** ['reɪn 'gʌn] *noun* one of the means of applying irrigation water; water is applied under pressure by a series of widely-spaced guns, which can be either fixed or mobile and are used for grassland or sugar beet irrigation

◊ **rainmaking** ['reɪnmeɪkɪŋ] *noun* attempting to create rain by scattering carbon dioxide on clouds

◊ **rainwater** ['reɪnwɔːtə] *noun* water which falls as rain and is collected and used for irrigation

> COMMENT: rain is normally slightly acid (about 5.6 pH) but becomes more acid when pollutants from burning fossil fuels are released into the atmosphere

raise [reɪz] *verb* (i) to make plants germinate and nurture them as seedlings;

(ii) to breed livestock; *the plants are raised from seed in special seedbeds*

rake [reɪk] **1** *noun* implement with a handle, a crossbar with several prongs, used for pulling hay together, or for smoothing loose soil to form a seedbed; **side rake** = machine which picks up two swaths and combines them into one before baling **2** *verb* **(a)** to pull hay or dead grass with a rake; to smooth loose soil to form an even seedbed **(b)** to move a flock of sheep from one pasture to another

ram [ræm] *noun* male sheep or goat, that has not been castrated (NOTE: a male sheep is also known as a **tup**)

rambutan [ræmbu'tæn] *noun* tropical fruit *(Nephelium lappaceum)* grown in South-East Asia; the fruit is similar to the litchi

ranch [rɑːnʃ] *noun* **(a)** large farm, specializing in raising cattle, sheep or horses; **cattle ranch** = very large grassland farm where cattle are raised; *governments have encouraged the conversion of rainforest to livestock ranches* (NOTE: in Australia, this is also called a **station**) **(b)** *US* large specialized farm

◊ **rancher** ['rɑːnʃə] *noun* owner of a ranch; worker on a cattle ranch

◊ **ranching** ['rɑːnʃɪŋ] *noun* agricultural system based on commercial grazing on ranches

QUOTE plantations and ranching have laid waste millions of hectares of forest. In Central America cattle ranching is responsible for the clearance of almost two-thirds of the forests
Ecologist

range [reɪnʒ] *noun* open space, particularly for poultry; eggs produced on a range are called 'freerange eggs'

rape [reɪp] *noun* **oilseed rape** = plant of the cabbage family *(Brassica napus)* closely related to the turnip, and usually grown as a catch crop for forage or for seed; the seed provides an edible oil, and the residual oilcake is used for animal feed; *see also* DOUBLE LOWS

◊ **rapeseed** ['reɪpsiːd] *noun* seed of the rape

rare [reə] *adjective* not common, (disease) of which there are very few cases; **Rare**

Breeds Survival Trust = a trust established in 1973 to foster interest in breeds which have historical importance and may prove useful in the future

◊ **rarity** ['reərɪti] *noun (of wild species)* state of being rare

COMMENT: species are classified internationally into several degrees of rarity: rare, means that a species is not very common and is restricted to small local populations, but is not necessarily likely to become extinct. Vulnerable means that a species has a small population and that population is declining. Endangered means that the species has such a small population that it is likely to become extinct

RAS = ROYAL AGRICULTURAL SOCIETIES (NOTE: qualified people may become associates or fellows of the RAS, and then can put the initials **ARAgS** or **FRAgS** after their name)

RASE = ROYAL AGRICULTURAL SOCIETY OF ENGLAND

raspberry ['rɑːzbəri] *noun* cane *(Rubus idaeus)* which provides a most important soft fruit, sold fresh, sent for freezing and also used for processing into jams (the main area for commercial raspberry growing in the UK is in the southern part of Scotland); **raspberry beetle** = a serious pest *(Byturus tomentosus)* whose larvae feed on young raspberry fruit

rat [ræt] *noun* **brown rat** *or* **Norway rat** = *Rattus norvegus* a pest which eats and damages growing and stored crops; rats also carry infection to cattle and pigs

ration ['ræʃən] *noun* amount of food given to an animal; **maintenance ration** = amount of food needed to keep a farm animal healthy and in good condition, and which must be met before any nutrients are used for production such as lactation; **production ration** = amount of food needed to make a farm animal produce meat, milk, eggs, etc.

ratoon *or* **ratoon crop** [rə'tuːn] *noun* the second and later crops taken from the regrowth of a crop that has been harvested once (for example, sugar cane plants can be harvested many times)

ray [reɪ] *noun* line of light *or* heat; **ray fungus**

= bacterium which affects grasses and cereals, and can cause actinomycosis in cattle

RCGM = RECTIFIED CONCENTRATED GRAPE MUST

RCVS = ROYAL COLLEGE OF VETERINARY SURGEONS

reafforestation [riːæfɒrɪsˈteɪʃn] *noun* planting trees again in an area which was once covered by forest

QUOTE the first priority in any programme of ecological recovery must be a worldwide programme of reafforestation. Even the most degraded lands in the dry tropics can be restored to forest, successful reafforestation schemes having been implemented in Costa Rica and India

Ecologist

reap [riːp] *verb* to cut a grain crop; **reaping hook** = short-handled semicircular implement with a sharp blade, formerly used for cutting corn by hand

◊ **reaper-binder** [ˈriːpəˈbaɪndə] *see* BINDER

rear [rɪə] *verb* to look after young animals until they are old enough to look after themselves

◊ **rearer** [ˈrɪərə] *noun* person who rears livestock

recessive [rɪˈsesɪv] *adjective* (trait) which is weaker than and hidden by a dominant gene

COMMENT: since each physical characteristic is governed by two genes, if one gene is dominant and the other recessive, the resulting trait will be that of the dominant gene. Traits governed by recessive genes will appear if genes from both parents are recessive

reclaim [rɪˈkleɪm] *verb* to take virgin land or land which has already been developed and make it available for agricultural or commercial purposes

◊ **reclamation** [rekləˈmeɪʃn] *noun* (i) act of reclaiming land; (ii) land which has been reclaimed; *the authority is studying the costs of the land reclamation scheme in the city centre*

COMMENT: reclamation includes the drainage of marshes and lakes, and the improvement of heathland and moorland

rectified concentrated grape must [ˈrektɪfaɪd ˈkɒnsəntreɪtɪd ˈgreɪp mʌst] *noun* form of grape sugar produced by distillation from surplus wine, used to add to new wine during chaptalization

recumbent [rɪˈkʌmbənt] *adjective* lying down (as in the case of cows after illness or injury); *reduced phosphorus levels may also play a part in keeping affected animals recumbent*

RDC = RED DEER COMMISSION

RDS = RESEARCH AND DEVELOPMENT SERVICE

Red [red] *noun* English name for the Rouge de l'Ouest breed of sheep

Red Bororo [ˈred ˈbɒrərəʊ] *noun* West African longhorn breed of cattle. Able to survive in hardy conditions and able to cover great distances at remarkable speed. They are poor milkers, however

red clover [ˈred ˈkləʊvə] *noun* short-lived deep-rooting species of clover *(Trifolium pratense)*

COMMENT: clovers are essential plants for the longer ley and permanent pasture: the following red clovers are important: *broad red clovers:* suitable as annual or biennial crops, producing good hay or silage; *late-flowering red clovers:* used in long leys and hardier than the broad red clovers *extra-late-flowering clovers:* these can support intensive grazing by sheep

red currant [ˈred ˈkʌrənt] *noun* soft fruit, growing on bushes, and used mainly for making jams

◊ **Red Data Book** [ˈred ˈdeɪtə ˈbʊk] *noun* world catalogue of species which are rare or in danger of becoming extinct

◊ **red deadnettle** [ˈred ˈdednetl] *noun* weed *(Lamium purpureum)* which is common in gardens, and now affects cereals and oilseed rape (NOTE: also called **French nettle**)

◊ **red fescue** [ˈred ˈfeskjuː] *noun* species of grass *(Festuca rubra),* used on hill and marginal land and in fine-leaved lawns

redlegs *or* **redshank** [ˈredlegz *or*

'redʃæŋk] *noun* common weed *(Polygonum persicaria)* which affects spring crops, and causes problems when harvesting

◊ **Red List** ['red 'lɪst] *noun* list of pesticides issued by the Department of Environment, which are due for review on grounds of stricter safety measures

◊ **Red Poll** ['red 'pəʊl] *noun* dual-purpose breed of cattle, which originated in East Anglia; deep red in colour, with a white swish at the end of the tail

◊ **Red Sindhi** ['red 'sɪndi] *noun* breed of dairy cattle found in Pakistan; the animals are small, hardy and deep red in colour. The breed is now found in Malaysia, Burma, Japan and Brazil

◊ **Red Sokoto** ['red sə'kəʊtəʊ] *noun* West African breed of goat

◊ **Red Steppe** ['red 'step] *noun* dual-purpose breed of cattle found in the Ukraine; the animals are red in colour with yellow-white horns

◊ **redwater** ['redwɔːtə] *noun* parasitic disease of cattle transmitted by the common tick; the affected animal becomes very dull, feverish, salivates freely and often staggers and falls; the acute form of the disease is often fatal

◊ **redwood** ['redwʊd] *noun* American conifer, often growing to a very large size

redistribution of land [riːdɪstrɪ'bjuːʃn əv 'lænd] *noun* taking areas of land from large landowners and splitting it into smaller plots for peasant owners

reeds [riːdz] *noun* aquatic plants growing near the shores of lakes, used to make thatched roofs

◊ **reedbed** ['riːdbed] *noun* mass of reeds growing together

reel [riːl] *noun* part of the mechanism of a combine harvester, which holds the crop against the cutter bar for cutting; the reel directs the crop after it has been cut onto the cutter bar table or platform; most combines have a pick-up reel which can be adjusted to deal with inlaid or tangled crops

refection [rɪ'fekʃən] *noun* consumption by an animal of its own faeces

reference price ['refrəns 'praɪs] *noun* the minimum price at which certain fruit and vegetables can be imported into the EU

refine [rɪ'faɪn] *verb* to make more pure; *juice from the sugar cane is refined by boiling*

QUOTE when palm oil was sold overseas without refining, the free fatty acid content of the crude oil as produced by the mills was a very important quality parameter

The Planter

refrigerate [rɪ'frɪdʒəreɪt] *verb* to cool produce and keep it at a cool temperature; **refrigerated lorry** = special lorry which carries produce under refrigeration

◊ **refrigeration** [rɪfrɪdʒə'reɪʃn] *noun* method of prolonging the life of various foods by storing them at very low temperatures

◊ **refrigerator** [rɪ'frɪdʒəreɪtə] *noun* device for cooling produce and keeping it cool; **refrigerator ship** = ship which carries produce under refrigerated conditions

COMMENT: low temperature retards the rate at which food spoils, because all the causes of deterioration proceed more slowly. In freeze drying, the food has to be quick-frozen and then dried by vacuum, so removing the moisture. Pre-cooked foods should be cooled rapidly down to $-3°C$ and eaten within five days of production. Certain high-risk chilled foods should be kept below $5°C$; these foods include soft cheese and various pre-cooked products. Eggs in shells can be chilled for short-term storage (i.e. up to one month) at temperatures between $-10°C$ and $-16°C$. Bakery products, including bread, have storage temperatures between $-18°C$ and $-40°C$; bread goes stale quickly at chill temperatures which are above these. Potatoes in the form of pre-cooked chips can be stored at $-18°C$ or colder, but ordinary potatoes must not be chilled at all. Apples and pears can be kept in air-cooled boxes at between $-1°C$ and $+4°C$ (this is known as 'controlled temperature storage'). Lettuces and strawberries (which normally must not be chilled) can be kept fresh by vacuum cooling; celery and carrots can be chilled by hydrocooling

refund ['riːfʌnd] *noun* money which is paid back; **export refunds** = refunds made by the EU to farmers to compensate for a lower export price for produce

regenerate [rɪ'dʒenəreɪt] *verb* to replace by new growth; *a forest takes about ten years to regenerate after a fire*

◊ **regenerative** [rɪ'dʒenərətɪv] *adjective* which allows new growth to replace damaged tissue; **regenerative properties** = ability of something, such as an ecosystem, to recover from pollution

◊ **regeneration** [rɪgenə'reɪʃn] *noun* growing again (of vegetation on land which has been cleared); *grazing by herbivores prevents the regeneration of forests destroyed by fire*

regrowth [rɪ'grauθ] *noun* new growth after a plant has been cut down or burnt

regulate ['regjuleɪt] *verb* to control how something is done; to control the growth of a plant

◊ **regulations** [regju'leɪʃnz] *noun* (i) rules laid down by a government or official body; (ii) rules laid down by the Council of Ministers or the Commission of the EU, which have legal force in all member countries; for example, European regulations require that animals being transported should be rested, fed and watered every 24 hours

◊ **regulator** ['regjuleɪtə] *noun* thing which controls development; **growth regulator** = a chemical used to control the growth of certain plants; mainly used for weed control in cereals and grassland

rein [reɪn] *noun* long narrow strap, each end of which is attached to the bit; used to control a horse

REIS = RURAL ENTERPRISE INFORMATION SERVICE

remembrement [rə'mɒmbrəmɒŋ] *noun* French term for the enlargement of agricultural holdings

rendzina [ren'ziːnə] *noun* soil developed on chalk and limestone rocks characterized by its shallowness and lack of true subsoil

rennet ['renɪt] *noun* extract from the stomach of a calf; it contains the enzyme rennin, which clots milk; it is used in the production of certain milk products such as cheese

rennin ['renɪn] *noun* enzyme which makes milk coagulate in the stomach, so as to slow down the passage of the milk through the digestive system

rent [rent] **1** *noun* money paid to use a farm

or land for a period of time **2** *verb* to pay money to hire a farm *or* land for a period of time

COMMENT: since 1950, there has been a decline in the area and number of farm holdings which are rented in Great Britain. In 1950, rented agricultural land in England, Wales and Scotland accounted for 60% of the holdings. In 1993 the figure was 46%

replacement [rɪ'pleɪsmənt] *noun* the action of putting something in the place of something else; **replacement milk** = milk which is used to feed young animals which cannot be fed by their mothers (such as 'lamb replacement milk'); **replacement rate** = the rate of introduction of heifers into a dairy herd to replace ageing cows or cows with low milk yields

replant [riː'plɑːnt] *verb* to plant again; *after the trees were felled, the land was cleared and replanted with mixed conifers and broadleaved species;* **replant disease** = condition affecting apple trees planted in an orchard which has been grubbed out

repot [rɪ'pɒt] *verb* to take a plant out of its pot and plant it in another, changing or adding to the soil at the same time

reproduce [rɪprə'djuːs] *verb* **(a)** to produce offspring; *(of bacteria, etc.)* to produce new cells **(b)** to do a test again in exactly the same way

◊ **reproduction** [riːprə'dʌkʃn] *noun* process of making offspring *or* derived cells, etc.; **organs of reproduction** = REPRODUCTIVE ORGANS **asexual reproduction** = reproduction by taking cuttings of plants *or* by cloning

◊ **reproductive** [riːprə'dʌktɪv] *adjective* referring to reproduction; **reproductive organs** = parts of the bodies of animals which are involved in the conception and development of a foetus; **reproductive system** = arrangement of organs and ducts in the bodies of animals which produces spermatozoa and ova; **reproductive tract** = series of tubes and ducts which carry spermatozoa and ova from one part of the body to another

COMMENT: service by the male is only allowed by the females of most animals during the heat period or oestrus. This acts as a natural check on the breeding rate of animals; the length of the oestrus varies with the animal

reseed [rɪ'siːd] *verb* to reestablish a ley by sowing seed again

COMMENT: reseeding is carried out to improve permanent pasture; this is done by direct reseeding which involves sowing again without a cover crop, or by undersowing, where the seed mixtures are sown with another crop, usually a cereal

QUOTE with the cost of reseeding a sward costing about £165/ha - plus the value of herbage lost during establishment - even partial failure is expensive
Smallholder

reserve [rɪ'zɜːv] *noun* area of unspoilt land where no commercial exploitation is allowed; **game reserve** = area of land kept for raising wild animals for shooting; **nature reserve** = special area where the wildlife is protected

reservoir ['rezəvwɑː] *noun* area of water made by man, used for storing water for domestic or industrial use

residual [re'zɪdjuəl] **1** *adjective* remaining *or* which is left behind; **residual herbicide** = herbicide applied to the surface of the soil which acts through the roots of the plants; not only growing weeds are killed but also new plants as they germinate; **residual spraying** = spraying insecticide onto walls so that it will stay there and kill insects which come into contact with it

◊ **residue** ['rezɪdjuː] *noun* material left after a process has taken place; **pesticide residues** = amounts of pesticide that remain in the environment after spraying

resin ['rezɪn] *noun* sticky oil which comes from some types of conifer

◊ **resinous** ['rezɪnəs] *adjective* referring to resin

resist [rɪ'zɪst] *verb* to be strong enough to avoid being killed *or* attacked by a disease *or* by a pesticide

◊ **resistance** [rɪ'zɪstəns] *noun* **(a)** (i) ability of an organism not to get a disease; (ii) ability of a germ not to be affected by antibiotics; (iii) ability of a pest *or* weed not to be affected by a pesticide *or* herbicide; *the bacteria have developed a resistance to certain antibiotics; after living in the tropics his resistance to colds was low; increasing insect resistance to chemical pesticides is a major problem* **(b)** environmental resistance

= all the external factors like predators, competition, weather, food availability, which inhibit the potential growth of a population

◊ **resistant** [rɪ'zɪstənt] *adjective* able not to be affected by something; *the bacteria are resistant to some antibiotics; fungus-resistant genetic material was found in genetic stocks from Mexico;* **resistant strain** = strain of an insect which is not affected by pesticides *or* of bacterium which is not affected by antibiotics

◊ **-resistant** [rɪ'zɪstənt] *suffix* meaning which can resist; *DDT-resistant strain of insects*

COMMENT: resistant strains develop quite rapidly after application of the treatment. Some strains of insect have developed which are resistant to DDT. The resistance develops as non-resistant strains die off, leaving only individuals which possess a slightly different and resistant chemical makeup. Hence a pesticide will select out only resistant individuals. This can be avoided by using pesticides in combination or by not using the same chemical (or chemicals with a similar mode of action) repeatedly

respiration [respɪ'reɪʃn] *noun* action of breathing; **external respiration** = part of respiration concerned with oxygen in the air being exchanged in the lungs for carbon dioxide from the blood; **internal respiration** = part of respiration concerned with the passage of oxygen from the blood to the tissues, and the passage of carbon dioxide from the tissues to the blood

◊ **respiratory** [rɪ'spɪrətəri] *adjective* referring to breathing; **respiratory pigment** = blood pigment which can carry oxygen collected in the lungs and release it in tissues; **respiratory quotient (RQ)** = ratio of the amount of carbon dioxide taken into the alveoli of the lungs from the blood to the amount of oxygen which the alveoli take from the air; **respiratory system** = series of organs and passages which take air into the lungs, and exchange oxygen for carbon dioxide

COMMENT: respiration includes two stages: breathing in (inhalation) and breathing out (exhalation). Air is taken into the respiratory system through the nose or mouth, and goes down into the lungs through the pharynx, larynx, and windpipe. In the lungs, the bronchi take the air to the alveoli (air sacs) where oxygen in the air is passed to the bloodstream in exchange for waste carbon dioxide which is then breathed out

response [rɪs'pɒns] *noun* the beneficial reaction of a growing crop to the application of fertilizer; **response curve** = graph showing the yield (or some associated factor) against fertilizer input, level of feed, antibiotics, etc.

rest [rest] *verb* **to rest land** = to let land lie fallow, without growing any crops and to remove grazing animals from pasture to allow it time to recover

restore [rɪ'stɔː] *verb* to give back; *by letting the land lie fallow for a couple of years, farmers hope to restore some of the natural nutrients which have been removed from the soil*

retained afterbirth [rɪ'teɪnd 'ɑːftəbɜːθ] *noun* disease of cattle caused by interference at calving, premature calving or milk fever. The afterbirth should be removed by a veterinary surgeon

retard [rɪ'tɑːd] *verb* to make something happen later

◊ **retardant** [rɪ'tɑːdənt] *noun* substance which retards; **growth retardant** = substance used to slow or stop the growth of a plant

reticulum [re'tɪkjʊləm] *noun* series of small fibres forming a network in the second stomach of ruminants, connected to the rumen; its acts as a sort of filter and holds water and hard objects which have been swallowed (also known as the honeycomb stomach); *see also* ABOMASUM, OMASUM, RUMEN

retinol ['retɪnɒl] *noun* = VITAMIN A

retting ['retɪŋ] *noun* process used in the preparation of flax, where flax is soaked in water and allowed to rot, so freeing the fibres from the plant stems

reversible plough [rɪ'vɜːsɪbl 'plaʊ] *noun* plough with left- and right-handed mouldboards, which make it possible to plough up and down the same furrow; this gives a level surface and reduces the headland; very little marking is needed before the start of ploughing

QUOTE the problem with reversible ploughs, especially on some Irish soils, is that their weight can cause compaction
Practical Farmer

revolution [revə'luːʃn] *see* GREEN REVOLUTION

rhizome ['raɪzəʊm] *noun* fleshy plant stem which lies under the ground, and contains leaf buds (as opposed to a root which lies under the ground but is not a stem)

◊ **rhizomania** [raɪzəʊ'meɪnɪə] *noun* notifiable virus disease affecting sugar beet; hairs grow on the roots and the leaves turn yellow (the disease is endemic in the Netherlands, and some cases have been reported in the UK)

rhizosphere ['raɪzəʊsfɪə] *noun* the region immediately surrounding the roots of a plant

Rhode Island Red (RIR) ['raʊd aɪlənd 'red] *noun* heavy breed of fowl, with red feathers on the body, and black tail and wing feathers; it produces large brown eggs

RHS = ROYAL HORTICULTURAL SOCIETY

rhubarb ['ruːbɑːb] *noun* perennial plant (*Rheum rhaponticum*), of which the leaf stalks are cooked and eaten as dessert; it has a high oxalate content, and the leaves are toxic

rib grass ['rɪb 'grɑːs] *noun* palatable deep-rooting herb with a high mineral content, which may benefit pasture

◊ **ribwort** ['rɪbwɜːt] *see* PLANTAIN

Ribes ['raɪbiːs] *Latin for* blackcurrant

rice [raɪs] *noun* cereal grass (*Oryza sativa*) the most important cereal crop, and the staple food of half the population of the world

◊ **ricefield** ['raɪsfiːld] *noun* field for growing rice

COMMENT: wet rice is by far the commonest method of cultivation: the paddies are enclosed by low banks and are kept flooded during the growing season. They are allowed to dry out before the crop is harvested. Dry land rice is cultivated in a similar way to wheat or barley. Rice is classified according to the length of the grains: long-grain rice is grown in tropical climates, such as India, while short-grain rice is grown in colder climates such as Japan. There are over 120,000 varieties of rice grown world-wide; in India alone, more than 40,000 varieties are cultivated. Rice is an important crop in most countries of Asia, and is becoming increasingly important in Africa, South America, the USA and Australia. In Europe, Italy, France and Hungary grow considerable amounts of rice. The world's leading rice exporters are the USA and Thailand

QUOTE average rice yields in Egypt have recently increased from some 6.35 tonnes/hectare to 7.04 tonnes/hectare in 1989

Middle East Agribusiness

rich [rɪtʃ] *adjective* very fertile soil, soil which is full of nutrients

rick [rɪk] *noun* a stack, usually of hay, with a sloping roof

rickets ['rɪkɪts] *noun* disease of young animals due to deficiency of Vitamin D. Bones fail to ossify and joints become swollen

rickettsia [rɪ'ketsiə] *noun* bacteria which causes typhus and tsutsugamushi disease

riddle ['rɪdl] **1** *noun* coarse sieve for sieving soil **2** *verb* to grade and sort produce according to size, using a sieve; *potatoes are riddled to separate the best potatoes, called 'wares' from the small potatoes, called 'chats'*

ridge [rɪdʒ] **1** *noun* long raised section of earth; a ridge is made by ploughing up and down on either side of the furrow. In systematic ploughing, ridges first mark out land in a field before the plough is reset for normal work and the field is ploughed **2** *verb* to make a ridge while ploughing

◊ **ridger** ['rɪdʒə] *noun* type of plough used to form ridges for earthing up crops such as potatoes

rig [rɪg] *noun* male animal in which one or both testicles have not descended into the scrotum at the usual time

right of way ['raɪt ʌv 'weɪ] *noun* legal title to go across someone else's property

rill [rɪl] *noun* small channel eroded in soil by rainwater; it can be removed during ordinary cultivation

ring [rɪŋ] **1** *noun* metal circle which goes through the nose of an animal; **ring-barking**

= the cutting of a strip of bark from a tree as a means of making the tree more productive; it restricts growth and encourages fruiting; **ring bone** = growth of bony tissue in the joints of a horse's foot **2** *verb* to attach a ring to an animal, such as to the nose of a bull

COMMENT: some animals can be ringed to allow them to be led, while others are ringed to prevent excessive grubbing in the ground

◊ **ring rot** ['rɪŋ 'rɒt] *noun* disease affecting potatoes

◊ **ringworm** ['rɪŋwɜːm] *noun* any of various infections of the skin by a fungus, in which the infection spreads out in a circle from a central point (ringworm is very contagious and very difficult to get rid of). In animals, it is most common in young store cattle; it also affects humans

riparian [rɪ'peəriən] *noun* referring to the bank of a river; **riparian fauna** = animals which live on the banks of rivers

ripe [raɪp] *adjective* ready for eating (used of fruit and grain); *when the corn is ripe the harvest can start; the early varieties of apple are ripe in August; ripe peaches cannot be kept very long; bananas should be packed before they are ripe, and allowed to ripen during transport and storage* (NOTE: the opposite is **unripe)**

◊ **ripen** ['raɪpən] *verb (of fruit)* to become ready for eating; *unripe bananas are shipped in special containers and will ripen in storage; tomatoes can be picked when still pink and allowed to ripen off the plant*

ripper ['rɪpə] *noun* heavy cultivator consisting of a strong frame with long tines attached to it; it is used to break up compacted soil to allow free passage of air and water; also called a 'subsoiler'

QUOTE another way to break up the tillage pan is by using subsoilers or rippers

Practical Farmer

RIR = RHODE ISLAND RED

river ['rɪvə] *noun* large flow of water, running from mountains *or* hills down to the sea; **river system** = series of small streams and rivers which connect with each other; **river terrace** = flat plain left when a river cuts more deeply into the bottom of a valley, usually fertile and used for cultivation

◊ **riverine** ['rɪvəraɪn] *adjective* referring to a river; *the dam has destroyed the riverine fauna and flora for hundreds of kilometres*

roan [rəʊn] *noun* coat of an animal in which the main colour is mixed with another, as for example red and white, or black and white

robot milkers ['rəʊbɒt 'mɪlkəz] *noun* system of completely automated milking parlour; lasers, mirrors and cameras are used to put all four caps on the teats simultaneously

robusta coffee [rə'bʌstə 'kɒfiː] *noun* the coffee *Coffea canephora*, grown chiefly in West Africa

roca ['rəʊkæ] *noun* forest clearing in Brazil

◊ **roceiro** [rə'seərəʊ] *noun (in Brazil)* a subsistence farmer who cultivates a 'roca'

rock [rɒk] *noun* solid mineral substance which forms the outside crust of the earth; **rock phosphate** = natural rock ground to a fine powder, used as a fertilizer; *see also* PHOSPHATE; **rock salt** = natural salt used in salt licks

rod [rɒd] *noun* old measurement of land, both a measurement of length (equals 5 metres) and of area (equals 25 square metres)

rodent ['rəʊdənt] *noun* order of mammals including rabbits, rats and mice, which have sharp incisor teeth for gnawing; with over 6500 species, this is the largest group of living animals

roe deer ['rəʊ 'dɪə] *noun Capreolus capreolus,* one of the breeds of deer which are found wild in the UK

rogue [rəʊg] **1** *noun* inferior plant; a plant of a different variety found growing in a crop **2** *verb* to remove unwanted plants from a crop, usually by hand; **roguing glove** = glove impregnated with herbicide which is used to destroy wild oats

roll *or* **roller** [rəʊl *or* 'rəʊlə] *noun* tractor-drawn implement used for breaking clods, firming the soil, pushing stones into the soil and providing a smooth firm surface for

drilling; the two main types are the Cambridge roll, with a number of cast iron rings on an axle (these leave a corrugated surface) and the flat roll, which leaves a smooth surface; **roller crusher** = machine used to condition freshly-cut grass; the swath of cut grass is picked up by the rolls and the stems are flattened as the grass is passed between them; with the sap removed from the stems, the drying process is much faster; **roller mill** = equipment used in the preparation of flour and animal feed; it has two smooth steel rollers which crush the grain; **roller table** = machine, consisting of a horizontal line of rotating rollers, used for removing stones and clods from a crop such as potatoes

◊ **rolled grain** ['rəʊld 'greɪn] *noun* grain which has been through a roller mill before it is fed to livestock; rolled grain (usually barley) is more easily digested

Romagnola [rɒmæ'njəʊlə] *noun* large docile hardy breed of beef cattle from north-east Italy; the animals are grey with a black muzzle and hooves

Roman ['rəʊmən] *noun* breed of white goose, now quite rare

Romney ['rɒmni] *noun* hardy breed of sheep found in large numbers on Romney Marsh; it has heavy fine-quality long wool fleece; the Romney half-breed has been developed by crossing Romney ewes with North Country Cheviot rams. The breed has been widely exported. It is also called the 'Kent'

rook [rʊk] *noun* crow-like bird which causes much damage to crops

◊ **rookery** ['rʊkri] *noun* breeding place for a colony of rooks

roost [ruːst] **1** *noun* place where birds rest at night **2** *verb (of birds)* to sleep on a perch at night

◊ **rooster** ['ruːstə] *noun (especially US)* cock, male domestic fowl

root [ruːt] **1** *noun* part of a plant which is normally under the ground and absorbs water and nutrients from the surrounding soil; **root crops** = plants which store edible material in a root, corm or tuber; root crops used as food vegetables or fodder include carrots, parsnips, swedes and turnips; starchy root crops include potatoes, cassavas and yams. The foliage of some root crops is valuable fodder, and sheep are often 'folded' on them. No continent depends as much on root crops and food legumes as does Africa; **root cutting** = cutting taken from a rhizome which can be planted and used to propagate a plant; **root harvester** = machine for lifting root crops out of the ground, such as a sugar beet harvester; **root system** = all the roots of a plant **2** *verb* to produce roots; *the cuttings root easily in moist sand;* **rooting compound** = powder containing plant hormones (auxins) into which cuttings can be dipped to encourage them to root; **rooting depth** = depth of soil from which plant roots take up water, or the depth of soil to which roots reach

◊ **rootlet** ['ruːtlət] *noun* little root which grows from a main root

◊ **rootstock** ['ruːtstɒk] *noun* **(a)** rhizome, plant stem which lies under the ground and contains buds, as opposed to a simple root which is not a stem **(b)** stock on which a scion is grafted; **dwarfing rootstock** = stock which is normally low-growing so making the grafted tree grow smaller than normal

rosewood ['rəʊzwʊd] *noun* a tropical hardwood *(Caesalpinia)* used in furniture making

rosemary ['rəʊzməri] *noun* aromatic herb *(Rosemarinus officinalis)* used for flavouring; also a source of oil used in soaps and cosmetics

rot [rɒt] **1** *noun* one of a number of plant diseases caused by bacteria or fungi (such as brown rot) **2** *verb (of organic tissue)* to decay *or* to become putrefied (because of microbial action)

rotary cultivator *or* **rotavator** ['rəʊtəri 'kʌltɪveɪtə *or* 'rəʊtəveɪtə] *noun* mounted or trailed machine with a shaft bearing a number of L-shaped blades; used for stubble-clearing, seedbed work and general land reclamation and cleaning; **rotary mower** = machine used for cutting grass and other upright crops. Rotary mowers have two or four rotors each with three or four swinging blades; the rotors contra-rotate and leave a single swath of cut grass; **rotary milking parlour** = the most expensive and complex of the four milking systems, where the cows stand on a rotating platform with the milker in the middle; the operator may work on the inner or outer side of the circle; **rotary sprinkler** = machine used for irrigation purposes; sprinklers can be fitted with fine spray nozzles for protection of fruit crops and potatoes against frost damage

rotate [rəʊ'teɪt] *verb* to move in a circle; **rotating flails** = parts used on manure spreaders to distribute materials and on machines for cutting crops or grass verges. Used also in mixing machines such as composters; **rotating tines** = spikes used on machines such as rotavators and power harrows for cultivation purposes. They are also used on machine pick-ups

rotation [rəʊ'teɪʃn] *noun* moving in a circle; **rotation of crops** *or* **crop rotation** = system of cultivation, where different crops are planted in consecutive growing seasons to prevent the nutrients in the soil being totally used up (each plant uses different nutrients). Rotation also reduces the effects of pests and diseases; **Norfolk rotation** = system for farming, using arable farming for fodder crops, and involving the temporary sowing of grass and clover; the system was introduced into England in the early 18th century; the rotation involved root crops (turnips or swedes), then cereal (barley), followed by ley (usually red clover), and ended with cereal (usually wheat)

◊ **rotational grazing** [rəʊ'teɪʃnəl 'greɪzɪŋ] *noun* movement of livestock around a number of fields or paddocks in an ordered sequence (NOTE: also called **on-off grazing)**

COMMENT: the advantages of rotating crops are firstly that pests particular to one crop are discouraged from spreading, and secondly that some crops actually benefit the soil. Legumes (peas and beans) increase the nitrogen content of the soil if their roots are left in the soil after harvesting. If the rotation includes a ley, the system is known as alternative husbandry or mixed farming. One of the best-known rotations was the Norfolk four-course system. A rotation should increase and maintain soil fertility, control weeds and pests, decrease the risk of crop failure and employ labour throughout the year

rotavator ['rəʊtəveɪtə] *noun* = ROTARY CULTIVATOR

rotenone ['rəʊtənəʊn] *noun* insecticide derived from derris

Rothamsted ['rɒθəmsted] site of the Agricultural Experimental station, established in 1843 by John Bennett Lawes; the station specialized in research into plant nutrition; it demonstrated the importance of nitrogen, phosphorus and potassium to plants. Today it is important for its research into biotechnology and is the largest centre of the AFRC Institute of Arable Crops Research

Rouen ['ruːɒŋ] noun breed of table duck; the drake has a green head and neck, rich claret-coloured breast and grey-black body. The female is mostly brown

Rouge de l'Ouest ['ruːʒ də 'lwest] noun breed of sheep originating in France (also called 'Red' in the UK)

rough [rʌf] adjective not smooth; **rough grazing** = unimproved grazing, found in mountain, heath and moorland areas; **rough stalked meadow grass** = type of grass, highly palatable but low in production compared to ryegrass; common in lowland pastures on rich moist soils. When found in cereal crops it is treated as a weed

◊ **roughage** ['rʌfɪdʒ] noun animal feedingstuffs with high fibre content, such as hay or straw

◊ **Rough Fell** ['rʌf 'fel] noun hardy moorland breed of horned sheep, closely related to the Swaledale; it has a dark-coloured face with irregular patterns. The wool is of coarse quality

round and round ploughing ['raʊnd nd 'raʊnd] noun system of ploughing in which fields are ploughed from the centre to the outside or from the edge to the centre

round baler ['raʊnd 'beɪlə] noun tractor-drawn machine which straddles the swath with a pickup cylinder; the crop is passed over a system of belts to form a round bale; when the bale is complete, twine is wrapped round it and it is thrown out of the machine

rounds [raʊndz] noun circular walls built to protect sheep from snow drifts

roundworm ['raʊndwɜːm] noun type of worm with a round body, some of which are parasites of animals, others of roots of plants; very common in pigs, it causes digestive problems and jaundice; in lambs, it causes Nematodirus disease; see also PARASITIC GASTRO-ENTERITIS

Roussin ['ruːsæŋ] noun breed of sheep imported into the UK from France

row [rəʊ] noun line of plants

◊ **row crop** ['rəʊ 'krɒp] noun crop planted in rows wide enough to allow cultivators between the rows (most farm crops are drilled in rows, in preference to broadcasting); **row crop tractor** = tractor used by market gardeners and farmers who grow row crops; a lightweight tractor with a narrow turning circle and adjustable wheel track widths

Royal Agricultural Society of England (RASE) organization whose main task is running the annual Royal Show held at The National Agricultural Centre, Stoneleigh, Kenilworth, Warwickshire

> QUOTE up to £6 million is to be spent on improvements to the layout and infrastructure of the Royal showground. Plans are to increase the area devoted to horses and horticulture in 1995
> *Farming News*

Royal College of Veterinary Surgeons (RCVS) body which organizes the examinations for veterinary surgeons and represents them. In 1994, the RCVS admitted 620 new members

Royal Horticultural Society (RHS) national society which organizes the Chelsea Flower Show and has permanent gardens at Wisley in Surrey

Royal Society for the Protection of Birds (RSPB) British society whose aim is the conservation of birds and their habitats

royalty ['rɔɪəlti] noun payments made to plant breeders for the use of seed of registered plant varieties

> QUOTE all seed growers have to pay plant royalties, but ware growers and merchants avoid this
> *Farmers Weekly*

RQ = RESPIRATORY QUOTIENT

RTV = ROUGH TERRAIN VEHICLE

rubbed seed ['rʌbd 'siːd] noun seed such as sugar beet which is formed of a cluster of seeds and can be separated out by rubbing (NOTE: also called **graded seed**)

rubber ['rʌbə] *noun* **(a)** material which can be stretched and compressed, made from a thick white liquid (latex) from a tropical tree; Malaysia is the leading world producer of rubber, followed by Indonesia and Sri Lanka **(b)** *Hevea brasiliensis,* the rubber tree, a tropical tree grown for its latex; in commercial practice, trees are grafted onto suitable rootstock

Rubus ['ru:bəs] *noun* genus of plants including cane fruits such as raspberries and blackberries

ruddle ['rʌdl] *noun* red colouring material on a harness worn by rams so that ewes which have been mated will be marked and identified

Rules of Good Husbandry ['ru:lz ʌv gʊd 'hʌzbəndri] *noun* unwritten set of 'rules' which, if they are deemed to have been broken by a tenant, can give a landlord the excuse to evict him

rumen ['ru:mən] *noun* the first stomach of ruminating animals such as cows, sheep or goats, all of which have four stomachs. It is used for storage of food after it has been partly digested and before it passes to the second stomach; *see also* ABOMASUM, OMASUM, RETICULUM

ruminant ['ru:mɪnənt] *noun* animal (such as cow, sheep or goat) which chews cud

◊ **rumination** [ru:mɪ'neɪʃn] *noun* process by which food taken to the stomach of a ruminant is returned to the mouth, chewed again and then swallowed

COMMENT: ruminants have stomachs with four sections, the rumen, the reticulum, the omasum and the abomasum. They take in foodstuffs into the upper chamber where it is acted upon by bacteria. The food is then regurgitated into their mouths where they chew it again before passing it to the last two sections where normal digestion takes place

run [rʌn] **1** *noun* **(a)** enclosure for animals, such as a chicken run **(b)** *(New Zealand & Australia)* extensive area of land used for sheep grazing **2** *verb* to keep animals; *we run 400 sheep on the common grazing land*

◊ **runholder** ['rʌnhəʊldə] *noun (in New Zealand & Australia)* farmer who owns a sheep run

runch [rʌnʃ] *noun* common weed *(Raphanus raphanistrum)* also known as the wild radish

runner ['rʌnə] *noun* long lateral shoot of a plant, such as a strawberry, ending in a tuft of leaves which will take root; **runner bean** = garden bean *(Phaseolus coccineus)* grown exclusively for the fresh trade

runoff ['rʌnɒf] *noun* **(a)** (i) the portion of rainfall which finally reaches a stream; (ii) the running of rainwater or melted snow from the surface of land into streams; *runoff is faster and greater during heavy rain than during a long drizzle and is greater on clay than on sandy soil; saturated soil can lead to surface runoff and flooding* **(b)** flowing of excess fertilizer or pesticide from farmland into rivers; *nitrate runoff causes pollution of lakes and rivers; fish are extremely susceptible to runoff of organophosphates; fertilizer runoff, such as runoff of phosphates, can cause problems of eutrophication*

runt [rʌnt] *noun* small individual animal, one that it is smaller than average for its kind; the smallest animal in a litter

RURAL = RESPONSIBLE USE OF RESOURCES IN AGRICULTURE AND ON THE LAND

rural ['ru:rəl] *adjective* referring to the country, as opposed to the town; *many rural areas have been cut off by floods; the government is planning an electrification project for rural areas;* **rural depopulation** = loss of population from the countryside due to various causes, including decline in agriculture and increased mechanization

COMMENT: it has been estimated (1992) that some 50,000-60,000 farming families will go out of business within the next decade

rush [rʌʃ] *noun* common weed *(Juncus)* growing near water, in moors and marshes, of little nutritional value

russet ['rʌsɪt] *noun* type of dessert apple with a rough brown skin

◊ **russeting** ['rʌsɪtɪŋ] *noun* formation of brown patches on the skin of an apple; *Cox's Orange often have some russeting on them, the amount depending on the weather conditions*

rust [rʌst] *noun* type of fungus disease

which gives plants a powdery surface; some types are particularly serious among cereal crops, such as yellow rust on barley; *see also* BROWN RUST

rustle ['rʌsl] *verb* to steal livestock (especially cows and horses)

◊ **rustler** ['rʌslə] *noun* person who steals livestock; *a cattle rustler*

◊ **rustling** ['rʌslɪŋ] *noun* crime of stealing cattle *or* horses

rut [rʌt] *noun* **(a)** period of intense sexual activity in males of various mammals, such as cattle, sheep and particularly deer **(b)** deep track left by wheels in soft ground

rye [raɪ] *noun* hardy annual grass *(Secale cereale)* a cereal crop grown in temperate areas (Poland, Germany, etc.) and used for making bread flour; rye flour is also used to make crispbreads. It is dark and the dough lacks elasticity. Another use for rye is for malting for whisky. Rye is also used for very early grazing; the long straw is good for thatching

ryegrass ['raɪgrɑːs] *noun* term for a most important group of grasses; **Italian ryegrass** = short lived ryegrass *(Lolium multiflorum);* it is sown in spring, and is very quick to establish; it produces good growth in its seeding year and early graze the following year; it is commonly used for short duration leys; **perennial ryegrass** = a grass *(Lolium perenne)* which forms the basis of the majority of long leys in the UK; the most important grass in good permanent pasture and often sown mixed with other grasses and clover; perennial ryegrass has a long growing season, is quick to become established and responds well to fertilizers. It is best suited to grazing and is highly palatable for animals

COMMENT: many varieties of hybrid ryegrass are now used; they are crosses between perennial and Italian ryegrasses, and often also tetraploids

Ryeland ['raɪlənd] *noun* rare breed of sheep; a medium-sized animal, white faced and without horns; the sheep has a very symmetrical shape and a thick growth of wool

Ss

S *chemical symbol for* sulphur

Saanen ['sɑːnən] *see* BRITISH SAANEN

sack [sæk] *noun* measure of capacity, particularly for cereals, equal to four bushels (NOTE: also called a **coomb**)

saddle ['sædl] *noun* **(a)** leather seat for a rider placed on the back of a horse; **saddle bow** = high part of a saddle in front of the rider **(b)** coloured patch on the back of an animal (such as a pig) which looks a little as if the animal is carrying a saddle **(c)** part of a shaft-horse's harness that bears the shafts

◊ **saddleback** ['sædlbæk] *noun* (i) a breed of pig now known as the British Saddleback; (ii) any pig with a white saddle, such as the American-bred Hampshire breed

◊ **British Saddleback** ['brɪtɪʃ 'sædlbæk] *noun* breed of pig, derived from the Wessex, Hampshire and Essex breeds; it is coloured black with a white band round the front of the body. Used for breeding hardy crosses

safety cab ['seɪfti 'kæb] *noun* protective cab fitted to a tractor to prevent injury to the driver if the tractor turns over; it also reduces the noise levels affecting the driver

safflower ['sæflaʊə] *noun* an oilseed crop (*Carthamus tinctorius*) grown mainly in India; the oil is used in the manufacture of margarine, and the residual oilseed cake has a limited use as a livestock feed

saffron ['sæfrən] *noun* spice obtained from the dried flowers of the crocus plant *Crocus sativus*

sage [seɪdʒ] *noun* aromatic herb (*Salvia officinalis*); the leaves are dried and used for flavouring

sago palm ['seɪgəʊ 'pɑːm] *noun* *Metroxylon saga*, tree native of South-East Asia; the seeds give sago flour and pearl sago used in desserts

Sahel [sə'hel] *noun* semi-desert region south of the Sahara in the process of being desertified through drought, overcropping and overgrazing of marginal land

Sahiwal ['sæhɪwæl] *noun* type of cattle found in Bangladesh

Saidi [sæ'iːdi] *noun* type of Egyptian cattle; *see also* EGYPTIAN

sainfoin ['sænfɔɪn] *noun* *Onobrychis sativa*, a forage legume very similar to lucerne, grown mainly in areas with calcareous soil

Saler ['seɪlə] *noun* hardy breed of French cattle, found in the Cantal department of central France; The animals are reddish in colour and are reared both for meat and for milk production; it is one of the best French suckler cows

saline ['seɪlaɪn] *adjective* referring to salt; *salt marshes have saline soil*

◊ **salinity** [sə'lɪnɪti] *noun* proportion of salt (sodium chloride) present in a given amount of water or soil; *rising salinity has now become a severe health hazard*

◊ **salination** *or* **salinization** [sælɪ'neɪʃn *or* sælɪnaɪ'zeɪʃn] *noun* process by which soil or water becomes more salty, found especially in hot countries where irrigation is practised

◊ **salinized** ['sælɪnaɪzd] *adjective* (land) where evaporation leaves salts as a crust on the dry surface

saliva [sə'laɪvə] *noun* fluid in the mouth, secreted by the salivary glands, which starts the process of digesting food

◊ **salivary** [sə'laɪvəri] *adjective* referring to saliva; **salivary digestion** = the first part of the digestive process, which is activated by the saliva in an animal's mouth; **salivary gland** = gland which secretes saliva

◊ **salivate** ['sælɪveɪt] *verb* to produce saliva

◊ **salivation** [sælɪ'veɪʃn] *noun* production of saliva

COMMENT: saliva is a mixture of a large quantity of water and a small amount of mucus, secreted by the salivary glands. Saliva acts to keep the mouth and throat moist, allowing food to be swallowed easily. It also contains the enzyme ptyalin, which begins the digestive process of converting starch into sugar while food is still in the mouth

Salmonella [sælmə'nelə] *noun* genus of bacteria in the intestines, which are acquired by eating contaminated food, and which cause typhoid fever and food poisoning; found in eggs from infected hens, but destroyed by heating. Many types of salmonella exist, some of which cause infectious diseases in livestock; *Salmonella pullorum* causes bacillary white diarrhoea in chicks

◊ **salmonellosis** [sælmənə'ləʊsɪs] *noun* disease caused by Salmonella bacteria

QUOTE Salmonellae are the most prevalent cause of bacterial food poisoning in this country
Farmers Weekly

salsify ['sælsɪfi] *noun* *Tragopogon porrifolius*, plant with a long white root which is used as a vegetable

salt [sɔːlt] **1** *noun* **common salt** = sodium chloride (NaCL), used as a cheap fertilizer for sugar beet and carrots; also used to preserve food; **salt lick** = block of salt given to cattle to lick in hot weather; **salt marsh** = marsh near the sea, formed with sea water; **salt pan** = area where salt from beneath the soil surface rises to form crystals on the surface **2** *verb* to preserve food by keeping it in salt or in salt water; cabbage, gherkins, ham and many types of fish are salted for preservation

◊ **salt poisoning** ['sɔːlt 'pɔɪzənɪŋ] *noun* disease of pigs usually caused by inadequate provision of water. May also be caused by increased salt in the ration. Pigs become constipated before twitching, fits and death

◊ **salty** ['sɔːlti] *adjective* full of salt; tasting of salt; *excess minerals in fertilizers combined with naturally saline ground to make the land so salty that it can no longer produce crops*

COMMENT: salt forms a necessary part of diet, as it replaces salt lost in sweating and helps to control the water balance in the

body. It also improves the working of the muscles and nerves. An adequate intake of sodium chloride is necessary to all animals. Grass is relatively poor in sodium and its high potassium content induces secretion of sodium in urine; this loss causes a craving for salt which may be supplied by salt licks

QUOTE vitamins and minerals are an essential part of a goat's diet. Usually a basic all-purpose mineral, plus a salt lick (goats have a high requirement for salt) is sufficient
Smallholder

sand [sænd] *noun* fine grains of weathered rock, usually round grains of quartz, and a constituent of soil

◊ **sandstone** ['sændstəʊn] *noun* type of sedimentary rock, formed of particles of quartz

◊ **sandy soils** ['sændi 'sɔɪlz] *noun* soil containing a high proportion of sand particles (approximately 50%). Sandy soil feels gritty. These soils drain easily and are naturally low in plant nutrients through leaching. They are often called 'light' soils, as they are easy to work and also 'hungry' soils since they need fertilizer. Market gardening is particularly well-suited to sandy soils

Santa Gertrudis ['sæntæ gɜː'truːdɪs] *noun* breed of beef cattle developed in Texas, USA. A cross between the Brahman and Shorthorn stock; the animals are red in colour. The breed is now found in many tropical countries

SAOS = SCOTTISH AGRICULTURAL ORGANISATION SOCIETY

sap [sæp] *noun* liquid which flows inside a plant carrying nutrients and water

◊ **sapling** ['sæplɪŋ] *noun* young tree

◊ **sappy** ['sæpi] *adjective* (wood) which is full of sap

◊ **sapwood** ['sæpwʊd] *noun* outer layer of wood on the trunk of a tree, which is younger than the heartwood inside and carries the sap; *see also* HEARTWOOD

sapele [sə'piːli] *noun* fine African hardwood, formerly widely exploited but now becoming rarer

sapodilla [sæpə'dɪlə] *noun* *Achas sapota*,

tropical fruit grown in most Central American countries. The tree also yields the chicle gum used in making chewing gum

sapro- ['sæprəʊ] *prefix* meaning decay *or* rotting

◊ **saprobe** ['sæprəʊb] *noun* microbe that lives in rotting matter

◊ **saprogenic** *or* **saprogenous** [sæprəʊ'dʒenɪk *or* sə'prɒdʒənəs] *adjective* (organism) which grows on decaying organic matter

◊ **saprophagous** [sə'prɒfəgəs] *adjective* (animal) which lives on decaying matter

◊ **saprophyte** ['sæprəfaɪt] *noun* organism (such as a fungus) which lives on dead *or* decaying tissue

◊ **saprophytic** [sæprə'fɪtɪk] *adjective* (organism) which lives on dead *or* decaying tissue and plays an essential part in helping to break down plant remains into humus

Sardinian [sɑː'dɪnɪən] *noun* important breed of Italian sheep which provides milk for the manufacture of cheese

sarson ['sɑːsən] *noun* type of oilseed rape grown in India

COMMENT: a new variety of yellow sarson has been developed at the Division of Genetics, Indian Agricultural Research Institute, New Delhi. This variety matures in 132 days

sarvah ['sɑːvə] *noun* intensive subsistence system of rice production found in south and south-east Asia

SAS = SET ASIDE SCHEME

satellite ['sætəlaɪt] *noun* man-made device that orbits the earth, receiving, processing and transmitting signals and generating images such as pictures of the earth and weather pictures. Information from satellites is being used to produce field yield maps and is also used to monitor farms during the growing season. The results can be matched to claims by farmers for crops and set aside. The move is part of an initiative to reduce farming fraud

sativus *or* **sativa** *or* **sativum** ['sætɪvəs *or* 'sætɪvə *or* 'sætɪvəm] *Latin word meaning*

'sown' *or* 'planted', used in the generic names of many plants

saturate ['sætʃəreɪt] *verb* take up *or* dissolve the maximum amount of something, e.g. to fill a substance so that no more liquid can be absorbed; *nitrates leach from forest soils, thus showing that the soils are saturated with nitrogen;* **saturated fat** = fat which has the largest amount of hydrogen possible

COMMENT: animal fats such as butter and fat meat are saturated fatty acids, and contain large amounts of hydrogen. It is known that increasing the amount of unsaturated and polyunsaturated fats (mainly vegetable fats and oils, and fish oil), and reducing saturated fats in the food intake helps reduce the level of cholesterol in the blood

saturation [sætʃə'reɪʃn] *noun* (i) result of saturating; (ii) point where air contains 100% humidity

COMMENT: when continuous rainfall occurs for long periods, the soil becomes saturated with water; no more water can enter the soil and it ponds on the surface or runs off down slopes

savanna *or* **savannah** [sə'vænə] *noun* dry grass-covered plain with few trees (usually referring to South America or Africa); growth is abundant during the rainy season but vegetation dies back during the dry season

savory ['seɪvəri] *noun* Mediterranean aromatic herb *(Satureia hortensis)*

savoy [sə'vɔɪ] *noun* type of winter cabbage with crinkly leaves

sawah ['sɑːwæ] *noun* system of irrigated rice farming on terraces, in Indonesia

sawdust ['sɔːdʌst] *noun* powder produced when sawing wood; sawdust is used both as a mulch for plants and as bedding for animals

sawfly ['sɔːflaɪ] *noun* family of insects, the larvae or caterpillars of which cause serious damage to fruit and crops; they include the Apple Sawfly, Gooseberry Sawfly, Pear and Cherry Sawfly and Rose Sawfly

saw-toothed beetle ['sɔːtuːθt 'biːtl]

noun dark brown beetle which lives in stored grain; the eggs are laid in the grain and the larvae feed on the grain, causing mould

scab [skæb] *noun* **(a)** disease of which the scab is a symptom; it affects the skin of animals; **sheep scab** = serious disease of sheep, caused by a parasitic mite, which results in intense irritation, skin ulcers, loss of wool and emaciation; a notifiable disease **(b)** fungal disease of fruit and vegetables, including potato scab and apple and pear scab

QUOTE after its elimination in the 1950s, sheep scab reappeared in 1973 and two compulsory dipping periods were brought in during 1976
Farmers Weekly

scabies ['skeɪbiːz] *noun* very irritating infection of the skin caused by a mite which lives under the skin

scald [skɔːld] *noun* **(a)** defect in stored apples, where brown patches appear on the skin and the tissue underneath becomes soft **(b)** bacterial disease of sheep; it causes lameness in lambs

scale [skeɪl] *noun* thin membranous leaf structure; **scale insect** = flat parasitic insect that secretes a protective scale around itself and which lives on plants

◊ **scaly** ['skeɪli] *adjective* covered in scales; **scaly leg** = disease of affecting the legs of poultry; caused by a mite which burrows under the leg scales causing considerable itching; large hard scales develop on the unfeathered parts of the legs

scar [skɑː] *noun* mark on the skin of animal or fruit, or on the bark of a tree, where a wound has healed

scarecrow ['skeəkrəʊ] *noun* figure shaped like a man, wearing old clothes, put in fields to keep birds way from growing crops

scarify ['skærɪfaɪ] *verb* **(a)** to stir the surface of the soil with an implement with tines, such as a wire rake, but without turning the soil over; lawns can be scarified to remove moss and matted grass **(b)** to slit the outer coat of seed in order to speed up germination

◊ **scarifier** ['skærɪfaɪə] *noun* machine with tines which stirs the soil surface without turning it over

Schistosoma [ʃɪstəˈsəʊmə] *noun* Bilharzia, a fluke which enters the blood stream and causes bilharziasis

Schleswig-Holstein system [ʃlezwɪɡˈhɒlsteɪn ˈsɪstəm] *noun* system of cereal cultivation practised in North Germany, giving high average yields

COMMENT: the system involves careful management of the crop and includes high seed rates and high amounts of fertilizer. Crops are carefully monitored and visited each day. Disease is controlled by spraying

Schwarzkopf ['ʃwɑːtskɒpf] *noun* breed of German sheep found mainly in Hesse and Westphalia

sciarid fly ['saɪərɪd 'flaɪ] *noun* pest *(Bradysia)* affecting greenhouse pot plants; the larvae feed on fine roots causing plants to wilt

scion ['saɪən] *noun* plant which is grafted onto a stock; the variety grafted or budded onto a rootstock to make a tree or bush

Scots pine ['skɒts 'paɪn] *noun Pinus sylvestris,* common commercially grown European conifer

Scottish Blackface ['skɒtɪʃ 'blækfeɪs] *noun* very hardy breed of small mountain sheep; the fleece gives a long coarse springy wool, valued for making carpets. Older ewes are crossed with Border Leicester rams to give Greyface hybrids

◊ **Scottish halfbreed** ['skɒtɪʃ 'hɑːfbriːd] *noun* crossbred type of sheep obtained by using a Border Leicester ram on a Cheviot ewe; they are used widely in lowland Britain

scour [skaʊə] *verb* to wash wool to remove grease and contaminants

◊ **scouring** *or* **scours** ['skaʊrɪŋ *or* 'skaʊəz] *noun* diarrhoea in livestock; it may be a symptom of other diseases such as Johne's disease, dysentery or coccidiosis; it may simply be due to a chill or to poor diet

SCPS = SUCKLER COW PREMIUM SCHEME

scraper ['skreɪpə] *noun* steel-framed attachment for a tractor; it has a rubber scraping edge, and is used for heavy duty work, clearing slurry from farmyards

scrapie ['skreɪpi] *noun* brain disease of sheep; animals twitch, followed by intense itching and thirst; the animal becomes extremely thin and death follows; *see note at* BOVINE SPONGIFORM ENCEPHALOPATHY

scratch plough ['skrætʃ 'plaʊ] *noun* ard, plough made of a sharpened piece of wood which is dragged across the surface of the soil; this was the first type of plough to be used for cultivating the land (first used in South-West Asia)

screen [skri:n] **1** *noun* hedge *or* row of trees grown to shelter other plants *or* to protect something from the wind *or* to prevent something from being seen **2** *verb* **(a)** to protect plants from wind (as by planting windbreaks) **(b)** to pass grain through a sieve to grade it

◊ **screenings** ['skri:nɪŋz] *noun* grains which are small and pass through the sieve when screening grain

screwworm ['skru:wɜ:m] *noun* *Cochliomya hominivorax*, a fly similar to the bluebottle, but dark green in colour. It is common in Central and South America; it devastated cattle in the USA in the 1950s, but has now been eradicated there. It is found in North Africa and is a particular threat to cattle in tropical regions. The fly lays its eggs in open wounds, where maggots develop very quickly. It can kill an animal within a few weeks if left untreated

scrub [skrʌb] *noun* vegetation consisting mainly of stunted trees and bushes

◊ **scrubland** ['skrʌblænd] *noun* land covered with small trees and bushes

◊ **scrub typhus** ['skrʌb 'taɪfəs] *noun* tsutsugamushi disease, a severe form of typhus found in Southeast Asia, caused by the Rickettsia bacterium and transmitted to humans by mites

QUOTE the vegetation is dominated by stands of pine with a thick understorey of evergreen scrub oaks and densely branched shrubs. This is the true scrub
Natural History

SCU = SCOTTISH CROFTERS UNION

scurs [skɜ:z] *noun* small horns that are not part of the animal's skull but are attached to the skin

scutch [skʌtʃ] *noun* common name for couch grass

scythe [saɪð] *noun* hand implement with a long slightly curved blade attached to a handle with two short projecting hand grips; now used for cutting grass, and formerly for reaping

SDA = SCOTTISH DEVELOPMENT AGENCY

Se *chemical symbol for* selenium

sealed silage ['si:ld 'saɪlɪdʒ] *noun* silage produced in airtight containers; silage is stored in wrapped big bales; it is also common practice to use a wall made of old railway sleepers with plastic sheeting as lining; old car tyres are often placed on top of the plastic sheeting to keep it weighed down

seakale ['si:keɪl] *noun* *Crambe maritima*, plant of the mustard family whose cabbage-like leaves are used as vegetable

season ['si:zn] *noun* **(a)** one of the four parts into which a year is divided (spring, summer, autumn, and winter) **(b)** time of year when something happens; **growing season** = time of year when a certain plant grows; **rainy season** *or* **wet season** = time of year when most rain falls **(c)** the oestrus period of a female animal

◊ **seasonal** ['si:zənəl] *adjective* referring to a season; occurring at a season; *seasonal changes in temperature; plants grow according to a seasonal pattern*

seaweed ['si:wi:d] *noun* general name for several species of large algae, growing in the sea and usually rooted to a surface; several species of seaweed are used as human food, particularly in the Far East and Wales (where it is called 'laver' and eaten at breakfast); seaweeds are also used as animal feeds, as soil conditioners and fertilizers. Seaweed contains nitrogen, phosphate and potash in addition to salt. The organic matter in seaweed breaks down very quickly. It is also a source of iodine and other trace elements, as well as of vitamins A,B,C and D

Secale [sɪ'keɪli] *Latin name for* rye

secateurs [sekə'tɜ:z] *noun* cutting tool, like small shears with sharp curved blades, used for pruning

second ['sekənd] **1** *verb* to hoe between rows of rootcrops that have previously been thinned out **2** *noun* **seconds** = grain of medium size **3** *adjective* coming after the first; **second cut** = grass cut a second time in the season for hay or silage; **second early potatoes** = crop of potatoes that follows the first early crop

◊ **secondary** ['sekəndrı] *adjective* **secondary host** = some parasites need two hosts: the secondary host is one in which the parasite completes its life cycle; *compare* PRIMARY HOST; **secondary substances** = chemical substances found in plant leaves, believed to be a form of defence against herbivores; **secondary succession** = succession that takes place in an area where a natural community existed and was removed

secrete [sɪ'kriːt] *verb* to produce as a liquid; *the cut branch will secrete a type of sticky sap*

◊ **secretion** [sɪ'kriːʃn] *noun* liquid which is secreted

sedentary agriculture ['sedəntrı 'ægrɪkʌltʃə] *noun* subsistence agriculture practised in the same place by a settled farmer

sedge [sedʒ] *noun* one of a number of grass or rushlike herbs of the family *Cyperaceae;* common in marshlands and poorly drained areas; they have minimal nutritional value

sedimentary rocks [sedɪ'mentərı 'rɒks] *noun* rocks which were formed by deposition of loose material, such as sand and gravel, mainly in water

seed [siːd] **1** *noun* fertilized ovule which forms a new plant on germination; **seed bank** = (i) all the seeds existing in the soil; (ii) collection of seeds from plants, kept for research purposes; **seed certification** = the testing, sealing and labelling of seed sold to farmers; this ensures that the seed is free from disease and from weeds; **seed coat** = testa, the outer layer which surrounds a seed and protects the embryo; **seed corn** = cereal grown to give grain which is used as seed; **seed dressing** = treatment of seeds with a fungicide and/or an insecticide to prevent certain soil and seed-borne diseases; **seed leaf** = cotyledon, first thick leaf on a plant as it germinates; **seed mixture** = seeds of different plants supplied by seed merchants to farmers to produce a new ley; it will include grasses and legumes;

seed onion *see* ONION SET **seed potato** = potato tuber which is sown to produce new plants; in the UK, these are grown mainly in Scotland, and produced under a certification scheme; **seed rate** = amount of seed sown per hectare shown as kilos per hectare (kg/ha); **seed ripeness** = stage at which the seed can be harvested successfully; **seed royalties** = money paid by seed growers to breeders of seeds; **seed weevil** = pest affecting brassica seed crops; seeds are destroyed in their pods by the larvae **2** *verb* **(a)** to sow seeds in an area; *the area of woodland was cut and then seeded with pines;* **seeding year** = year during which a ley is sown with grass; *(of garden plant)* **to seed itself** = to sow seeds naturally and grow the following year; *the poppies seeded themselves all over the garden; several self-seeded poppies have come up in the vegetable garden* **(b)** to encourage rain to fall by flying over a cloud and dropping crystals onto it

COMMENT: EU regulations require all seed sold to farmers to be tested and to be guaranteed to certain standards of purity and free from pests and diseases

seedbed ['siːdbed] *noun* area of land tilled to produce a fine tilth, firm and level, into which seeds will be sown; some crops, such as potatoes, do not need a fine tilth and a rough damp bed is preferable; **seedbed wheels** = set of wheels bolted onto the front of a tractor which will give even compaction and a uniform sowing depth

seed-borne ['siːdbɔːn] *adjective* carried by seeds; **seed-borne disease** = disease which is carried in the seed of a plant

seedbox ['siːdbɒks] *noun* **(a)** box in which seeds can be planted for cultivation in a greenhouse **(b)** part of the plant head which contains the seeds

seedcase ['siːdkeɪs] *noun* the hard shell round certain seeds

seed drill ['siːd 'drɪl] *noun* machine consisting of a hopper carried on wheels with a feed mechanism which delivers grain to seed tubes

COMMENT: most seed drills are designed to plant seed in prepared seedbeds; drills for cereals and grasses sow the seed at random while precision drills, used mainly for sugar beet and vegetable crops, place seed at preset intervals in the rows; these are also called 'seeder units'. Grain drill feed mechanisms may be internal force feed, external force feed or studded roller. Some drills have a hopper divided into two parts: one contains the seed, the other fertilizer; this is the combine drill which drills grain and fertilizer at the same time

seeder ['si:də] *noun* machine for sowing seeds; **seeder unit** = precision drill, a seed drill which sows the seed separately at set intervals in the soil

seedhead ['si:dhed] *noun* top of a stalk with seeds, either in a seedcase or separately attached to the stem

seeding year ['si:dɪŋ 'jɜ:] *noun* the calendar year in which the seed is sown

seedless ['si:dləs] *adjective* with no seeds; **seedless hay** = hay obtained from a grass seed crop after threshing out the seedheads

◊ **seedling** ['si:dlɪŋ] *noun* very small plant which has just sprouted from a seed

◊ **seed tree** ['si:d 'tri:] *noun* tree left standing when others are cut down, to allow it to seed the cleared land

seep [si:p] *verb (of a liquid)* to flow slowly through a substance; *water seeped through the rock; silage liquor seeped out of the container*

◊ **seepage** ['si:pɪdʒ] *noun* slow oozing out of ground water from the soil surface; **seepage tank** = tank attached to a septic tank, into which the liquids from the septic tank are drained

select [sɪ'lekt] *verb* to make a choice *or* to choose some things but not others; *the strongest plants are selected for further breeding*

◊ **selection** [sɪ'lekʃn] *noun* **artificial selection** = selection by people of individual animals or plants from which to breed further generations, because the animals or plants have useful characteristics; **natural selection** = evolution whereby the characteristics which help individuals to survive and reproduce are passed on to their offspring

◊ **selective** [sɪ'lektɪv] *adjective* which chooses only certain things and not others; **selective pesticide** = pesticide which takes toxic action against specific pests without affecting the growing crop; **selective weedkiller** = weedkiller which is supposed to kill only certain plants and not others

QUOTE a species of small mammal has evolved selective resistance to venom
BBC Wildlife

selenium [sə'li:niəm] *noun* trace element, an essential part of the diet for all animals; white muscle disease is the symptom of selenium deficiency (NOTE: chemical symbol is **Se**; atomic number is **34**)

self- [self] *prefix* referring to the subject itself; **self-blanching celery** = variety of celery where the stalks are naturally white, and do not need to be earthed up; **self-contained herd** = dairy herd which breeds its own replacements, the calves being kept and reared; **self-raising flour** = flour with baking powder added to it; **self-sown plant** = plant which grows from seeds which have been naturally dispersed

self-feed ['self'fi:d] *verb* to take a controlled amount of feed from a large container as required; **self-feed silage** = feeding system where stock feed from silage, the amount of silage available being limited by a form of control

self-fertile ['self'fɜ:taɪl] *adjective* (plant) which fertilizes itself by male gametes from its own flowers

◊ **self-fertilization** ['selffɜ:tɪlaɪ'zeɪʃn] *noun* action of fertilizing itself

◊ **selfing** ['selfɪŋ] *noun* = SELF-FERTILIZATION

◊ **self-pollination** ['selfpɒlɪ'neɪʃn] *noun* pollination of a plant by pollen from its own flowers

◊ **self-purification** ['selfpju:rɪfɪ'keɪʃn] *noun* ability of water to clean itself of polluting substances

◊ **self-regulating** ['self'regju:leɪtɪŋ] *adjective* (ecosystem) which keeps itself in balance; *most tropical rainforests are self-regulating environments*

◊ **self-sterile** ['self'steraɪl] *adjective* (plant) that cannot fertilize itself from its own flowers

◊ **self-sufficient** ['selfsʌ'fɪʃənt] *adjective* being able to supply all one's needs of a commodity, such as food. British farmers have increased their market share of the food bought in Britain. In 1976 Britain was 63% self-sufficient in homegrown produce and 50% self-sufficient in all foods, including tropical produce. In 1994 the figures were 73% and 56% respectively

sell-by date ['selbaɪ 'deɪt] *noun* date on the label of a food product, which is the last date on which the product should be sold and be guaranteed as of good quality

selva ['selvə] *noun* evergreen rain forest of the Amazon basin in South America

semen ['siːmən] *noun* thick pale fluid containing spermatozoa, produced by the testes and seminal vesicles of the male

semi- ['semi] *prefix* meaning partly; **semi-digger** = type of mouldboard on a plough

◊ **semi-mounted**   [semɪ'maʊntɪd] *adjective* (implement) which is supported by a tractor but also has its own wheels (NOTE: an implement without wheels is said to be **mounted;** one which is not supported by a tractor is said to be **trailed)**

QUOTE semi-mounted ploughs have been surpassed by fully trailed models capable of turning several furrows at a pass, pulled by very powerful tractors
Practical Farmer

seminal ['semɪnəl] *adjective* referring to semen or to seed; **seminal propagation** = growing new plants from seed or from tubers (such as potatoes); **seminal roots** = secondary roots of a plant which support the primary root; this root system is then replaced by adventitious roots

semolina [semə'liːnə] *noun* coarse flour made from wheat after the fine flour has been ground; used to make puddings

sepal ['sepəl] *noun* green leaf-like outer part which with others forms the calyx surrounding a flower

separated milk [sepə'reɪtɪd 'mɪlk] *noun* milk from which the cream has been removed; also called skimmed milk

septic tank ['septɪk 'tæŋk] *noun* waste tank in which sewage is collected to begin the decomposition process

sericulture ['serɪkʌltʃə] *noun* raising silkworms for the production of silk

serve [sɜːv] *verb (of a male animal)* to copulate with *or* to mate with (a female)

sesame ['sesəmi] *noun Sesamum indicum,* a minor oilseed cultivated throughout the drier tropics and subtropics; the plant is an

erect herb which grows to two metres tall; in West Africa it is called 'benniseed' and in East Africa 'sim-sim'

COMMENT: the crop is of African origin, but is now grown in tropical, subtropical and Mediterranean areas. It is important in India, China, Sudan and Nigeria. The seeds are crushed to make an oil used in cooking, and the sesame cake which remains after the oil has been extracted is an important protein-rich stock feed

sessile ['sesaɪl] *adjective* (i) attached directly to a branch *or* stem (not by a stalk); (ii) permanently attached to a surface

set [set] **1** *noun* **(a)** (i) a seed potato; (ii) a seed onion **(b)** badger's burrow **2** *verb* **(a)** to form fruit or seed **(b)** to plant **(c)** *(of skin of a newly dug potato)* to harden

set aside ['set ə'saɪd] *verb* to take a piece of arable land and allow it to lie fallow *or* to use it for other types of use (for example as woodland *or* for recreational use)

◊ **set-aside** ['setəsaɪd] *noun* using a piece of formerly arable land for another use (such as for recreational use *or* for woodland *or* to lie fallow) (NOTE: the US equivalent is **the acreage reduction program)**

QUOTE the scheme will offer subsidies to farmers if they set aside - i.e. leave fallow - 20 per cent of their land for five years
Sunday Times

QUOTE agriculture is changing, in the form of set-aside policies for the reduction of overproduction and conservation of important habitats and landscape. Set-aside land must either be managed with green cover crops like clover and kept in good agricultural condition, or put to forestry, or used for non-farming purposes
Environment Now

QUOTE he will be setting aside some whole fields on poorer-yielding chalk soils, but mainly 15m headland strips adjacent to woodland which suffer from shading
Farmers Weekly

QUOTE a total of 1820 farmers are to take out 58,000 hectares (143,000 acres) in the first year of land set-aside scheme. Most producers had opted for permanent fallow covering 75% of the total set aside
Ministry of Agriculture

QUOTE in 1994 around 1.4 million acres on an estimated 30,000 farms will be set aside at a cost of over £200 million.
Guardian

QUOTE Britain is to be allowed to introduce a non-rotational set-aside programme at a rate of 18% compared with 20% for other EU member states. This is due to the larger average size of farms in the UK and a greater proportion of arable land being set aside
British Farmer

set on ['set 'ɒn] *verb* to foster an orphaned animal to another female (as a lamb onto a ewe)

◊ **set stocking** ['set 'stɒkɪŋ] *noun* grazing system associated with extensive grazing; livestock graze an area where they remain for an indefinite period; the traditional practice in Britain

◊ **set to** ['set 'tʊ] *noun* orphan lamb given to a foster mother

sewage ['suːɪdʒ] *noun* waste water and other refuse such as faeces, carried away in sewers; **sewage farm** = place where sewage is treated chemically to make it safe to be used as fertilizer; **sewage farming** = using sewage as a fertilizer

sex [seks] **1** *noun* one of two groups (male and female) into which animals and plants can be divided; *the relative numbers of the two sexes in the population are not equal, more males being born than females;* **sex chromosome** = chromosome which determines if an organism is male or female; **sex determination** = way in which the sex of an individual organism is determined by the number of chromosomes which make up its cell structure; **sex organs** = organs which are associated with reproduction and sexual intercourse **2** *verb* to identify the sex of an animal, egg or carcass

COMMENT: in mammals, individuals have either a pair of identical XX chromosomes (and so are female) or have one X and one Y chromosome and are male. Out of the pairs of chromosomes in each animal cell, only two are sex chromosomes. Females have a pair of X chromosomes and males have a pair consisting of one X and one Y chromosome. The sex of the offspring is determined by the father's sperm. While the mother's ovum only carries X chromosomes, the father's sperm can carry either an X or a Y chromosome. If the ovum is fertilized by a sperm carrying an X chromosome, the embryo will contain the XX pair and so be female

sex-linkage [seks'lɪŋkɪdʒ] *noun* existence of characteristics which are transmitted through the X chromosomes

◊ **sex-linked** [seks'lɪŋkt] *adjective* (i) (genes) which are linked to X chromosomes; (ii) (characteristics, such as colour of the eyes in fruit flies) which are transmitted through the X chromosomes

SFA = SMALL FARMERS ASSOCIATION

SFW = STANDARD FEED WHEAT

shade [ʃeɪd] *noun* place sheltered from direct sunlight; **shade plants** = plants which prefer to grow in the shade; **shade-tolerant tree** = tree (such as beech) which will grow in the shade of a larger tree; **shade intolerant tree** = tree (such as Douglas fir) which does not grow in the shade of other trees

◊ **shading** ['ʃeɪdɪŋ] *noun* action of cutting off the light of the sun; *parts of the field near tall trees suffer from shading*

shadoof *or* **shaduf** [ʃə'duːf] *noun* primitive device for raising water from a well, consisting of a horizontal beam or branch with a bucket at one end, supported by a forked post. Still widely used in Asia and North Africa

shafts [ʃɑːfts] *noun* two pieces of wood reaching in front of a cart, between which the horse stands

◊ **shaft-horse** ['ʃɑːft 'hɔːs] *noun* horse which pulls a cart, standing between two shafts

shaggy ['ʃægi] *adjective (of animals)* with long hair (as Highland cattle)

shank [ʃæŋk] *noun* lower part of a horse's leg between the knee and the foot

share [ʃeə] *noun* ploughshare, the heavy metal blade which cuts the bottom of the furrow slice; it can be in one piece or in three sections, the point, wing and shin

sharecropper ['ʃeəkrɒpə] *noun* tenant

farmer who pays a part of his crop to the landlord as a form of rent

◊ **sharecropping** ['ʃeəkrɒpɪŋ] *noun* system of land tenure, whereby tenants pay an agreed share of the crop to the landlord as a form of rent

sharefarming ['ʃeəfɑːmɪŋ] *noun (in Britain)* a joint enterprise between a party with an interest in the land and another party involved in farming operations. Usually one party provides the capital and the other the farm management inputs such as labour and equipment

sharp eyespot ['ʃɑːp 'aɪspɒt] *noun* soilborne fungus *(Rhizoctonia solani)* affecting cereals, which can cause lodging and shrivelled grain

shavings ['ʃeɪvɪŋz] *noun* thin curled pieces of wood removed when planing, used as litter for animals

shea butter ['ʃiː 'bʌtə] *noun* kernel oil from the shea butter-nut tree, found in West Africa

sheaf [ʃiːf] *noun* bundle of corn stalks tied together after reaping (NOTE: plural is **sheaves)**

shear [ʃɪə] *verb* to clip the fleece from a sheep (NOTE: **shearing - shorn)**

◊ **shears** [ʃɪəz] *noun* large clippers used to shear a sheep

◊ **shearer** ['ʃɪərə] *noun* person who clips the fleece off a sheep

◊ **shearling** ['ʃɪəlɪŋ] *noun* young sheep which has been shorn for the first time

COMMENT: shearing nowadays is done with electric shears; using these clippers, a skilled shearer can fleece a ewe in under two minutes. Shearers often work in gangs. The fleece is clipped off in one piece; that of a longwool breed may weigh over 10kg

QUOTE scientists are experimenting with an organic compound which will defoliate sheep, literally making their fleeces fall off
Sunday Times

shed [ʃed] **1** *noun* wooden farm building used to house livestock or farm implements **2** *verb* **(a)** to separate one or more animals from a flock or herd **(b)** to let leaves or grain fall; *cereals may shed grain as a result*

of strong winds or heavy rain; deciduous trees shed their leaves in winter

sheen [ʃiːn] *noun* bright shiny appearance of a surface (used of fruit, animals' coats, meat, etc.)

sheep [ʃiːp] *noun* ruminant of the genus Ovis, family Bovidae; one of many domesticated varieties, farmed for their wool, meat and milk; **sheep farming** = raising sheep on a farm; most sheep are raised under extensive grazing conditions; **sheep ked** = SHEEP TICK; **sheep maggot fly** = type of fly that lays its eggs on the wool of sheep; the eggs hatch into maggots that burrow into the flesh causing a condition known as 'strike'; **sheep pox** = highly contagious viral disease; symptoms include fever, loss of appetite, difficulty in breathing and in the final stages scabs and ulcers appear; a notifiable disease; **sheep run** = extensive area used for sheep grazing (especially in New Zealand and Australia); **sheep scab** = serious disease of sheep, caused by a parasitic mite, which results in intense irritation, ulceration of the skin, loss of wool and emaciation; a notifiable disease; **sheep's fescue** = *Festuca ovina,* species of grass useful under hill and marginal conditions; **sheep's sorrel** = common weed *(Rumex acetosella)* **sheep station** = very large sheep farm in Australia; **sheep tick** = small wingless dipterous insect, parasitic on sheep; **sheep walk** = area of land on which sheep are pastured

COMMENT: most sheep in the UK are kept for meat, and milk production is relatively unimportant. Wool is the main product of sheep in some other countries, such as Australia; in the UK, however, it is an important by-product of sheep farming. In 1994 the total sheep population of the UK was approximately 40 million, accounting for 30% of all sheep in the EU member states. Home production was worth over £1 bn or about 8% of total agricultural output value. Sheep are kept under a wide range of environmental and management conditions, from coastal lowland areas such as Romney Marsh to the upland areas of Wales, Scotland and the North of England. Lambs from the upland areas are moved to lowland farms for fattening. In the UK, a great many breeds of sheep have survived and there are some 50 recognized breeds as well as a variety of local types and many crossbreds. More recently, the introduction of continental breeds has increased the variety. A broad classification into three main categories may be made: *the long-woolled breeds* which include the Romney, Lincoln and Leicester; *the short-woolled breeds* including the Southdown, Dorset Down and Suffolk, and the *mountain, moorland and hill breeds* which include the Cheviot, Radnor, Scottish Blackface, Swaledale and Welsh Mountain

sheep-dip [ˈʃiːpdɪp] *noun* chemical preparation used in a dipping bath to disinfect sheep to control diseases such as sheep scab

COMMENT: all sheep in Britain are dipped for scale once a year, following the ministerial decision to have a single national dip. Dipping ceased to be compulsory in the UK in 1992. It is illegal to buy organophosphorous sheep-dip without a certificate of competence. To obtain a certificate farmers must sit a two-hour exam on ways of limiting exposure to pesticides, fewer than 10% of farmers have complied with these regulations

QUOTE toxic sheep dip chemicals have been found in water supplies in concentrations 23 times higher than the government recommended limits. The chemicals are organophosphorous insecticides which can accumulate in the body and cause nervous disorders. Hundreds of farmers claim they have been harmed after dipping sheep against scab and blowfly
Guardian

sheepdog [ˈʃiːpdɒg] *noun* breed of dog trained and used by shepherds in controlling sheep

sheepman [ˈʃiːpmən] *noun* shepherd, a farm worker who looks after sheep

sheepmeat [ˈʃiːpmiːt] *noun* term used in the EU for meat from a sheep or lamb

sheep premium quotas [ˈʃiːp ˈpriːmiəm ˈkwəʊtəz] *noun* introduced under 1992 CAP reforms and allocated to existing producers to enable them to claim EU support payments. Quotas were based on stock held in 1992

sheet erosion [ˈʃiːt ɪˈrəʊʒn] *noun* during heavy rainfall, the saturation of the soil may result in surface runoff down slopes; water flow is even over the whole surface and does not create channels as in rill erosion, but the soil is eroded over the whole area

shelf-life [ˈʃelflaɪf] *noun* number of days or weeks which a product can stay on the shelf of a shop and still be good to use

shell lime [ˈʃel ˈlaɪm] *noun* another name for burnt lime or quicklime

shelter [ˈʃeltə] **1** *noun* protection (from wind *or* sun, etc.); **shelter belt** = row of trees planted to shelter smaller crops from wind; it may also check soil erosion by wind; *see also* CHILLSHELTER **2** *verb* to protect something from sun *or* wind

◊ **shelterwood** [ˈʃeltəwʊd] *noun* large trees left standing when others are cut, to act as shelter for seedling trees

shepherd [ˈʃepəd] *noun* person who looks after sheep; **shepherd's purse** = common weed *(Capsella bursa-pastoris)* in gardens and market gardens, found particularly among vegetables and root crops (NOTE: also called **pepper and salt)**

Shetland [ˈʃetlənd] *noun* **(a)** rare breed of cattle, native to the Shetland Isles; medium sized, black and white, with short legs, short horns and a bulky body **(b)** breed of sheep, native to the Shetland Isles; the colour varies from white, through grey and black to light brown; the ewes are polled and the rams horned; it produces fine soft wool of high quality, used in the Shetland wool industry. A small Shetland ewe yields a fleece 1.5 - 2 kilos in weight **(c)** breed of pony, used as a riding horse for children

shifting cultivation [ˈʃɪftɪŋ ˈkʌltɪˈveɪʃn] *noun* agricultural practice using the rotation of fields rather than crops, short cropping periods followed by long fallows and the maintenance of fertility by the regeneration of vegetation; *see also* FALLOW

shigella [ʃɪˈgelə] *noun* bacillus which causes dysentery

shin [ʃɪn] *noun* **(a)** lower part of the foreleg (of cattle) **(b)** upper part of a ploughshare

shire horse [ˈʃaɪə ˈhɔːs] *noun* tall heavy breed of draught horse; the coat may be of various colours, but there is always a mass of feather at the feet

shivering [ˈʃɪvərɪŋ] *noun* an affliction of the nervous system with involuntary muscular contractions, usually of the hind legs. It is a progressive condition found in horses

shock [ʃɒk] *noun* stook, sheaves of corn standing up against each other to dry

shoddy ['ʃɒdi] *noun* waste product of the wool industry; it contains up to 15% nitrogen and is used as a fertilizer, particularly in market gardens

shoe [ʃuː] *verb* to make, fit and fix horseshoes to the feet of a horse

shoot [ʃuːt] **1** *noun* **(a)** part of a young seed plant, the stem and first leaves which show above the surface of the soil **(b)** new growth from the stem of a plant; *after pruning the plant will send out new shoots* **2** *verb* to kill with a gun

> COMMENT: shooting is the only effective way of reducing common pests such as the wood pigeon; most birds are more useful than harmful, though

short day plant [ʃɔːt 'deɪ 'plɑːnt] *noun* plant (such as the chrysanthemum) which flowers only if the daily period of light is less than a particular critical length, as in the autumn

◊ **short duration ley** ['ʃɔːt dju'reɪʃn leɪ] *noun* ley which is kept only for a short time; **short duration ryegrass** = class of grasses which are important to the farmer, including Westerwolds, Italian and Hybrid. These grasses are quick to establish and give early grazing; they are used where persistency is not important

short grain ['ʃɔːt 'greɪn] *see* RICE

shorthorn ['ʃɔːthɔːn] *noun* breed of cattle, with short horns

> COMMENT: in the 18th century, Charles Colling used many of he breeding principles established by Robert Bakewell to develop the shorthorn breed; the breed became the most common in Britain and remained so for over a hundred years. It has later developed into three different strains: the Beef Shorthorn, the Dairy Shorthorn and the Lincoln Shorthorn

short ton ['ʃɔːt 'tʌn] *noun* American ton (equal to 907kg)

shred [ʃred] *verb* to tear into tiny pieces; *farmyard manure is shredded before being spread on fields*

◊ **shredder** ['ʃredə] *noun* machine for shredding waste vegetable matter before composting

shrivel ['ʃrɪvl] *verb* to become dry and wrinkled; *in the drought the fruit shrivelled and fell off the stems*

Shropshire ['ʃrɒpʃə] *noun* medium-sized breed of sheep with a black face and heavy fleece, now rare

shrub [ʃrʌb] *noun* a bush, a low-growing perennial plant; *shrubs may provide useful grazing, especially for goats*

◊ **shrubby** ['ʃrʌbi] *adjective* (plant) that grows like a shrub

Shuwa Arab ['ʃuːwə 'ærəb] *noun* type of Nigerian cattle with a hump on the shoulders

sickle ['sɪkl] *noun* curved knife-edged metal tool with a wooden handle, used for harvesting cereals

side rake ['saɪd 'reɪk] *noun* machine which picks up two swaths and combines them into one before baling

◊ **sideland** ['saɪdlænd] *noun* strip of land left at the side of a field during ploughing; it may be ploughed up with the headlands

◊ **sidewalk farmer** ['saɪdwɔːk 'fɑːmə] *noun US* farmer who cultivates land some way away from his house in a town

sieve [sɪv] **1** *noun* garden implement with a base made of mesh or with perforations through which fine particles can pass while coarse material is retained; *compare* RIDDLE **2** *verb* to pass soil, etc. through a sieve to produce a fine tilth, or to remove the soil from root crops such as potatoes

silage ['saɪlɪdʒ] *noun* food, chiefly used for cattle, but also fed to other ruminants and horses; it is formed of grass and other green plants, cut and stored in silos, big bales or clamps; **big bale silage** = silage stored in big bales; **silage additive** = additive containing bacteria and/or chemicals which will enhance the ensilage process; **silage effluent** = acidic liquid produced by the silage process which can be a serious pollutant, especially if it drains into a watercourse; legislation controlling the storage and disposal of silage effluent is becoming very strict; **silage tower** = container used for making and storing silage

> COMMENT: silage is an animal feedstuff made by fermenting a crop with a high moisture content under anaerobic

conditions; silage may be made from a variety of crops, the most common being grass and maize, but also including grass and clover mixtures, green cereals, kale, root tops, sugar beet pulp and potatoes. Trials indicate that very high quality grass silage can be fed to adult pigs

QUOTE big bale silage is much easier to handle or to feed than either hay or conventional silage, and much more palatable. Because of its higher DM, intakes are much higher and this helps milk yields or gives excellent growth in beef cattle and store lambs
Practical Farmer

silk [sɪlk] *noun* **(a)** fibre produced by silkworms when preparing a cocoon or a web, used for weaving into high-quality fabric **(b) silks** = the long thin styles of the female flowers of the maize plant

◊ **silkworm** ['sɪlkwɜːm] *noun* larva of a moth *(Bombyx mori)* which feeds on the leaves of the white mulberry, and spin cocoons of silk

silo ['saɪləʊ] *noun* large container for storing grain or silage

◊ **silopress** ['saɪləʊpres] *noun* polythene 'sack' into which silage is forced; as the sack fills up, it gradually grows longer and when completely full is sealed. A 'sack' may contain up to 80 tonnes of silage

COMMENT: there are many different types of silo: some are pits dug into the ground; others are forms of surface clamp; built silos are towers which may be either top- or bottom-loaded, and are built of wood, concrete or steel

silt [sɪlt] *noun* **(a)** particles of mineral intermediate between clay and sand (with a diameter of between 0.002 and 0.06mm in the UK), generally formed of fine quartz **(b)** *(in general)* soft mud which settles at the bottom of water

◊ **silty soil** ['sɪlti 'sɔɪl] *noun* soil containing a high proportion of silt; such soils are difficult to work and drainage is a problem

silver-laced Wyandotte ['sɪlvəleɪst 'waɪəndɒt] *noun* dual-purpose breed of poultry; the feathers are silvery, with black edges, especially on the tail

silvi- ['sɪlvi] *prefix* referring to trees

◊ **silvicide** ['sɪlvɪsaɪd] *noun* substance which kills trees

◊ **silviculture** ['sɪlvɪkʌltʃə] *noun* growing of trees, mainly for timber

sim-sim ['sɪmsɪm] *noun* East African name for sesame; *see also* SESAME

Simmental ['sɪməntɑːl] *noun* breed of cattle originating in Switzerland; the colour is yellowish-brown or red; it is a dual-purpose breed, with a high growth rate potential and good carcass quality

single flower ['sɪŋgl 'flaʊə] *noun* flower with only one series of petals (as opposed to a double flower)

single-suckling ['sɪŋglsʌklɪŋ] *verb* natural method of rearing beef cattle, where calves are permitted to suckle their own mothers

◊ **singleton** ['sɪŋgltən] *noun* single offspring

◊ **singling** ['sɪŋglɪŋ] *noun* (i) reducing the number of plants in a row; (ii) reducing the number of plants from a multigerm seed to a single plant

sire [saɪə] *noun* male parent of an animal; male animal selected for breeding

QUOTE AI centres not only supply semen from high-ranking sires, but generally offer a high degree of choice. The infusion of genes from a range of sires helps to maintain genetic variability within a herd
Queensland Agricultural Journal

sisal ['saɪzəl] *noun* tropical plant *(Agave rigida)* which yields a hard fibre used for making binder twine and mats

site [saɪt] *noun* place *or* position of something *or* place where something happened; **Site of Special Scientific Interest (SSSI)** = small area of land which has been noted as particularly important by the Nature Conservancy Council, and which is preserved for its fauna, flora or geology; **nesting site** = place where a bird may build a nest

Sitka spruce ['sɪtkə 'spruːs] *noun Picea sitchensis,* a fast-growing temperate softwood tree, used for making paper

six-tooth sheep ['sɪkstuːθ 'ʃiːp] *noun* sheep between two and three years old

skim coulter ['skɪm 'kuːltə] *noun* part of a

plough which turns a small slice off the corner of the furrow about to be turned and throws it into the bottom of the one before; it is attached to the beam behind the disc coulters

skim milk *or* **skimmed milk** ['skɪm *or* 'skɪmd 'mɪlk] *noun* milk which has had both fat and fat-soluble vitamins removed; used as a milk substitute for calves and lambs

skin [skɪn] *noun* outer layer on an animal, fruit or vegetable; **skin spot** = potato disease causing pimple-like dark brown spots which can harm the buds in the eyes of seed tubers

slag [slæg] *noun* waste matter which floats on top of the molten metal during smelting, used to lighten heavy soils, such as clay, and also for making cement; **basic slag** = calcium phosphate, waste slag from blast furnaces, used as a fertilizer because of its phosphate content; now less valuable than previously because of improved steel-making processes

slaked lime ['sleɪkt 'laɪm] *noun* calcium hydroxide, mixture of calcium oxide and water, used to improve soil quality; the lime is in powder form, having been treated to break it down from large lumps

slash [slæʃ] *verb* to make a long cut with a knife; **slash and burn** = swidden farming, form of agriculture where forest is cut down and burnt to create open space for growing crops; the space is abandoned after several crops have been grown and further forest is cut down

slat [slæt] *noun* thin flat piece of wood; **slatted floor** = floor covering made of slats of wood, allowing chicken droppings to fall through; **slatted mouldboard** = type of mouldboard which breaks up the soil as it is being ploughed

slaughter ['slɔːtə] **1** *noun* killing of animals **2** *verb* to kill animals

◊ **slaughterhouse** ['slɔːtəhaʊs] *noun* building where animals are slaughtered and the carcasses prepared for sale for human consumption; slaughterhouses are mainly owned by local authorities and are subject to strict control and frequent inspection

sledge [sledʒ] *noun* vehicle without wheels, which is dragged along the ground;

bale sledge = sledge pulled behind a baler to collect the finished bales

SLF = SCOTTISH LANDOWNERS FEDERATION

slings [slɪŋz] *noun* device for supporting an animal with some form of disability; it is an elaborate type of harness and supports the weight of the animal

slink calf ['slɪŋk 'kɑːf] *noun* calf born early, before the normal period of gestation is complete

slip [slɪp] **1** *noun* small piece of plant stem, used to root as a cutting, or in budding **2** *verb* (*of an animal*) to miscarry

sloe [sləʊ] *noun* Prunus spinosa, or blackthorn, the wild plum

sludge [slʌdʒ] *noun* **raw sludge** = solid sewage which falls to the bottom of a sedimentation tank; **sludge digestion** = final treatment of sewage when it is digested anaerobically by bacteria; **sludge gas** = sewage gas, methane mixed with carbon dioxide which is given off by sewage

slug [slʌg] *noun* kind of mollusc without a shell; it causes damage to crops especially in wet conditions; under the surface of the soil, it can damage bulbs, roots and tubers; **slug pellets** = small amounts of poisonous mixture containing substances such as metaldehyde, used to kill slugs

QUOTE slugs have undergone a population explosion over the past two years due to a combination of factors including prolonged moist conditions during the summers and autumns. A big reduction in straw burning and the large area of oilseed rape also have helped the pest to thrive

QUOTE this year no slug pellets were used with corn drilled into fields previously burnt and there is no sign of slug damage
Farmers Weekly

sluice [sluːs] *noun* channel for water, especially channel for water through a dam; the flow is controlled by a sluicegate; sluices are used in the distribution of irrigation water

slurry ['slʌri] *noun* **(a)** liquid waste from animals, stored in tanks and treated to be used as fertilizer; it may also be stored in a lagoon, from which it can either be piped to the fields or transferred to tankers and then distributed **(b) lime slurry** = mixture of lime and water added to hard water to make it softer

◊ **slurry injector** ['slʌri ɪn'dʒektə] *noun* tractor-hauled machine which injects slurry into the soil

◊ **slurry spreader** ['slʌri 'spredə] *noun* machine which spreads slurry

COMMENT: slurry can be spread on the land using pumps, a pipeline system and slurry guns; more often, slurry spreaders are used, which can load, transport and spread the material. Spreading may be by gravity feed onto a spinning disc which spreads it on the land, or by a pressure system which directs the slurry onto a deflector plate which spreads the slurry in a wide belt. New regulations to control the pollution from slurry were introduced in 1989. These require the base and walls of silage clamps to be impervious. An artificial embankment must be constructed around slurry tanks. Silage clamps in the middle of fields with covers held down with rubber tyres will be banned

QUOTE during the survey, the farmers revealed that utilization of wet slurry with irrigation water is more effective than dried slurry, relatively free from weeds and produces no ill-effects on the crops
Indian Farming

QUOTE feed stances and cubicle passages are cleaned by automatic scraper every one and a half hours. Slurry is scraped towards the centre of the building where it passes through steel gratings into shallow liquid-filled chambers
Farm Building Progress

smallholding ['smɔːlhəʊldɪŋ] *noun* small agricultural unit under 20 hectares in area; often run as a family concern, and with some form of specialization, such as market gardening, raising goats, etc.

◊ **smallholder** ['smɔːlhəʊldə] *noun* person who farms a smallholding

small nettle ['smɔːl 'netl] *noun* weed (*Urtica urens*) which is common on rich friable soils; it affects vegetables and other row crops (NOTE: also called **annual nettle** or **burning nettle)**

SMD = STANDARD MAN DAY

smoke [sməʊk] *verb* to preserve food by hanging it in the smoke from a fire (used now mainly for fish, but also for some bacon and cheese)

smooth-stalked meadowgrass ['smuːðstɔːkt 'medəʊgrɑːs] *noun Poa pratensis*, species of grass which can withstand quite dry conditions; a perennial grass with smooth greyish-green leaves and green purplish flowers

smudging ['smʌdʒɪŋ] *noun* burning of oil to produce smoke to prevent loss of heat from the ground and so to minimize or prevent frost damage to crops and orchards

smut [smʌt] *noun* disease of cereal plants, caused by a group of fungi, which covers the plant with black spots; varieties include common smut, loose smut and bunt

snail [sneɪl] *noun* mollusc of the class Gastropoda; a pest which can be controlled by molluscicides; **edible snails** = snails reared for human consumption

QUOTE there are 48 snail farms in operation in the UK, with 26 actively selling snails. Some were producing more than 2 tons per year
Farmers Weekly

snap beans ['snæp 'biːnz] *noun US* beans which are eaten in the pod (such as green beans, French beans), or of which the seed is eaten after drying (such as haricot beans); as opposed to broad beans or Lima beans, of which the seed is eaten fresh

SNF = SOLID-NOT-FAT

snout [snaʊt] *noun* nose and mouth of some animals, including the pig

snow mould ['snəʊ 'məʊld] *noun* fungal pre-emergent blight and root rot of cereals (*Micronectriella nivalis*)

snow rot ['snəʊ 'rɒt] *noun* white mould growth affecting wheat, causing leaves to turn brown and shrivel (*Typhula incarnata*)

soakaway ['səʊkəweɪ] *noun* hole in the ground filled with gravel, which takes rainwater from a downpipe and allows it to soak into the surrounding soil

Soay ['səʊeɪ] *noun* rare horned breed of sheep which sheds its fleece naturally, thought to be the link between wild and domesticated breeds; the short hairy fleece is tan or dark brown (the breed originally came from the island of Soay in the Outer Hebrides)

sodium ['səʊdiəm] *noun* chemical element which is the basic substance in salt; sodium is essential for animals and sodium deficiency can reduce performance; **sodium balance** = balance maintained in the body between salt lost in sweat and urine and salt taken in from food; the balance is regulated by aldosterone; **sodium chlorate** = total herbicide which is highly toxic to plants and is no longer used; **sodium chloride** = common salt (NOTE: chemical symbol is **Na**; atomic number is **11**)

soft [sɒft] *adjective* **soft fruit** = general term for all fruits and berries that grow on bushes and canes, have a relatively soft flesh, and so cannot be kept, except in some cases by freezing; typical soft fruit are raspberries, strawberries, blueberries and blackberries, and the various currants; **soft rot** = bacterium which affects stored potatoes and carrots; the cell walls of the plant dissolve causing mushy soft rot *(Erwinia carotovora)* **soft water** = water which does not contain calcium and other minerals which are found in hard water; **soft wheat** = wheat containing grains which, when milled, break down in a random manner; soft wheats have less protein than hard wheats and have poor milling qualities

◊ **softwood** ['sɒftwʊd] *noun* white wood from pine trees and other conifers; **softwoods** = (forests of) pine trees (as opposed to hardwoods)

soil [sɔɪl] *noun* loose material on the surface of the earth in which plants grow; it consists of disintegrated rock, organic material, water and living organisms; **rich soil** = soil with many useful nutrients in which plants grow well; **poor soil** = soil with few useful nutrients and so less suitable for plants; **soil air** *or* **soil atmosphere** = air content of the soil, made up of the gases of the atmosphere but in different amounts, because it is modified by the constituent parts of the soil; aeration is the replacement of stale air in the soil with fresh; **soil association** = group of soils associated with one area and which occur in a predictable pattern; **the Soil Association** = organization whose main aims are to promote organic farming and gardening, and to protect the environment; **soil ball** = the rooting system of a plant, complete with the soil attached to it; as when a plant is lifted from a pot or seedbed; **soil capping** = hard crust on the surface of the soil which can be caused by heavy rain drops or the passage of heavy farm machinery; **soil compaction** = firming of the soil into a condensed state, often brought about by the use of heavy machinery; compaction lowers the infiltration rate and increases the likelihood of runoff; **soil conservation** = maintenance of soil structure, moisture levels and nutrient status, by careful management such as by irrigation, planting and mulching, and preventing it being eroded; **soil creep** = slow movement of soil material down a slope under the force of gravity; **soil drainage** = drainage of water from soil (either naturally or by putting pipes and drainage channels into the soil); **soil erosion** = removal of soil by natural means (such as being removed by water *or* by wind); the main types of water erosion are gully erosion, in which deep channels are cut into the land, sheet erosion, which is the gradual uniform removal of surface soil by water, and rill erosion, which is the formation of small channels by the removal of soil by running water; **soil fertility** = potential capacity of soil to grow plants; **soil flora** = minute plants (such as fungi and algae) which live in the soil; **soil horizons** = different layers present in soil (the topsoil (the 'A' horizon) containing humus, the subsoil ('B' horizon) containing minerals leached from the topsoil, and the unweathered parent rock or 'C' horizon); **soil loosener** = trailed implement which loosens the surface of the soil; **soil map** = map showing the different types of soil found in an area; **soil pan** = hard layer in the soil; **soil parent material** = material from which soil is formed; **soil profile** = the various layers of soil particles, as seen in a vertical section obtained by making a pit; **soil salinity** = measurement of the quantity of mineral salts found in a soil; **soil series** = classification of soils based on their similarities, used in soil mapping; soil series are defined using a combination of three main properties: the parent material, the texture of the soil material and the presence or absence of material with a distinctive mineralogy and the presence or absence of distinctive horizons; **soil sterilization** =

treatment of glasshouse, greenhouse and other horticultural soils in order to kill weed seeds, plant disease organisms and pests; **soil structure** = the arrangement of soil particles in groups or individually; the importance of soil structure lies in the fact that it largely controls the spaces within the soil, and it is along the structure faces that air and water move through the soil; **soil survey** = mapping of soil types using a soil classification system; **soil texture** = relative proportions of sand, silt and clay particles in soil; *see also* CHERNOZEM, LOESS, PODZOL, SUBSOIL, TOPSOIL

COMMENT: soil is a mixture of mineral particles, decayed organic matter and water. Topsoil contains chemical substances which are leached through into the subsoil where they are retained. Soils are classified either according to the areas of the world in which they are found, or according to the types of minerals and stage of development which a soil has reached. Without care, soils easily degrade, losing the few nutrients they possess and become increasingly acid or sour

◊ **soilage** ['sɔɪlɪdʒ] *noun* green forage crops which are cut and carried to feed animals grazing on unproductive pastures, in order to supplement their diets. Crops commonly used for soilage are clovers and lucerne

◊ **soil-borne fungus** ['sɔɪlbɔːn 'fʌŋgəs] *noun* fungus of which the spores are carried in the soil

◊ **soil classification** ['sɔɪl klæsɪfɪ'keɪʃn] *noun* the classification of soils into groups with broadly similar characteristics (used in soil surveys)

COMMENT: all forms of soil classification are artificial as soils vary in three dimensions and in time; therefore no clear boundaries exist between soil types

◊ **soilless gardening** ['sɔɪləs 'gɑːdənɪŋ] *noun* hydroponics, growing plants in vermiculite or other substances which are not soil

◊ **soil moisture deficit** ['sɔɪl 'mɔɪstʃə 'defɪsɪt] *noun* difference between the amount of water a soil has and its higher field capacity

QUOTE soil moisture deficits around the country continue to remain high at about twice the level expected at this time of the year, making drilling almost impossible, giving concern for oilseed rape already sown, and making mature beet plants and maincrop potatoes wilt in the fields
Farmers Weekly

Sokoto [sə'kəutəu] *noun* Nigerian breed of cattle, grey in colour. A dairy breed, docile and easily trained for farm work

Solanum [sə'leɪnəm] *Latin name for* the potato family

Solari piggery [sə'lɑːri 'pɪgəri] *noun* type of housing for pigs, with fattening pens on each side of a central feeding passage, housed in an open-sided Dutch barn

sole [səul] *adjective* alone; **sole crop** = crop grown in a pure stand, with no other crops; **sole furrow** = the last slice cut during ploughing

solid-not-fat (SNF) percentage ['sɒlɪdnɒt'fæt pə'sentɪdʒ] *noun* measure of milk quality, showing the percentages of all substances other than fat in the milk

solstice ['sɒlstɪs] *noun* **summer solstice** = 21st June, the longest day in the Northern Hemisphere, when the sun is as its furthest point south of the equator; **winter solstice** = 21st December, the shortest day in the Northern Hemisphere, when the sun is furthest north of the equator

solum ['səuləm] *noun* technical word for the weathered material above the parent rock, including both topsoil and subsoil

somatotropin [səumətə'trəupɪn] *see* BOVINE SOMATOTROPIN, PORCINE SOMATOTROPIN

soot [sut] *noun* black deposit of carbon formed by smoke; after it has been weathered it is a valuable fertilizer and provides nitrogen

sooty mould ['suti 'məuld] *noun* fungal disease of wheat *(Cladosporium)*

sorghum ['sɔːgəm] *noun* drought-resistant cereal plant *(Sorghum vulgare)* grown in the semi-arid tropical regions, such as Mexico, Nigeria and Sudan; it is a staple food in the drier regions of Africa and is grown as animal feed in the USA; also called 'Kaffir corn' in South Africa

sorrel ['sɒrəl] *noun Rumex acetosa,* plant with a sour juice sometimes eaten as a salad; varieties include the sheep sorrel and the wood sorrel

sour soil ['sauə 'sɔɪl] *noun* soil which is excessively acid and hence needs liming to restore the correct balance between acidity and alkalinity

South Devon ['sauθ 'devən] *noun* heaviest breed of British cattle; its colour is light brownish-red. It was originally a dual-purpose breed, but now is mainly raised for beef

◊ **Southdown** ['sauθdaun] *noun* smallest of the Down breeds of sheep; compact body and very well fleeced, with a dense fleece of high-quality short wool.The Southdown is an early maturing breed and produces meat of high quality

sovkhoz ['sɒvhɒz] *noun* formerly, a Soviet state-run farm, usually a very large enterprise. They are now being divided into smaller farms

sow 1 [sau] *noun* mature female pig, after she has had and weaned her first litter **2** [sau] *verb* to put seeds into soil so that they will germinate and become plants; *peas should be sown in April or May; newly-sown seed can be attacked by mice; sow the seed thinly in shallow drills*

soya ['sɔɪə] *noun Glycine max.,* a plant which produces edible beans which have a high protein and fat content and very little starch

◊ **soybean** *or* **soyabean** ['sɔɪbiːn *or* 'sɔɪə 'biːn] *noun* bean from a soya plant

◊ **soyoil** ['sɔɪɔɪl] *noun* oil extracted from the soybean

COMMENT: soybeans are very rich in protein and, apart from direct human consumption, are used as a livestock feedingstuff and for their oil; after the oil has been extracted, the residue is used as a high protein feedingstuff. Other by-products are soybean milk and soy sauce, both widely used in China and Japan. Soybeans are widely grown in China

(where they are the most important food legume) Brazil and in the USA

QUOTE soybean is the cheapest source of protein and the beans contain about 40% high quality protein and 20% excellent edible oil. Over 300 food preparations can be made from it
Indian Farming

sp. = SPECIES

spaced plant ['speist 'plɑːnt] *noun* a plant grown in a row so that its canopy does not touch or overlap that of any other plant

◊ **spacing** ['speisɪŋ] *noun* making places between things, as between plants in a row; *the correct spacing of underground drains depends on the permeability of the soil and subsoil;* **spacing drill** = a precision seed drill

spade [speid] *noun* common garden tool, with a wide square blade at the end of a strong stick; used for making holes or digging by hand

◊ **spading machine** ['speidɪŋ mə'ʃiːn] *noun* machine which uses rotating digger blades to cultivate compacted topsoil and dig out pans created by other cultivators

spawn [spɔːn] *noun* **mushroom spawn** = spores of the edible mushroom which are sold to be used to propagate mushrooms

spay [spei] *verb* to remove the ovaries of a female animal

SPCS = SEED POTATO CERTIFICATION SCHEME

spear [spiə] *noun* shoot of a green plant, such as asparagus or broccoli

species ['spiːʃiːz] *noun* division of a genus, a group of living things which can interbreed (NOTE: plural is also **species.** Note that when referring to a species, it is usual to abbreviate it to **sp.** after the species name)

◊ **speciation** [spiːsi'eiʃn] *noun* process of forming a new species

specific [spə'sɪfɪk] *adjective* referring to a species; **specific name** = name by which a species is shown to be different from other members of the genus (it is the second name in the binomial classification system, the first being the generic name which identifies the genus); **specific weight** = bulk density of a grain sample measured in hectolitres or bushels

◊ **specificity** [spesɪˈfɪsɪti] *noun* being specific; *parasites show specificity in that they live on only a certain limited number of hosts*

specimen tree [ˈspesɪmən ˈtriː] *noun* tree planted in a garden in a position by itself, where it can be seen to advantage

speckle [ˈspekl] *noun* small spot; *see also* BEULAH

◊ **speckled yellowing** [ˈspekld ˈjeləʊɪŋ] *noun* disease of sugar beet caused by a deficiency of manganese

speedwell [ˈspiːdwel] *noun* widespread weed *(Veronica persica)* found in cereal crops and oilseed rape; because it spreads rapidly it is a hazard in row crops

spermatozoon *or* **sperm** [spɜːmætəˈzəʊɒn *or* spɜːm] *noun* mature male sex cell, which is capable of fertilizing an ovum (NOTE: plural is **spermatozoa**)

◊ **spermicide** [ˈspɜːmɪsaɪd] *noun* substance which kills sperm

COMMENT: a spermatozoon is very small, and comprises a head, neck and very long tail. It can swim by moving its tail from side to side

sphagnum [ˈsfægnəm] *noun* type of moss which grows in bogs, noted for its ability when dry to absorb large quantities of water; **sphagnum peat** = peaty soil made up of dead sphagnum moss

spice [spaɪs] *noun* pungent or aromatic parts of plants used a flavouring in cooking; spices are obtained from seeds, fruit, flowers, roots, bark or buds of plants; the commonest are pepper, mustard, ginger, cloves and nutmeg

spider [ˈspaɪdə] *noun* one of a large group of animals, with two parts to their bodies and eight legs

spike [spaɪk] *noun* **(a)** pointed end of a pole or piece of metal; **spike tooth harrow** = tractor-trailed implement consisting of a simple frame with tines attached where the frame members cross **(b)** flower cluster or tall pointed inflorescence

◊ **spiked** [spaɪkt] *adjective* (farm implement) with spikes, such as a spiked chain harrow

◊ **spikelet** [ˈspaɪklət] *noun* part of the flower head of plants such as grass, attached to the main stem without a stalk

spin [spɪn] *verb* **(a)** to turn round very fast **(b)** *(of spiders, silkworms)* to make thin thread for a web or cocoon

◊ **spinner** [ˈspɪnə] *noun* device used for harvesting potatoes; the potatoes are left on the surface of the soil for picking later

spinach [ˈspɪnɪtʃ] *noun* annual plant *(Spinacia oleracea)* grown for its succulent green leaves and eaten as a vegetable; there are several varieties including perpetual spinach, summer or round spinach and prickly or winter spinach. Quantities of spinach are grown commercially both for the fresh trade and for freezing

◊ **spinach beet** [ˈspɪnɪtʃ ˈbiːt] *noun Beta vulgaris,* plant similar to sugar beet, but grown for its leaves which are cooked in the same way as spinach

spine [spaɪn] *noun* sharp pointed part of a plant, such as blackthorn

◊ **spinous** [ˈspaɪnəs] *adjective* (plant) with spines

spinney [ˈspɪni] *noun* small wood with undergrowth

spirochaete [ˈspaɪərəʊkiːt] *noun* type of bacteria shaped like a spiral

spit [spɪt] *noun* depth of soil which is dug with a spade

splay leg [ˈspleɪ ˈleg] *noun* disorder in piglets, which are born unable to stand properly

spoil [spɔɪl] *verb* to go bad

◊ **spoilage** [ˈspɔɪlɪdʒ] *noun* making food bad and inedible (by rotting, damp, etc.)

sponge gourd [ˈspʌnʒ ˈgʊəd] *noun* vegetable sponge which produces loofahs, which are the fibrous insides of the ripe fruits *(Luffa cylindrica)*

spongiform [ˈspʌnʒɪfɔːm] *see* BSE

spore [spɔː] *noun* reproductive body of certain plants and bacteria, which can survive in extremely hot or cold conditions for a long time

◊ **sporicidal** [spɒrɪ'saɪdəl] *adjective* which kills spores

◊ **sporicide** ['spɒrɪsaɪd] *noun* substance which kills spores

◊ **Sporozoa** [spɔːrə'zəʊə] *noun* type of parasitic Protozoa which includes Plasmodium, the cause of malaria

COMMENT: spores are produced by plants such as ferns or by algae and fungi. They are microscopic, and float in the air or water until they find a resting place where they can germinate. Other spores are produced by bacilli and cause food poisoning

spot price ['spɒt 'praɪs] *noun* price for a commodity which is actually ready for sale (as opposed to 'futures')

QUOTE merchants were quoting between £100/t and £104/t ex-farm for spot wheat
Farmers Weekly

spp = SPECIES (plural)

spraing [spreɪŋ] *noun* disease of potatoes spread by nematodes in the soil

spray [spreɪ] **1** *noun* **(a)** mass of tiny drops of liquid applied under pressure, often used to apply insecticides, herbicides, and liquid fertilizers; **spray drift** = blowing away of spray in windy conditions; vibrojets are used where spray drift is a problem; otherwise low pressure or rapidly vibrating sprayers may help to cut down drift; **spray irrigation** = system of irrigation using sprinklers which are located along a boom. Some booms rotate and can distribute water over a large circular area; **spray lines** = method of distributing irrigation water using flexible hose (mainly used for horticultural crops) **(b)** special liquid for spraying onto a plant to prevent insect infestation *or* disease **2** *verb* to send out a liquid in fine drops; *they sprayed the room with disinfectant; apple trees must be sprayed twice a year to kill aphids*

◊ **spraybar** ['spreɪbɑː] *noun* attachment consisting of a horizontal tube with nozzles or jets, used for spraying over a wide area

◊ **sprayer** ['spreɪə] *noun* machine which forces a liquid through a nozzle under pressure, used to distribute liquid herbicides, fungicides, insecticides and fertilizers; a sprayer consists of a tank, a power take-off driven pump, a control valve and a spraybar or boom, with a number of nozzles or jets

QUOTE most trailed sprayer booms in France are now fitted with 24m (80ft) booms. At the Paris show a trailed 36m (120ft) boom was introduced
Farmers Weekly

spread [spred] *verb* to put manure *or* fertilizer *or* mulch out over a large area

◊ **spreader** ['spredə] *noun* **(a)** device used for spreading, such as one for spreading granules of fertilizer evenly over a lawn **(b)** agent added to an insect spray in order to make sure that the foliage is covered uniformly

sprig [sprɪg] *noun* small shoot or twig

spring [sprɪŋ] *noun* **(a)** place where water comes naturally out of the ground; *the first settlers found springs of fresh water in the mountains* **(b)** device like a coiled wire, used to support; **spring-tined harrow** *or* **harrow with spring-loaded tines** = cultivator which has tines of spring steel which vibrate in the soil; this gives fast seedbed preparation **(c)** the first season in the year (around April in the Northern Hemisphere); **spring wheat** = wheat which is sown in the spring and harvested towards the end of the summer (the wheat is softer than winter wheat)

◊ **springer** ['sprɪŋə] *noun* cow almost ready to calve; also called a 'down-calver'

springtail ['sprɪŋteɪl] *noun* primitive wingless insect very common in soils, where they may do damage to fine roots

sprinkle ['sprɪŋkl] *verb* to water with drops in a light shower

◊ **sprinkler** ['sprɪŋklə] *noun* hose which sends out a shower of drops; **rotary sprinkler** = sprinkler system where the water comes out of a central rotating nozzle, thus watering a circular area

sprout [spraʊt] **1** *noun* little shoot from a plant, with a stem and small leaves; *see also* BRUSSELS **2** *verb* (*of plants*) to send out new growth; *about 50% of the UK potato crop is grown from sprouted tubers*

spruce [spruːs] *noun* temperate softwood tree of the genus Picea; **spruce-larch adelgid** = relative of the aphid which may cause serious damage on spruce grown for Christmas trees (*Adelges viridis*)

spur [spɜː] *noun* **(a)** short spinelike

projection on the leg of a domestic cock **(b)** side shoot of a tree, with clusters of fruit buds

spurrey ['spʌri] *noun Spergula arvensis*, a common weed on arable land; also called 'corn spurrey'

square ploughing ['skweə 'plauɪŋ] *noun* method of ploughing suitable for large areas. A piece of land is ploughed in the centre of a field and then the field is ploughed in a clockwise direction starting from this central point

squash [skwɔʃ] *noun* plants of the *Cucurbitaceae* family, such as marrows, cucumbers

squirrel ['skwɪrəl] *noun* medium-sized rodent living in trees; squirrels are harmless as far as crops are concerned; there are two types of squirrel in the UK: the grey squirrel *(Sciurus carolinensis)* and the red squirrel *(Sciurus vulgaris)*

SSEW = SOIL SURVEY OF ENGLAND AND WALES

SSLRC = SOIL SURVEY AND LAND RESEARCH CENTRE

SSPDC = SCOTTISH SEED POTATO DEVELOPMENT COUNCIL

SSSI = SITE OF SPECIAL SCIENTIFIC INTEREST

stable ['steɪbl] **1** *noun* building used for lodging and feeding horses; **stable fly** = fly which is like the house fly, but with a distinct proboscis which can pierce the skin; it breeds in stable manure and is a serious pest to animals as the bites cause irritation **2** *adjective* not changing; **stable climax** = more or less stable community of plants and animals in equilibrium with its environment, the final stage of an ecological succession; **stable population** = population which remains at the same level, where births and deaths are equal

◊ **stability** [stə'bɪlɪti] *noun* being stable *or* resisting any change

◊ **stabilization lagoon** [steɪbɪlaɪ'zeɪʃn lə'guːn] *noun* pond used to purify sewage in warm countries by allowing sunlight to fall on the mixture of sewage and water

◊ **stabilize** ['steɪbəlaɪz] *verb* to make something stable; to become stable;

marram grass was planted on the sand dunes to stabilize them

◊ **stabilizer** ['steɪbəlaɪzə] *noun* artificial substance added to processed food to keep the mixture stable (as in sauces containing water and fat): in the EU emulsifiers and stabilizers have E numbers E322 - E495

stack [stæk] *noun* pile of sheaves of grain, hay or straw; stacks can be round or square

stag [stæg] *noun* **(a)** male deer **(b)** male of various animals castrated after maturity **(c)** male turkey

staggers ['stægəz] *noun* condition of animals in which they stagger about, as in looping-ill and swayback disease; grass staggers in cattle is caused by hypomagnesaemia

stake [steɪk] **1** *noun* thick wooden post, to which a tree or shrub is attached to keep it upright; *a newly planted tree needs to have a stake* **2** *verb* to attach a plant to a stake

stale seedbed ['steɪl 'siːdbed] *noun* method of killing weeds by using a contact herbicide at the time of drilling

stalk [stɔːk] *noun* stem which attaches a leaf *or* flower *or* fruit to the main part of a plant; **stalk rot** = disease of maize caused by a fungus

stall [stɔːl] *noun* single compartment in a stable or cowshed, where an animal can stand or lie down

stallion ['stæljən] *noun* uncastrated full-grown male horse, especially one kept for breeding

stamen ['steɪmən] *noun* male part of a flower consisting of a stalk (the filament) bearing a container (the anther) in which pollen is produced

stance [stɑːns] *noun* **feed stance** = open stall in a building where animals can feed

stand [stænd] *noun* group of plants growing together; *(in a forest)* group of standing trees; *softwoods from the vast conifer stands in Scandinavia could be widely used in place of hardwood;* **cereal stand** = field of standing cereals

◊ **standing** ['stændɪŋ] *adjective* growing upright; **standing crop** = (i) crop (such as corn) which is growing upright in a field; (ii) the numbers and weight of the living vegetation of an area, calculated by weighing the vegetation growing in a sample section; **standing timber** = trees which are growing in a wood, ready to be felled

standard ['stændəd] **1** *adjective* normal; *it is the standard practice to monitor the pollution levels twice a day* **2** *noun* **(a)** something which has been agreed upon and is used to measure other things by; **effluent standard** = amount of sewage which is allowed to be discharged into a river *or* the sea; **emission standard** = amount of a pollutant which is allowed to be released into the environment **(b)** type of fruit tree (or rose tree) where the stem is about two metres high, on top of which the head is developed; **standard apple** *or* **pear** = apple or pear tree grown as a standard; *see also* HALF-STANDARD

COMMENT: in UK orchards, standard trees have been replaced by bush varieties which are easier and cheaper to prune, spray and harvest

standard man day (SMD) ['stændəd 'mæn deɪ] *noun* eight hours of work, used as a measure for calculating labour costs on a farm

COMMENT: one milking cow is calculated as equal to ten SMDs, and one hectare of barley or wheat is equal to 50 SMDs; if a farm has less than 275 SMDs it is not counted as a full-time farm

staphylococcus [stæfɪlə'kɒkəs] *noun* bacterium which grows in a bunch like a bunch of grapes, and causes boils and food poisoning. *Staphylococcus aureus* causes mastitis in cows, sheep and pigs (NOTE: plural is **staphylococci,** sometimes shortened to **staphs)**

◊ **staphylococcal** [stæfɪlə'kɒkəl] *adjective* referring to Staphylococci; **staphylococcal mastitis** = condition of cows caused by several types of staphylococci especially when accompanied by stress resulting from liver fluke or cold conditions. Milk becomes watery and the cow has a high temperature

COMMENT: staphylococcal infections are treated with antibiotics such as penicillin, or broad-spectrum antibiotics such as tetracycline

staple ['steɪpl] **1** *adjective* **staple**

commodity = basic food or raw material **2** *noun* wool fibres used in determining quality; **staple length** = the length of the wool fibre

starch [stɑːtʃ] *noun* substance found in green plants, the usual form in which carbohydrates exist in food, especially in bread, rice and potatoes

◊ **starchy** ['stɑːtʃi] *adjective* (food) which contains a lot of starch

COMMENT: starch is present in common foods and is broken down by the digestive process into forms of sugar. Carbohydrate is not stored in the bodies of animals in the form of starch, but as glycogen

starter ['stɑːtə] *noun* culture of bacteria, used to inoculate animals, or to start growth in milk used in cheese production

station ['steɪʃn] *noun (in Australia)* very large farm, specializing in raising sheep or cattle

steam up ['stiːm 'ʌp] *verb* to feed a cow before it calves, to prepare it for the next lactation

stecklings ['steklɪŋz] *noun* young sugar beet plants grown in seedbeds in summer, to be transplanted in the autumn or following spring

steer [stɪə] *noun* castrated male bovine over one year old; also called a 'bullock'

steerage hoe ['stiːrɪdʒ 'həʊ] *noun* hoe mounted behind a tractor and steered by the driver to avoid crop damage

stell [stel] *noun* stone shelter for sheep and cattle in upland areas

stem [stem] *noun* piece of plant tissue which holds the plant upright *or* which bears the leaves and flowers; **stem eelworm** = pest affecting cereals, in particular oats; the plant stem swells and is prevented from growing and producing any ears; **stem rot** = disease caused by deficiency of nutrients; **stem rust** = *Puccinia graminis,* a disease of wheat, infecting the stem

◊ **stemmed** ['stemd] *adjective* with a stem; *a short-stemmed variety of rose*

steppe [step] *noun* wide grass-covered plain, with very few trees, found in central Asia and Russia (NOTE: the North American equivalent is **prairie)**

sterile ['sterail] *adjective* **(a)** with no microbes *or* infectious organisms **(b)** infertile *or* not able to produce offspring *or* incapable of sexual reproduction

◊ **sterility** [ste'rɪlɪti] *noun* (i) being free from germs; (ii) infertility, being unable to produce offspring (NOTE: the opposite is **fertile, fertility**)

◊ **sterilization** [sterɪlaɪ'zeɪʃn] *noun* (i) action of making soil, etc., free from bacteria; (ii) action of making an organism unable to produce offspring

◊ **sterilize** ['sterɪlaɪz] *verb* **(a)** to make something sterile (by killing microbes *or* bacteria); *the soil needs to be sterilized before being used for intensive greenhouse cultivation;* **sterilized milk** = milk prepared for human consumption by heating in sealed airtight containers to kill all bacteria **(b)** to make an organism unable to have offspring

sticker ['stɪkə] *noun* substance added to a fungicide or bactericide preparation to help it to stick to the sprayed surface

stigma ['stɪgmə] *noun* part of the female organ of a flower which receives the pollen, and where the pollen grains are stimulated to germinate

stile [staɪl] *noun* steps arranged so that one can climb over a wall or fence

stillbirth ['stɪlbɜːθ] *noun* birth of a dead animal or abortion at a late stage of pregnancy

◊ **stillborn** ['stɪlbɔːn] *adjective* (animal) which is born dead

Stilton ['stɪltən] *noun* English blue cheese

stipula ['stɪpjulə] *noun* newly sprouted feather

stipule ['stɪpjuːl] *noun* one of the pair of wing-like growths at the base of a leaf stalk

stirk [stɜːk] *noun* Scottish term for cattle, both male and female, under two years old

stock [stɒk] **1** *noun* **(a)** animals *or* plants which are derived from an ancestor; **stock bull** = bull kept for breeding purposes in a pedigree herd; **stock breeder** = farmer who specializes in breeding livestock; **stock farming** = breeding livestock for sale; *see also* LIVESTOCK, ROOTSTOCK **(b)** plant with roots on which a piece of another plant (the scion) is grafted; **stock culture** =

basic culture of bacteria from which other cultures can be taken **(c)** quantity of something held for future use **2** *verb* to introduce livestock into an area or into a farm; **stocking rate** = measure of the carrying capacity of an area in terms of the number of livestock in it at a given time, such as the number of animals per hectare; **set stocking** = grazing system associated with extensive grazing; livestock graze an area where they remain for an indefinite period; the traditional practice in Britain

◊ **stockman** ['stɒkmən] *noun* farm worker who looks after animals, especially cattle

◊ **stockproof** ['stɒkpruːf] *adjective* (fence) which livestock cannot get through

stocky ['stɒki] *adjective* (animal) with short strong legs

stolon ['stəʊlɒn] *noun* thin horizontal shoot at the base of a plant, which gives rise to new shoots

stoma ['stəʊmə] *noun* pore in a plant, especially in the leaves (NOTE: plural is **stomata**)

COMMENT: the stomata are the holes through which a plant takes in carbon dioxide and sends out oxygen

stone [stəʊn] *noun* **(a)** hard seed inside a fruit such as a date or peach **(b)** small piece of rock; *(in a combine harvester)* **stone trap** = trough, with a trap door, which prevents stones passing between the cylinder and the concave; **stone-ground flour** = flour made by grinding with millstones

Stoneleigh ['stəʊnliː] home of the National Agricultural Centre and proposed site for the National Museum of Food and Farming

stook [stʊk] *noun* several corn sheaves (usually twelve) gathered together in a field to form a small pyramid; also called a 'shock'

stop [stɒp] *verb* to remove the growing tip of a shoot, to encourage lateral growths

store [stɔː] *verb* to keep something until it is needed; **store cattle** *or* **stores** *or* **store lambs** = cattle *or* lambs bred or bought for fattening; animals are usually reared on one farm and then sold on to dealers or other farmers; *we usually buy store lambs in August for finishing*

◊ **storage** ['stɔ:rɪdʒ] *noun* keeping something until it is needed; **storage drying** = method of drying bales of hay by blowing air through them; there are several methods of storage drying: in a building with airtight sides, air is forced up through ventilation holes in the floor; in open barns, by radical drying or by a centre duct system

stover ['stəuvə] *noun* inferior type of fodder

straight [streɪt] *noun* animal foodstuff composed of one type of food; **straight fertilizer** = inorganic fertilizer containing one essential mineral, as opposed to compound fertilizers which contain more than one mineral; these fertilizers are also simply called 'straights'

strain [streɪn] *noun* distinct variety of a species, which will breed true, usually referring to cultivated plants; *they have developed a new strain of virus-resistant rice*

strake [streɪk] *noun* attachment bolted onto the rear wheels of a tractor to improve wheel grip; the strake has spikes which can be extended beyond the tyre and which dig into the soil

strangles ['stræŋgəlz] *noun* disease of mangolds and sugar beet. It occurs in fairly large seedlings after singling. The stem is severely damaged

straw [strɔ:] *noun* dry stems and leaves of cereal crops left after the grains have been removed by threshing; usually obtained from cereals, but sometimes from grain legumes; barley straw may be used as a low quality feed for livestock; wheat straw is used as litter; **straw burning** = cheap method of disposal of straw, which helps to control diseases. A ban on straw burning came into force in the UK in 1993. In addition to stubble burning it prohibits burning residues from oilseed rape, peas and beans. Broken bales and old straw stacks may still be burned; **straw chopper** = device fitted to the back of a combine which chops straw into short lengths and drops it on the stubble; chopped straw is easier to plough in; **straw spreader** = device attached to the back of a combine when the straw is not wanted; the straw is spread over the ground and then ploughed in; **straw walker** = part of a combine harvester where straw is carried away from the threshed grain after it has been separated from the stalks

COMMENT: straw can be ploughed back into the soil. Straw is often mixed with animal dung to make manure. Non-agricultural uses are varied and include thatching, making paper and bricks. It can be compressed into bundles to act as fuel and in this way can be used for heating farms and small local industrial buildings

strawberry ['strɔ:bəri] *noun* soft fruit of the *Fragaria* species, used as a dessert fruit, but also preserved as jam; **strawberry foot rot** = bacterial disease affecting sheep, causing ulcers

stray [streɪ] **1** *adjective (of an animal)* lost; *the farmers are looking for stray sheep in the hills* **2** *verb (of an animal)* to wander away from an enclosure

strength [streŋθ] *noun* ability of wheat flour to produce a yeasted dough capable of retaining carbon dioxide bubbles until the proteins in the bubble walls become relatively rigid (at about 75°C); the milling quality of wheat is measured by the Hagberg test

streptococcus [streptə'kɒkəs] *noun* genus of bacteria which grows in long chains, and causes fevers. *Streptococcus uberis* (informally 'strep uberis') affects the udders of cows and causes mastitis; streptococci are treated with antibiotics such as streptomycin (NOTE: plural is **streptococci** sometimes shortened to **streps)**

◊ **streptococcal** [streptə'kɒkl] *adjective* (infection) caused by a streptococcus

◊ **streptomycin** [streptə'maɪsɪn] *noun* antibiotic used against many types of infection, especially when streptococcal

stress [stres] **1** *noun* condition where an outside influence changes the working of the metabolism of an animal or plant; **drought stress** = lack of growth caused by drought **2** *verb* to worry an animal so that it becomes ill

strike [straɪk] **1** *noun* infestation of the flesh of sheep by the larvae of blowflies; it causes extreme irritation and death can occur in a short time **2** *verb (of cuttings)* to take root

string beans ['strɪŋ 'bi:nz] *noun* French beans

strip [strɪp] **1** *noun* long thin piece of land;

strip cropping = method of cultivation in which ploughing follows the contours; this is to check soil erosion on slopes; **strip cultivation** = method of communal farming where each family has a long strip or strips of land to cultivate; **strip farming** = method of farming where strips of land across the contours are planted with different crops; **strip grazing** = system of grazing which allows animals access to a small part of the field; the rest of the field is protected by a temporary fence, usually electric **2** *verb* to remove a covering from something; *spraying with defoliant strips the leaves off all plants;* **nutrient stripping** = removing nutrients from sewage to prevent eutrophication of water in reservoirs

◊ **strip cup** ['strɪp 'kʌp] *noun* container into which the first drops of milk are drawn by hand from the teats of the cow before the milking machine is attached to the udder

◊ **stripper-header** ['strɪpə'hedə] *noun* machine which harvests a crop such as linseed, and strips off the seedheads

◊ **strippings** ['strɪpɪŋz] *noun* last drops of milk from a cow's teats at the end of milking session

stripe rust ['straɪp 'rʌst] *see* YELLOW RUST

struck [strʌk] *noun* acute disease of sheep; a form of entero-toxaemia; it affects sheep which are one to two years old and is very localized (in Britain it occurs only in the Romney Marsh and in some Welsh valleys)

stubble ['stʌbl] *noun* short stems left in the ground after harvesting crops of cereals, peas and beans; **stubble burning** = straw burning, method of removing dry stubble by burning it before ploughing (now banned); **stubble cleaning** = working the stubble after harvest, using ploughs, cultivators, and harrows to free the weeds from the soil; **stubble field** = field in which the crop has been harvested and the stubble left in the ground; *the animals are driven into the stubble fields to graze after the harvest*

COMMENT: stubble burning has the advantage of removing weed seeds and creating a certain amount of natural fertilizer which can be ploughed into the soil. The disadvantage is that it pollutes the atmosphere with smoke, releasing large amounts of carbon dioxide. This, together with the possible danger that the fire may

get out of control, killing small animals and burning trees and crops, means that it is not recommended as a means of dealing with the stalks of harvested plants unless on heavy clay soil. Modern byelaws require ash to be incorporated into the soil so that true direct drilling is not possible

stud [stʌd] *noun* **(a)** stud farm, an establishment where horses are kept for breeding **(b)** *US* male horse kept for breeding **(c)** metal nail with a head projecting above the surface

◊ **studded roller feed drill** ['stʌdɪd 'rəʊlə 'fiːd drɪl] *noun* type of external force feed seed drill; fluted rollers are replaced by rolls with studs or pegs; it is suitable for drilling most types of seed

◊ **stud farm** ['stʌd 'fɑːm] *noun* farm where horses are bred

stump [stʌmp] *noun* **tree stump** = short section of the trunk of a tree left in the ground with the roots after a tree has been cut down

stunt [stʌnt] **1** *noun* plant disease characterized by retarded growth **2** *verb* to reduce the growth of something; *the poor mountain soil supports a few stunted trees*

sty [staɪ] *noun* **sty** *or* **pigsty** = housing for pigs, usually with a small run next to it

style [staɪl] *noun* stalk-like structure at the top of the carpel of a flower, bearing the stigma

sub- [sʌb] *prefix* meaning (i) less important; (ii) underneath

suboestrus [sʌb'iːstrəs] *noun* situation where a female animal comes on heat, but does not show any of the usual signs

subsistence [sʌb'sɪstəns] *noun* existing *or* surviving, with very little food; **subsistence agriculture** *or* **subsistence farming** = growing products which are not for sale but provide just enough to feed the farmer and his family; **subsistence food** = food which is needed to keep a person alive

subsoil ['sʌbsɔɪl] *noun* layer of soil under the topsoil, which contains little organic matter and into which leached chemical substances from the topsoil are deposited; **subsoil water** = water held in the subsoil

◊ **subsoiler** ['sʌbsɔɪlə] *noun* heavy

cultivator consisting of a strong frame with long tines attached to it; it is used to break up compacted soil to allow free passage of air and water, a process called 'subsoiling'

substandard [sʌb'stændəd] *adjective* not up to standard quality; *a load of substandard potatoes was rejected by the buyer*

substratum ['sʌbstrɑːtəm] *noun* layer of rock beneath the topsoil and subsoil (NOTE: plural is **substrata)**

subtropics [sʌb'trɒpɪks] *noun* area between the tropics and the temperate zone

◊ **subtropical** [sʌb'trɒpɪkl] *adjective* referring to the subtropics; *the islands enjoy a subtropical climate; subtropical plants grow in the sheltered parts of the coast*

succession [sək'seʃn] *noun* series of stages, one after the other by which a plant community reaches its final stable state (or climax); **ecological succession** = series of communities of organisms which follow one another in an area until a stable climax community is formed

◊ **successional cropping** [sək'seʃənəl 'krɒpɪŋ] *noun* (i) growing several crops one after the other during the same growing season; (ii) sowing a crop (such as lettuce) over a long period, so that harvesting takes place over a similarly long period

succulent ['səkjuːlənt] *adjective & noun* (plant, such as a cactus) which has fleshy leaves or stems in which it stores water; **succulent foods** = feedingstuffs which contain a lot of water; they are palatable and filling, and usually have a laxative effect. Most root crops (for example swedes and turnips) are succulents

sucker ['sʌkə] *noun* shoot which develops from the roots or lower part of a stem; typically, suckers may form from the stem or roots of a stock (as on a rose) and must be removed; they sometimes appear some distance from the main stem. Some plants and trees (such as raspberries or elms) are propagated by means of suckers

suckle ['sʌkl] *verb* to feed with milk from the udder

◊ **suckler** ['sʌklə] *noun* **(a)** calf or other young animal which is suckling; **suckler herd** = herd of beef cattle, where each dam suckles its own calf or calves; **suckler cow** =

cow which rears its own calf and is later used for beef production **(b) sucklers** = flowers of clover

◊ **suckling** ['sʌklɪŋ] *noun* nursing a calf at the udder; **single suckling** or **double sucking** or **multiple suckling** = nursing one calf, two or several calves; **suckling pig** = unweaned piglet

sucrose ['suːkrəʊz] *noun* sugar, sweet substance found in plants

Sudan Dwarf [suː'dɑːn 'dwɔːf] *noun* breed of very small goat, found in southern Sudan

sudd [sʌd] *noun* floating mass of dense vegetation, sometimes preventing boats from passing, found on tropical rivers, such as the Upper Nile

Suffolk (Down) ['sʌfək 'daʊn] *noun* breed of sheep developed from crosses between the now extinct Norfolk Horn ewes and the Southdown ram; a large quick-growing animal with a close short fleece and a black face which has no wool on it. Suffolk crosses perform well under a broad range of farming systems, being equally effective for over-winter storing and for intensive early lamb production

◊ **Suffolk Punch** ['sʌfək 'pʌnʃ] *noun* heavy draught horse; coloured chestnut; it is shorter and more stocky than the shire and lacks feathers on the fetlocks

suffrutescent or **suffruticose** [səfruː'tesənt or sə'fruːtɪkəʊz] *adjective* (perennial plant) which has a wooden base to its stem and does not die down to ground level in winter

sugar ['ʃʊgə] *noun* sucrose, sweet substance $(C_{12}H_{22}O_{11})$ obtained from sugar cane or sugar beet

◊ **sugar beet** ['ʃʊgə 'biːt] *noun* Beta vulgaris, specialized type of beet grown for the high sugar content of its roots; it is cultivated in temperate regions, and in Britain is an especially important crop in East Anglia. The crowns and leaves of the crop are used for feedingstuff, as is also the residue after the sugar content has been extracted from the roots; **sugar beet harvester** = machine for harvesting sugar beet; it may be trailed or self-propelled; the machine cuts off the beet tops, lifts the root, cleans off the soil and conveys the beet to a hopper which is then emptied by a second elevator onto a trailer; **sugar beet topper** = attachment to a sugar beet harvester which collects the sugar beet tops; some have choppers and blower units, which chop up the tops and then blow them into a trailer

◊ **sugar cane** ['ʃʊgə 'keɪn] *noun* *Saccharum officinarum* a large perennial grass, whose stems contain a sweet sap

COMMENT: sugar cane is rich in sucrose which is extracted and used for making sugar; rum is a by-product of sugar cane. Cane sugar is now one of the most scientifically produced tropical products, although cutting is still often done by hand. Cane is grown in many tropical and subtropical regions, in particular in the Caribbean. The principal sugar producers are Cuba, India, Brazil, China, Puerto Rico and Hawaii for cane sugar, and Russia, the Ukraine, France and Germany for beet sugar

suitcase farmer ['suːtkeɪs 'faːmə] *noun* US farmer who lives some distance from his holding (i.e. more than 30 miles)

sulphate *or* US **sulfate** ['sʌlfeɪt] *noun* salt of sulphuric acid and a metal; **sulphate of ammonia** = nitrogenous fertilizer available in granular or liquid form; **sulphate of potash** = a fertilizer (potassium sulphate) made from the muriate of potash; it contains about 50% potash and is used by potato growers and market gardeners

◊ **sulphite** *or* US **sulfite** ['sʌlfaɪt] *noun* salt of sulphuric acid which forms part of several chemical compounds which are used in processing paper

◊ **sulphur** *or* US **sulfur** ['sʌlfə] *noun* yellow non-metallic chemical element which is essential to biological life (it is found in volcanoes, in limestone, and is contained in some amino acids widely present in fossil fuels, which when burnt release sulphur dioxide and sulphur trioxide into the atmosphere); **powdered sulphur** *or* **flowers of sulphur** = sulphur in powdered form, used to dust on plants to prevent mildew (NOTE: chemical symbol for sulphur is **S**; atomic number is **16**. Note also that words beginning **sulph-** are spelt **sulf-** in US English)

sulphuric acid (H_2SO_4) [sʌl'fjuːrɪk 'æsɪd] *noun* acid formed from sulphur trioxide and water, which forms in the atmosphere and is a nutrient for some types of plant. It is used to make chemical fertilizers and also in the plastics industry

summer ['sʌmə] *noun* season of the year, when the weather is warmest, when the sun is highest in the sky and when plants flower and set seed; **summer feeding** = feeding of cattle on permanent pastures in the summer months; **summer mastitis** = infection of the udder thought to be spread by biting flies. Cows become very ill, lameness may occur and milk is watery and later bloody; **summer solstice** = 21st June, the longest day in the Northern Hemisphere, when the sun is as its furthest point south of the equator

sunflower ['sʌnflaʊə] *noun* *Helianthus annuus,* an important oilseed crop grown in temperate areas; **sunflower oil** = oil extracted from sunflower seeds: seeds contain between 25 and 45% of semi-drying oil which is high in unsaturated fats

COMMENT: the oil extracted from the seeds is used for cooking and for margarine production; the residual cake after pressing is a high-protein livestock feed, and the whole plant can be fed to cattle. It is also useful as a green manure plant. Birds can cause serious damage to sunflower crops by feeding on the ripening seeds. The main producing countries are Russia, the Ukraine, Argentina and Romania

superlevy ['suːpəlevi] = SUPPLEMENTARY LEVY

superphosphate [suːpə'fɒsfeɪt] *noun* chemical compound formed by treating calcium phosphate (basic slag) with sulphuric acid, used as a fertilizer; superphosphates are either sold and used straight, or as a constituent of compound fertilizer; they are suitable for all crops and all soil conditions

supplement ['sʌplɪmənt] **1** *noun* material which is added; *the feed gives a useful protein supplement to the diet* **2** *verb* to add to make something enough; *in the winter it is necessary to supplement the diet with extra protein*

◊ **supplementary** [sʌplɪ'mentəri] *adjective* which is added; **supplementary levy** = in the EU, payment introduced to penalize milk production over the quota level; **supplementary ration** = concentrates fed to livestock to supplement feeds of hay and roots

support [sə'pɔːt] **1** *noun* thing which holds something upright; *plants with weak stems, such as tomatoes, need supports to stop them from falling to the ground* **2** *verb* **(a)** to hold something upright; *a tree with a heavy crop of fruit needs to be supported with forked stakes* **(b)** to provide enough food to live; *the desert supports only a few stunted trees*

◊ **support buying** [sə'pɔːt 'baɪɪŋ] *noun* intervention buying, a system used under the Common Agricultural Policy, whereby governments or their agents offer to buy surplus agricultural produce at a predetermined price; **support energy** = total energy expenditure necessary for the production of plant and animal agricultural foodstuffs; **support price** *or* **intervention price** = price at which the EU will buy farm produce which farmers cannot sell, in order to store it

surface drainage ['sɜːfəs 'dreɪnɪdʒ] *noun* removing of surplus water from an area of land by means of ditches and channels; **surface water** = water (after rain) which lies on the surface of the soil and does not drain into the soil, but flows across the surface as a stream and drains into rivers; **surface wind** = wind which blows across the land surface (as opposed to winds higher in the atmosphere)

surgeon ['sɜːdʒən] *noun* person who has qualified in the treatment of disease by cutting the diseased part; **tree surgeon** = person who specializes in the treatment of diseased or old trees, by cutting or lopping branches; **veterinary surgeon** = person who is qualified in veterinary science, person who is qualified to treat animals

surplus ['sɜːpləs] *noun* extra produce, produce which is more than is needed; *surplus corn and butter are stored until they can be sold on the world markets; after the storm, surplus water will drain away in storm gullies*

survive [sə'vaɪv] *verb* to continue to live, even in bad conditions; *not all the litter will survive the first few days; only some of the saplings planted survived the dry summer*

Sussex ['sʌsɪks] *noun* beef breed of cattle, similar to the North Devon. Dark cherry red in colour, they were originally used as draught animals in preference to draught horses; a hardy and adaptable breed

SVD = SWINE VESICULAR DISEASE

SVS = STATE VETERINARY SERVICE

Swaledale ['sweɪldeɪl] *noun* very hardy breed of sheep, with distinctive twisting horns and a black face with a white nose; the breed originated in the North Pennines of Yorkshire; the fleece has an outer layer of long coarse wool and an inner layer of fine dense wool. The Swaledale ewe is the mother of the popular lowland 'mule' ewe when mated to the Blue-faced (Hexham) Leicester ram

swamp [swɒmp] *noun* area of very wet land overgrown with vegetation; **mangrove swamp** = swamp covered with mangroves

◊ **swampland** ['swɒmplænd] *noun* area of land covered with swamp

◊ **swampy** ['swɒmpi] *adjective* (land) which is waterlogged and supports a natural community of plants and animals

sward [swɔːd] *noun* cover of grasses and clovers which makes a pasture; **sward height record pad** = notebook in which the height of a sward is recorded

◊ **swardsman** ['swɔːdzmən] *noun* farm worker who looks after or grows pasture

swarm [swɔːm] **1** *noun* large number of insects (such as bees or locusts) travelling as a group **2** *verb* (of insects) to travel as a large group

swath [swɒθ] *noun* row of grass or other plants lying on the ground after being cut; row of potatoes which have been lifted and left lying on the ground; **swath turner** = haymaking machine used to move individual swaths sideways and turn them over at the same time, so making the drying process faster; also used in wet conditions to scatter a swath to dry it more quickly; **finger wheel swath turner** = machine used for raking

swayback disease ['sweɪbæk dɪ'ziːz] *noun* disease of lambs caused by copper deficiency in the ewe's diet; lambs become unsteady and unable to walk; the disease is often fatal. It is often a problem when there has been no snow during the winter

swede [swiːd] *noun* vegetable *(Brassica rutabaga)* with an swollen root; an important forage crop, it is grown for feeding sheep and cattle, either in the field or as winter feed for housed livestock. Swedes have a slightly higher feeding value and keep better than turnips, so they are often lifted and clamped

Swedish Red and White ['swiːdɪʃ 'red nd 'waɪt] *noun* dual-purpose breed of cattle found in Central and Southern Sweden; the animals are cherry red in colour with white markings

sweeper bull ['swiːpə 'bʊl] *noun* bull used to serve cows not caught up by Artificial Insemination

sweet corn ['swiːt 'kɔːn] *noun* type of maize in which the grains contain a large amount of sugar rather than starch; grown for human consumption, and also known as corn-on-the-cob

sweetener ['swiːtnə] *noun* material, such as high fructose corn syrup, used to sweeten a manufactured product

> QUOTE world sugar stocks do not reflect potential supplies of sweeteners. High sugar prices will lead in due course to an increase in other sweeteners
> *Farmers Weekly*

sweet potato ['swiːt pə'teɪtəʊ] *noun Ipomoea batatas,* a starchy root crop grown in tropical and subtropical regions

> COMMENT: the sweet potato is valuable as famine food in parts of Africa and South America. The main producing countries are Indonesia, Vietnam and Japan. In the Southern USA, the tubers are called 'yams'. The plant has no connection with the ordinary potato

swell [swel] *verb* to grow fat; *the rain will help the grapes to swell;* **swollen shoot** = viral disease affecting the cocoa tree (NOTE: **swelled - swollen)**

swidden farming ['swɪdən 'fɑːmɪŋ] *noun* slash and burn, a form of agriculture where forest is cut down and burnt to create open space for growing crops; the space is abandoned after several crops have been grown and further forest is cut down

swill [swɪl] *noun* waste food from kitchens used for pig feeding. Swill must, by law, be boiled for at least an hour before feeding; any excess fat should be skimmed off before use. A licence must be obtained for premises where swill is processed and fed. It is illegal to feed kitchen scraps to pigs

swine [swaɪn] *noun* collective term for pigs; **swine erysipelas** = infectious disease of pigs caused by bacteria; symptoms include inflammation and skin pustules, the red marks on the skin which are diamond-shaped, and from which the disease gets its common name of 'diamonds'; it occurs especially in hot muggy weather and in its acute form can be fatal; **swine fever** *or* **pig typhoid** = notifiable disease of pigs; the symptoms are fever, loss of appetite and general weakness; it can be fatal. The disease was eradicated in Britain, but some further cases have occurred; **African swine fever** = a virus disease which is highly contagious among pigs. Animals suffer fever and high temperature followed by death; in Europe, it occurs in parts of Spain; **swine vesicular disease** = virus infection similar to foot and mouth disease, but which only attacks pigs; a notifiable disease; infected animals are compulsorily slaughtered and movement licences apply

SWT = SCOTTISH WILDLIFE TRUST

sycamore ['sɪkəmɔː] *noun Acer pseudoplatanus,* large hardwood tree of the maple family

symbiosis [sɪmbaɪ'əʊsɪs] *noun* condition where two or more organisms exist together and help each other to survive

◊ **symbiont** ['sɪmbaɪɒnt] *noun* one of the organisms living in symbiosis; *compare* COMMENSAL

◊ **symbiotic** [sɪmbaɪ'ɒtɪk] *adjective* referring to symbiosis; living in symbiosis; *the rainforest has evolved symbiotic mechanisms to recycle minerals*

◊ **symbiotically** [sɪmbaɪ'ɒtɪkli] *adverb* in symbiosis

symptom ['sɪmptəm] *noun* change in the way the body works *or* change in the body's appearance, which shows that a disease *or* disorder is present

◊ **symptomatic** [sɪmptə'mætɪk] *adjective* which is a symptom

syndicate ['sɪndɪkət] *noun* group of people *or* companies working together to make money; **machinery syndicate** = group of farmers who join together to buy very large items of equipment, which they can use in turn

syndrome ['sɪndrəʊm] *noun* group of symptoms and other changes in the body's functions which, when taken together, show that a particular disease is present

system ['sɪstəm] *noun* arrangement of things *or* phenomena which act together;

ecosystem = grouping of living organisms and the environment in which they live; **the Linnaean system** = method of naming organisms worked out by the Swedish scientist Carolus Linnaeus

◊ **systematic** [sɪstɪ'mætɪk] *adjective* organized in a planned way; part of a system; **systematic ploughing** = ploughing a field by sections, each of which is called a land

◊ **systemic** [sɪs'temɪk] *adjective* which affects the whole organism; **systemic compound** = compound fertilizer which reaches all parts of a plant; **systemic fungicide** = fungicide which acts by entering the plant and therefore is not removed by washing; **systemic herbicide** = herbicide which is absorbed into the plant's sap system through its leaves; **systemic insecticide** = insecticide which is taken up by a plant and enters the sap stream so that biting insects take the insecticide when they suck the sap

Tt

t = TONNE

tabanidae *or* **tabanids** [tə'bænidiː *or* 'tæbənɪdz] *noun* family of horse flies which are often large and fly fast; they have strong antennae. Most females suck blood, and attack large mammals such as cattle with their blade-like jaws

◊ **Tabanus** [tə'beɪnəs] *noun* any of the tabanidae

table ['teɪbl] *noun* **table bird** = poultry bird reared for meat; **water table** = top level of water in the ground that occupies spaces in rock *or* soil and is above a layer of impermeable rock

tack [tæk] *noun* **(a)** harness equipment **(b) on tack** = fields supplied for other farmers' animals to graze on; *he takes 360 store lambs on tack in the autumn to remove surplus grass*

◊ **tackroom** ['tækrʊm] *noun* room for storing harness equipment; the tack should be kept in damp-free conditions

Taenia ['tiːniə] *noun* genus of tapeworm

tag [tæg] *noun* small piece of metal attached the ear of an animal to identify it

taiga ['taɪɡə] *noun* forest region between the Arctic tundra and the steppe

Taihu ['taɪhuː] *noun* group of very prolific and fecund Chinese pig breeds, including the Meishan

tail corn *or* **tailings** ['teɪl 'kɔːn *or* 'teɪlɪŋz] *noun* grains of corn of inferior size

◊ **tailbiting** ['teɪlbaɪtɪŋ] *noun* biting the tail of another animal, a vice associated with pigs; the cause is not known, but could be due to bad housing

◊ **tailpiece** ['teɪlbaɪtɪŋ] *noun* extension of the mouldboard of a plough which helps to press down the furrow slice

taint [teɪnt] *verb* to give an unpleasant taste to food

take [teɪk] *verb* (*of grafted or transplanted plants*) to grow successfully

◊ **take-all** ['teɪkɔːl] *noun Gaumannomyces graminis,* disease of wheat and barley, causing black discoloration at the base of the stem, premature ripening and white ears containing little or no grain (also called 'whiteheads', although whiteheads can be caused by other diseases)

tallow ['tæləʊ] *noun* cattle by-product produced by rendering down all the inedible waste, used in the manufacture of soap and candles and also incorporated into animal feeds

tall fescue ['tɔːl 'feskjuː] *noun Festuca arundinacea,* very hardy perennial grass often used for winter grazing in hilly or less fertile areas

Tamworth ['tæmwɜːθ] *noun* breed of pig, red-gold in colour, which makes the animal almost immune to sunburn

COMMENT: Tamworths are widely exported to the USA and Australia because of their ability to stand hot sunshine; they are hardy and can thrive on the roughest land

tan [tæn] **1** *noun* an artificial riding surface consisting principally of wood bark **2** *adjective* light brown colour **3** *verb* to convert animal skins to leather

◊ **tanning** ['tænɪŋ] *noun* the process of converting animal skins to leather

COMMENT: hides are soaked in a mixture of chromium salts and after a certain period of soaking they become leather

tandem parlour ['tændəm 'pɑːlə] *noun* milking parlour where the cows stand in line with their sides to the milker

tank [tæŋk] *noun* large container for liquid, part of a spraying machine

◊ **tank mix** ['tæŋk 'mɪks] *noun* mixing several herbicides or pesticides into one mixture to be used as a spray

COMMENT: a variety of mixes are available from manufacturers, and farmers also prepare their own tank mixes. The most dangerous operations are the handling of concentrated pesticides during mixing and dilution. FEPA will make only recommended mixes legal in the UK

tankard ['tæŋkəd] *noun* variety of mangel

tanker ['tæŋkə] *noun* container on wheels, used to carry liquids (such as a 'milk tanker' or 'slurry tanker')

tannia ['tæniə] *noun* *Xanthosoma sagittifolium,* a West African root crop similar to taro

tannin ['tænɪn] *noun* any substance capable of precipitating the gelatin in animal hides as an insoluble compound, and so changing the hides to leather which is resistant to putrefaction. All tannins are obtained from plants, including tea, and the bark and galls of oak

tap [tæp] *verb* to cut a thin ring round the bark of a rubber tree to let the sap run out

◊ **tappable girth** ['tæpəbl 'gɜːθ] *noun* measurement round the stem of a rubber tree, showing when tapping can be started (about 45cm)

◊ **tapper** ['tæpə] *noun* person employed on a rubber plantation to tap trees

◊ **taproot** ['tæprʊt] *noun* thick main root of a plant which grows straight down into the soil

COMMENT: the taproot system is the primary root system of a plant; it has a main root with smaller roots branching off it, as opposed to a fibrous root system which has no main root

tapeworm ['teɪpwɜːm] *noun* parasitic worm with a small head and a long body like a ribbon

COMMENT: tapeworms enter the intestine when a person eats raw fish or meat. The worms attach themselves with hooks to the side of the intestine and grow longer by adding sections to their bodies. Tapeworm larvae do not normally develop in humans, with the exception of the pork and dog tapeworms, *Taenia solium.* Tapeworms

seldom need treatment in livestock, although sheepdogs should be wormed regularly

tapioca [tæpɪ'əʊkə] *noun* flour of the cassava

tar oil ['tɑː 'ɔɪl] *noun* winter wash used to control aphis and scale insects on fruit trees; it also clears moss and lichen

tare [teə] *noun* **dirt tare** = percentage of dirt and waste material lifted with a crop (such as sugar beet) when it is harvested (NOTE: no plural)

Tarentaise [tærən'teɪz] *noun* breed of dairy cattle from the Savoie region of France; the animals are yellowish fawn in colour, with black muzzle, ears and tail

tares [teəz] *see* VETCHES

target price ['tɑːgɪt 'praɪs] *noun* wholesale target price within the EU for certain products, such as wheat, which market management is intended to achieve; it is linked to the intervention price

COMMENT: target prices are set in terms of fixed agricultural units of account, which are converted into different national currencies using adjusted exchange rates known as 'green rates' (in the UK, the 'green pound'). A system of levies on non-EU agricultural imports is used to protect target prices when they are set above the general level of world prices. In addition, the EU has established an internal price support system based on a set of intervention prices set slightly below the target price. If the level of supply is in excess of what is needed to clear the market at the target price, the excess supply is bought by the Community at the intervention price, thereby preventing overproduction from depressing the common price level

taro ['tærəʊ] *noun* root crop *(Colacasia esculenta)* grown in tropical regions; it is the staple diet in the Pacific islands where the food made from it is called 'poi'; in West Africa it is called the 'cocoyam'. The corms are used as a source of starch

tarragon ['tærəgən] *noun* aromatic plant *(Artemisia dracunculus)* of which the leaves are used for seasoning

tassel ['tæsl] **1** *noun* **(a)** male flower of the maize plant **(b)** appendage of hair hanging from the neck of male turkeys **2** *verb (of maize)* to produce a tassel; *the crop tasselled early*

taungya ['taʊŋɡiə] *noun* tropical agricultural system where arable farming and forestry are practised together

TBC = TOTAL BACTERIAL COUNT

TCDD = TETRACHLORO-DIBENZOPARADIOXIN highly toxic environmentally persistent by-product present in 2,4,5-T

tea [ti:] *noun* dried leaves of one or more shrubs of the Camellia family (the commonest is *Camelia sinensis*) **black tea** = tea where the leaves wither and are then allowed to ferment in a dry place before drying and crushing; **green tea** = tea where the leaves are heated to prevent fermentation

COMMENT: tea is one of the hardiest of all subtropical plants. The plants are pruned to form low bushes and to encourage the production of leaves and help plucking. The principal tea-producing countries are India, Bangladesh, Sri Lanka, China, Indonesia and Japan

teak [ti:k] *noun* tropical hardwood which is resistant to water

teaser ram ['ti:zə 'ræm] *noun* vasectomized ram, used to stimulate ewes by 'non-fertile' mating prior to the introduction of fertile rams

teat [ti:t] *noun* nipple on an udder (in cattle there are four quarters to the udder, each drained by a teat); **teat chaps** = sores on the teat, probably due to abrasions caused by the milking machine; **teat cup** = part of a milking machine, a tube which fits over the teat of the cow; **teat dipping** = measure for control of mastitis in cattle; the teats are dipped in a cup containing an iodophor disinfectant

technology [tek'nɒledʒi] *noun* applying scientific knowledge to industrial processes; **appropriate technology** = technology that is suitable for the local environment, usually involving skills *or* materials that are easily available locally; **intermediate technology** = technology that is half way between the advanced technology of developed countries and the lack of technology in underdeveloped countries

tedding ['tedɪŋ] *noun* spreading by lifting the swaths of new-mown grass in haymaking, so as to expose more grass to the sun and air and make it dry more quickly

◊ **tedder** ['tedə] *noun* machine used to lift and loosen a swath, enabling air to circulate through the cut crop

COMMENT: tedders may be mounted or trailed, and some are of the rotary type with spring tines which rotate on a vertical axis and move the crop sideways; others (overshot hay tedders) have a rotor with a series of spring tines which lift the crop over the rotor and back to the ground

Teeswater ['ti:zwɔ:tə] *noun* breed of sheep; the original Teeswater was a very big animal, but the introduction of Leicester blood in the late 18th century reduced the size. The modern Teeswater is a longwool breed with a dark muzzle. It is used to provide rams for cross-breeding with Swaledale rams to produce the hybrid Masham

teff [tef] *noun* type of millet grown widely in Ethiopia

teg [teg] *noun* sheep in its second year

Telemark ['telɪmɑ:k] *noun* Norwegian breed of dairy cattle; the animals are red with white patches

tel quel ['tel 'kel] *noun* referring to the weight of any type of sugar in tonnes

temperate ['temprət] *adjective* neither very hot nor very cold; **temperate climate** *or* **temperate region** = climate *or* region which is neither very hot in summer nor very cold in winter; **temperate forest** = forest in a temperate region

temperature ['temprətʃə] *noun* measurement of heat in degrees; **temperature inversion** = atmospheric phenomenon where cold air is nearer the ground than warm air, making the temperature of the air rises as one gets further from the ground; **mean temperature** = average temperature; *the mean temperature for July is 25°C*

COMMENT: seeds need to reach certain temperatures before they can germinate

temporary ['temprəri] *adjective* which is not permanent *or* which does not last a long time; **temporary grassland** = arable land sown to ley for a limited period

tenant ['tenənt] *noun* person who pays rent for the use of a farm and land owned by a landlord

tender ['tendə] *adjective* **(a)** food which is soft and easy to eat **(b)** (plant) which cannot stand frost

◊ **tenderize** ['tendəraɪz] *verb* to make meat tender (by keeping it for a certain time in cold conditions, by applying substances such as papain, by injecting with enzymes, etc.)

◊ **tenderometer** [tendə'rɒmɪtə] *noun* device used for testing vining peas to see how firm they are, so allowing harvesting to take place at the right time

tendon ['tendən] *noun* strip of connective tissue which attaches a muscle to a bone; **tendon sheath** = tube of membrane which covers and protects a membrane

tendril ['tendrɪl] *noun* part of a climbing plant which holds onto a surface, so allowing the plant to climb; members of the pea family climb by twining their tendrils round convenient supports

terminal ['tɜːmɪnəl] *adjective* **(a)** referring to a shoot or bud at the end of a shoot **(b)** **terminal sire** = sire used in cross-breeding, whose progeny will possess a high rate of growth and good carcass quality, but will not be suitable for breeding themselves

terrace ['terəs] **1** *noun* flat strip of land across a sloping side of a hill, lying level along the contours; **terrace cultivation** = hill slopes cut to form terraced fields which rise in steps one above the other and are cultivated, often with the aid of irrigation **2** *verb* to build terraces on the side of a mountain; *the hills are covered with terraced rice fields*

COMMENT: terracing is widely used to create small flat fields on steeply sloping land, so as to bring more land into productive use, and also to prevent soil erosion

terrain [tə'reɪn] *noun* type of land surface; **all terrain vehicle** = vehicle which can be driven over all types of land surface

terra rossa ['teræ 'rɒsæ] *noun* red soils that develop over limestone; found in Spain, Southern France and Southern Italy

terricolous [tə'rɪkələs] *adjective* (animal) which lives in soil

terrier ['teriə] *noun* record of land held and its occupation and use

Teschen disease ['teʃən dɪ'ziːz] *noun* virus disease of pigs; a notifiable disease which causes fever, paralysis and often death. It is caused by an Enterovirus

testa ['testə] *noun* the seed coat, an outer layer surrounding a seed and protecting the embryo inside

testicle *or* **testis** ['testɪkl *or* 'testɪs] *noun* one of two male sex glands in the scrotum, producing sperm (NOTE: plural of **testis** is **testes)**

tetanus ['tetənəs] *noun* infection caused by *Clostridium tetani* in the soil, which affects the spinal cord and causes spasms which occur first in the jaw (also called lockjaw)

COMMENT: livestock can be affected by the disease. People who work on the land or with soil (such as farm workers or construction workers) should be immunized against tetanus

tether ['teðə] **1** *noun* rope, chain or halter, used to tie up animals **2** *verb* to tie up an animal with a rope or chain, so that it cannot move away

COMMENT: according to EU regulations the construction of sow tethers and their use will be banned in the UK from 1999

tetrachlorodibenzoparadioxin (TCDD) [tetrəklɔːrəʊdaɪbenzəʊpærədaɪ'ɒksɪn] *noun* dangerous gas made as a by-product of the manufacture of 2,4,5-T (this was the gas which escaped during the Seveso disaster in 1976)

tetracycline [tetrə'saɪklɪn] *noun* antibiotic used against various bacterial diseases

tetraploid ['tetrəplɔɪd] *noun* forms of grass and clover with larger seeds and a larger plant than ordinary grass and clover; deep green in colour, they are lower in dry matter, palatable and digestible; there are also tetraploidal varieties of wheat. The number of chromosomes has been doubled from the normal diploid state in ryegrasses

Texas Longhorn ['teksəs 'lɒŋhɔːn] *noun* rare breed of long-horned cattle found in Texas; the animals are mainly red in colour

Texel ['teksel] *noun* breed of sheep from the North of Holland; used to cross-breed as a flock sire

texture ['tekstʃə] *noun* **fine texture** = substance with a smooth surface *or* with very small particles; **coarse texture** *or* **rough texture** = substance with a rough surface *or* with large particles; **soil texture** = description of the size of particles that make up a soil

TFA = TENANT FARMERS ASSOCIATION

TGE = TRANSMISSIBLE GASTRO-ENTERITIS a serious disease in pig production

TGUK = TIMBER GROWERS (UK) LIMITED

t/ha = TONNES PER HECTARE

Tharparkar ['θɑːpɑːkæ] *noun* Zebu breed of cattle found in Pakistan; the animals are used for draught purposes and also for milk; they are a variety of colours, including blue-grey and red

thatch [θætʃ] **1** *noun* roofing material made of straw and reeds **2** *verb* to cover a roof with reeds *or* straw; *reeds provide the best material for thatching*

theaves [θiːvz] *noun* female sheep between the first and second shearing

Theobroma cacao [θiːəʊ'brəʊmə kə'kæəʊ] the cocoa tree

theoretical field capacity [θiːə'retɪkl 'fiːld kə'pæsɪti] *noun* rate of work that would be achieved if a machine were performing its function at its full-rated forward speed for 100% of the time

therophyte ['θerəʊfaɪt] *noun* annual, a plant which grows from a seed in spring, flowers, sets seed and dies within one season, leaving the seed to remain dormant in the soil during the winter

thiabendazole [θaɪə'bendəzəʊl] *noun* substance used to worm cattle

thiamine ['θaɪəmiːn] *noun* vitamin in the vitamin B complex

thicket ['θɪkɪt] *noun* wood of saplings and bushes growing close together

thin [θɪn] *verb* to remove a number of small plants from a crop, so allowing the remaining plants to grow more strongly

◊ **thinnings** ['θɪnɪŋz] *noun* small plants removed to let others grow more strongly

thistle ['θɪsl] *noun* perennial weed *(Cirsium arvense, Cirsium vulgarae)* with spiny or prickly leaves. An erect plant with large purple or white flower heads

thorax ['θɔːræks] *noun (in insects)* middle section of the body, between the head and the abdomen; *(in animals)* the chest, the space in the top part of the body above the abdomen, containing the diaphragm, heart and lungs, all surrounded by the rib cage

thorn [θɔːn] *noun* sharp spike which grows on certain plants, as a protection against browsing animals; **thorn forest** = tropical vegetation with thorny shrubs and bushy trees, with little grassland

◊ **thornless** ['θɔːnləs] *adjective* (plant) without thorns; *he developed a new thornless variety of blackberry*

◊ **thorny** ['θɔːnɪ] *adjective* with thorns

thoroughbred ['θʌrəbred] *adjective & noun* horse which is bred for particular characteristics, in particular, a horse bred for racing

thousand grain weight ['θaʊzənd greɪn 'weɪt] *noun* weight of a thousand grains, used as an indicator of grain quality

thousand-headed *or* **thousand-head kale** ['θaʊzəndhedɪd 'keɪl] *noun* variety of kale grown for feeding to livestock, usually in the winter months; it has many branches and small leaves; the dwarf thousand-head produces a large number of new shoots during the winter

threadworm ['θredwɜːm] *noun* pinworm, a thin parasitic worm *Enterobius* which infests the large intestine

three-point linkage ['θriːpɔɪnt 'lɪŋkɪdʒ] *noun* method of coupling implements to a tractor; automatic couplers for three-point linkage permit implements to be attached rapidly and safely

three-times-a-day ['θriːtaɪmzə'deɪ] *noun*

milking system, where cows are milked three times a day; using this system can increase milk yields

thresh [θreʃ] *verb* to separate grains from stalks and the seedheads of plants; in wheat and rye, the chaff is easily removed from the grain; in barley, only the awns are removed; **threshing machine** = machine formerly used to thresh cereals, now replaced by the combine harvester

threshold price ['θreʃhəʊld 'praɪs] *noun* in the EU, the lowest price at which farm produce imported into the EU can be sold; the price in the home market below which the government or its agencies must buy all the produce offered by producers for sale at that price

thrift [θrɪft] *noun* good health in an animal

◊ **thrifty** ['θrɪfti] *adjective* (animal) which is developing well

thrips [θrɪps] *noun* insect pest of vegetables; thrips operate on the underside of leaves, leaving spots of sap or other liquid which are red or blackish-brown; typical examples are the onion thrips, grain thrips and pea thrips

thrive [θraɪv] *verb (of animal or plant)* to do well

throw [θrəʊ] *verb (of animals)* to give birth to young

thrunter ['θrʌntə] *noun* three year old ewe

thrush [θrʌʃ] *noun* disease affecting the frog of a horse's hoof

Thuya ['θuːjæ] *Latin for* cedar

thyme [taɪm] *noun* common aromatic Mediterranean plant *(Thymus)* used for flavouring soups, stuffings and sauces

tick [tɪk] *noun* tiny parasite of the order Acarida, which sucks blood from the skin, one of the most serious parasites of livestock, it causes worrying and lack of thrift; **tick bean** = BROAD BEAN; **tick-borne fever** = infectious disease transmitted by bites from ticks; in cattle, the disease causes loss of milk yield and a lower resistance to other diseases. In sheep it causes fever, listlessness and loss of weight. Abortions may occur as a result of tick-borne fever; **tick pyaemia** = disease

affecting young lambs resulting in limb joint and internal abscesses

tied cottage ['taɪd 'kɒtɪdʒ] *noun* house which can be occupied by the tenant as long as the tenant remains an employee of the landlord

tier [tɪə] *noun* range of things placed one row above another, such as the arrangement of cages in a battery

tile draining ['taɪl 'dreɪnɪŋ] *noun* means of draining land using underground drains made of clay, plastic or concrete; special machines (called 'tile-laying machines') are available for this work

till [tɪl] **1** *noun* boulder clay, clay soil mixed with rocks of different sizes, found in glacial deposits **2** *verb* to plough *or* to cultivate *or* to harrow the soil, to make it ready for the cultivation of crops

◊ **tillage** ['tɪlɪdʒ] *noun* husbandry *or* action of tilling the soil

◊ **tiller** ['tɪlə] *noun* shoot of a grass or cereal plant at ground level; tillers form at ground level in the angle between a leaf and the main shoot; true stems are only produced from the tillers at a later stage in the plant's development

◊ **tillering** ['tɪlərɪŋ] *noun* developing several seedheads in a plant of wheat, barley or oats. Tillering leads to the production of a heavier yield, and can be induced by rolling the young crop in the spring when it begins to grow; tillering is used to compensate for poor establishment. It is also important in grasses

tilth [tɪlθ] *noun* condition of the soil surface as a result of cultivation; *work the soil into a fine tilth before sowing seeds; small seeds need a finer tilth than large seeds; too fine a tilth should be avoided as it will cap after rain*

timber ['tɪmbə] *noun* trees which have been felled and cut into logs

◊ **timberline** ['tɪmbəlaɪn] *noun* line across the side of a mountain marking the highest point at which trees will grow; *the slopes above the timberline were covered with boulders, rocks and pebbles*

Timothy ['tɪməθi] *noun* palatable tufted perennial grass *(Phleum pratense)*. It grows on a wide range of soils and is winter hardy. It is used in grazing mixtures and as a hay plant in conjunction with ryegrass

tine [taɪn] *noun* **(a)** pointed spike on a cultivator or harrow; types of tine include rigid, spring-loaded and spring; **tine harrows** = sets of curved tines sometimes used when the soil surface is caked or compacted; a tine harrow will break up the soil to depth of several inches **(b)** sharp prong of a fork or rake

◊ **tined** ['taɪnd] *adjective* with tines; *a spring-tined harrow*

Tipulidae [tɪ'pjuːlɪdiː] *noun* family of insects including crane flies (and their larvae, leatherjackets) which destroy plant roots

tissue culture ['tɪʃuː 'kʌltʃə] *noun* method of plant propagation which reproduces clones of the original plant by growing plant material on nutrient media containing plant hormones

title ['taɪtl] *noun* (i) right to hold goods or property; (ii) document proving a right to hold property; **title deeds** = document showing who is the owner of a property

toadstool ['təʊdstuːl] *noun* general name for a poisonous fungus which looks like a mushroom

tobacco [tə'bækəʊ] *noun* leaves of a plant *(Nicotiana)* which are dried and smoked, either in a pipe or as cigarettes or cigars; **tobacco mosaic virus** = virus affecting both tobacco plants and tomatoes

COMMENT: The tobacco plant is a coarse fast-growing plant with a simple cylindrical stem from 1.2 to 2.4 metres in height. Each leaf is broad, often 30cm in width, and 90cm in length. Tobacco contains nicotine, which is an addictive stimulant. Nicotine is a poisonous alkaloid obtained mainly from tobacco and is used in agriculture as an insecticide, usually in a 40% solution of nicotine sulphate

toe-in ['təʊɪn] *noun* the shorter distance between the bases of the front wheels of a tractor, compared to their tops. Toe-in improves steering performance and reduces wear on the front tyres

Toggenburg ['tɒgənbɜːg] *noun* small Swiss breed of goat; its colour is pale brown with white markings on face, legs and rump. In Britain it has been developed into a larger, darker animal, which is a good milker with a long lactation period

tolerance ['tɒlərəns] *noun* ability of an animal or plant to withstand adverse conditions

◊ **tolerant** ['tɒlərənt] *adjective* which does not react to something; **tolerant variety** = variety of crop which has been developed to withstand a disease or attacks by certain pests

◊ **-tolerant** ['tɒlərənt] *suffix* meaning which tolerates; *a salt-tolerant plant*

◊ **tolerate** ['tɒləreɪt] *verb* to accept *or* not to react to (something)

◊ **toleration** [tɒlə'reɪʃn] *noun* not reacting to something; **toleration level** = level at which an organism will tolerate something

tom turkey ['tɒm 'tɜːki] *noun* male turkey

tomato [tə'mɑːtəʊ] *noun Lycopersicon esculentum,* an important food crop; the ripe fruit are used in salads and many cooked dishes; also pressed to make juice and sauces; large quantities are canned. In Britain, tomatoes are cultivated in heated greenhouses to provide out-of-season crops. In Spain, Italy, and other countries with a Mediterranean climate, tomatoes are cultivated out of doors

ton [tʌn] *noun* measure of weight; *GB* **(long) ton** = unit of measurement of weight (= 1,016 kilograms); **(metric) ton** = tonne *or* unit of measurement of weight (= 1,000 kilograms); *US* **(short) ton** = unit of measurement of weight (= 907 kilograms)

◊ **tonne** [tʌn] *noun* metric ton *or* 1,000 kilos (NOTE: **tonne** is shortened to **t** after figures: **10,000t of wheat; the price is £159/t**)

QUOTE Saudi Arabia has become not only self-sufficient in wheat, but also exports substantial quantities. Total production of wheat rose to 2.6m tonnes in 1987, 3m tonnes in 1988, and an estimated 3.2m tonnes in 1989. This compares with the annual Saudi consumption of about 800,000 tonnes
Middle East Agribusiness

top dressing ['tɒp 'dresɪŋ] *noun* fertilizer applied to a growing crop

◊ **topknot** ['tɒpnɒt] *noun* tuft of hair on the top of an animal's head, found in certain breeds of cattle and sheep

◊ **top link sensor** ['tɒp lɪŋk 'sensə] *noun* the mechanism by which most draught controls sense the draught on a tractor implement (it uses the top link of the three-point linkage)

◊ **topper** ['tɒpə] *noun* machine used to cut the tops off sugar beet; also used in pasture; **topper unit** = unit forming part of a two- or three-stage system, with a chopper and blower unit

◊ **topping** ['tɒpɪŋ] *noun* cutting the leaves and stems from the sugar beet root; it must be done accurately, as overtopping reduces yield

◊ **tops** [tɒps] *noun* leaves and stems of plants, such as sugar beet, cut off and used as fodder for cattle and sheep or made into silage

◊ **top-saving attachment** ['tɒpseɪvɪŋ ə'tætʃmənt] *noun* attachment to a topper unit which collects the tops of sugar beet after they have been cut off

◊ **topsoil** ['tɒpsɔɪl] *noun* top layer of soil, often containing organic material, from which chemical substances are leached into the subsoil below. The topsoil is the part of the soil which is cultivated and in which most plant roots are formed; *see also* SOIL HORIZON

total bacterial count (TBC) ['təʊtəl bæ'tɪərɪəl 'kaʊnt] *noun* system of calculating the strength of an infection by counting the number of bacteria present in a sample quantity of liquid taken from the animal

Toulouse ['tuːluːz] *noun* medium-large grey and white breed of goose, which originates in France

tow [təʊ] *verb (of a vehicle)* to pull another wheeled vehicle or implement

◊ **towbar** ['təʊbɑː] *noun* strong bar at the back of a car or tractor, to which another vehicle can be attached to be pulled along

tower silo ['taʊə 'saɪləʊ] *noun* tall circular tower used for storing silage

Townshend ['taʊnzend] Viscount Townshend, (1674 - 1738) an 18th century Norfolk landowner; nicknamed 'Turnip' Townshend, he did much to make the Norfolk four-course rotation system popular

tox- *or* **toxo-** ['tɒksəʊ] *prefix* meaning poison

◊ **toxaemia** [tɒk'siːmɪə] *noun* blood poisoning; *see also* PREGNANCY TOXAEMIA

◊ **toxic** ['tɒksɪk] *adjective* poisonous *or* harmful; **toxic agent** = substance which is harmful; **toxic substances** = substances, such as lead, which are harmful to living organisms; **toxic waste** = waste which is poisonous

◊ **toxicity** [tɒk'sɪsɪti] *noun* level to which a substance is poisonous *or* amount of poisonous material in a substance; **acute toxicity** = level of concentration of a toxic substance, which makes people seriously ill, or can even cause death; **chronic toxicity** = exposure to harmful levels of a toxic substance over a long period of time

◊ **toxico-** ['tɒksɪkəʊ] *prefix* meaning poison

◊ **toxicological** [tɒksɪkə'lɒdʒɪkl] *adjective* referring to toxicology; *irradiated food presents no toxicological hazard to humans*

◊ **toxicologist** [tɒksɪ'kɒlədʒɪst] *noun* scientist who specializes in the study of poisons

◊ **toxicology** [tɒksɪ'kɒlədʒɪ] *noun* scientific study of poisons and their effects on the human body

◊ **toxicosis** [tɒksɪ'kəʊsɪs] *noun* poisoning

◊ **toxin** ['tɒksɪn] *noun* poisonous substance produced by an animal or plant cell, which induces the formation of antibodies within an animal

◊ **toxoplasmosis** [tɒksəʊplæz'məʊsɪs] *noun* infectious disease affecting ewes, which causes pregnant animals to abort

t/pa = TONNES PER ANNUM

TPO = TREE PRESERVATION ORDER

trace [treɪs] *noun* **(a)** very small amount; *there are traces of radioactivity in the blood sample* **(b) traces** = side-straps or chains by which a horse pulls a cart or implement

◊ **trace element** ['treɪs 'elɪmənt] *noun* a micronutrient, an element which is essential to organic growth, but only in very small quantities

COMMENT: plants need traces of iron, manganese, copper, zinc; human beings need the trace elements cobalt, chromium, copper, magnesium, manganese, molybdenum, selenium and zinc

tracklayer *or* **tracklaying tractor** ['trækleɪə *or* 'trækleɪɪŋ 'træktə] *noun* heavy-duty caterpillar tractor, used mainly for earthmoving and drainage work

tractor ['træktə] *noun* heavy vehicle, with large wheels, which provides the main source of power in modern farms; **tractor-mounted loader** = loader which is mounted on a tractor, and not trailed; **tractor vaporizing oil (TVO)** = fuel formerly used in many tractors, but now replaced by diesel oil

COMMENT: by 1920, the tractor was established as an alternative to the horse. In its early days, the tractor was used to pull trailed implements and to drive stationary equipment with a belt drive. Today, the general purpose tractor does most of the work on arable and livestock farms and may be either two- or four-wheel drive. Lighter tractors, usually two-wheel drive models, are used by market gardeners and farmers who grow row crops. More powerful four-wheel drive tractors are needed for ploughing and heavy cultivation. One type has four large wheels, all the same size; another type has smaller front wheels and larger back wheels. The heaviest tractors are tracklayers, or 'crawlers', which are used for very heavy work and on heavy soils. Besides pulling trailed implements such as balers, forage harvesters and drills, the tractor's hydraulic system can be used to raise and lower mounted implements and operate lifting and loading equipment. The hydraulic system provides the power for fertilizer spreaders, hedge trimmers and a variety of other implements. Medium-sized tractors develop 40-60hp, while large models can develop as much as 200hp. Very modern tractors have in-cab computers which can tell the driver how much ground he has covered, and how much fuel has been used, and even can advise on the gear which will give the most economic use of fuel. In the UK, since 1970, all new tractors must be fitted with an approved protective cab or frame to prevent the driver being crushed in accidents where the tractor rolls over

trafficability [træfɪkə'bɪlɪti] *noun* ability of soil to take machinery or stock without significant soil damage; it is related to the soil water content

trail [treɪl] **1** *noun* path; **nature trail** = footpath with guideposts and explanatory notices, designed to explain the fauna and flora of a piece of countryside to the general public; some farms have designed trails as a means of supplementing farm income **2** *verb* to pull; **trailed implements** = implements such as harrows which are pulled behind a tractor, as opposed to mounted implements which are fixed to the tractor and take their power from it

◊ **trailer** ['treɪlə] *noun* machine used for carrying purposes; trailers are of the two-wheel or four-wheel types, and are used for carrying cereal and root crops, and for general use on the farm

◊ **trailing** ['treɪlɪŋ] *adjective* (plant) whose shoots lie on the ground

train [treɪn] *verb* to make plants (especially fruit trees and climbing plants) become a certain shape, by attaching shoots to supports or by pruning

tramline ['træmlaɪn] *noun* paths left clear for the wheels of tractors to drive over; they are used as guidemarks for spraying and when applying fertilizer so that damage to crops is kept to a minimum

trans- [trænz] *prefix* meaning through *or* across

transhumance [træns'hjuːməns] *noun* practice of moving flocks and herds up to summer pastures and bringing them down to the valley again to be stall-fed in winter (transhumance is practised in the Alps and other mountainous regions such as the Pyrenees)

translocate [trænslə'keɪt] *verb* to move substances through the tissues of a plant; **translocated herbicides** = herbicides (such as glyphosate) which kill a plant after being absorbed through its leaves or through its roots if applied to the soil

◊ **translocation** [trænslə'keɪʃn] *noun* movement of substances through the tissues of a plant

transmissible gastro-enteritis (TGE) [træns'mɪsɪbl gæstrəʊentɪ'raɪtɪs] *noun* very infectious disease, which mainly affects very young pigs

transmit [trænz'mɪt] *verb* to pass on (a disease) to another animal or plant; *some diseases are transmitted by insect bites*

transpire [træn'spaɪə] *verb* (*of plants*) to lose water through stomata; *in tropical* *rainforests, up to 75% of rainfall will evaporate and transpire into the atmosphere*

◊ **transpiration** [trænspɪ'reɪʃn] *noun* removing moisture from the soil by plant roots, passing up the stem to the leaves and its loss from the plant through its stomata into the atmosphere

COMMENT: transpiration accounts for a large amount of water vapour in the atmosphere. A tropical rainforest will transpire more water per square kilometre than is evaporated from the same area of sea. Clearance of forest has the effect of reducing transpiration, with the accompanying change in climate: less rain, leading in the end to desertification

transplant [træns'plɑ:nt] *verb* to take a growing plant from one place and plant it in the soil in another place either to give it more growing space or to check its root growth

◊ **transplanter** [træns'plɑ:ntə] *noun* machine for transplanting seedlings (especially used for planting brassicas)

COMMENT: there are two types of transplanter: the hand-fed machine, where a worker feeds seedlings into the machine as it passes over the field. The other is the automatic transplanter, where seedlings are raised in special containers and placed in the machine before transplanting starts

tray [treɪ] *noun* flat shallow container (usually made of plastic) in which seeds can be sown in a greenhouse

tree [tri:] *noun* very large plant with a wooden stem; **tree ferns** = very large ferns found in Australasia, growing like trees with a single large stem; **tree surgeon** = person who specializes in the treatment of diseased trees, especially by pruning or lopping

◊ **tree farming** ['tri: 'fɑ:mɪŋ] *noun* growing of trees for commercial purposes

COMMENT: schemes have been started to grow trees on surplus agricultural land to provide fuel for wood-burning power stations. Farmers will be paid to plant genetically engineered poplars and willows. 90,000 tonnes of dry wood are expected each year. In 1995 farmers choosing to grow trees on set-aside land would get a first year grant of £562 an acre followed by an annual subsidy of £104 an acre

◊ **treeline** ['tri:laɪn] *noun* (i) line at a certain altitude, above which trees will not grow; (ii) line in the Northern or Southern hemisphere, north or south of which trees will not grow

◊ **tree preservation order (TPO)** ['tri: presə'veɪʃn 'ɔːdə] *noun* official ruling by a local authority forbidding the cutting down of an important tree

trefoil ['tri:fɔɪl] *noun Lotus uliginaosus*, a leguminous plant, the thin wiry form of a small-flowered yellow clover, sometimes grown in pasture mixtures; it is a useful catch crop, and thrives in marshy acid soils

trematode ['tremətəud] *noun* fluke, a parasitic flatworm

trench [trenʃ] **1** *noun* long narrow hole in the ground; **drainage trench** = long hole cut in the ground to allow water to run away **2** *verb* to dig trenches

◊ **trenching** ['trenʃɪŋ] *noun* method of double digging which loosens the soil to a depth of two feet, twice as deep as in plain digging

trial ['traɪəl] *noun* test carried out to see if something works well; **seed trials** = tests of new seeds to see if they germinate correctly

triangle ['traɪæŋgl] *see* GOLDEN TRIANGLE

triazine ['traɪəziːn] *noun* one of a group of soil-acting herbicides, such as Atrazine and Simazine; maize is tolerant to these substances

trichinosis *or* **trichiniasis** [trɪkɪ'nəusɪs *or* trɪkɪ'naɪəsɪs] *noun* disease caused by infestation of the intestine by larvae of roundworms or nematodes, which pass round the body in the bloodstream and settle in muscles. Pigs are usually infected after eating raw swill

trichlorophenoxyacetic acid [traɪklɔːrəufenɒksɪə'setɪk 'æsɪd] *noun* 2,4,5-T, herbicide which forms dioxin as a by-product during the manufacturing process. (This was the gas which escaped during the Seveso disaster in 1976)

Trichomonas [trɪkə'mɒnəs] *noun* species of long thin parasite which infests the intestines; *Trichomonas foetus* is a cause of infertility in cattle

trickle ['trɪkl] *verb (of liquid)* to flow gently; **trickle system** = irrigation system where water is brought to the base of each plant and drips slowly into the soil

trifolium [traɪ'fəʊlɪəm] *noun* the crimson clover *(Trifolium incarnatum)*, a plant which does best on calcareous loams and is grown after cereals as a catch crop; it is planted in mixed herbage as a winter annual for forage, particularly for sheep

trim [trɪm] *verb* to cut neatly

◊ **trimmer** ['trɪmə] *noun* implement for trimming hedges

◊ **trimmings** ['trɪmɪŋz] *noun* small pieces of vegetation which have been cut off a hedge when trimming

trip device ['trɪp dɪ'vaɪs] *noun* device used to sense when a person is too close to a hazard and isolate the hazard before contact can occur. Can take the form of trip bars as used on some rotating arm bale wrappers

triple-purpose animal ['trɪplpɜːpəs 'ænɪml] *noun* breed of animal (usually of cattle) which is used for three purposes (milk, meat and as a draught animal)

tripoding ['traɪpɒdɪŋ] *noun* drying hay on a wooden frame in the field; rarely practised in the UK, but still common in some parts of Europe. Also used as a means of drying out peas

triticale [trɪtɪ'keɪli] *noun Triticum x Secale*, a new cereal hybrid of wheat and rye; it combines the yield potential of wheat with the winter hardiness and resistance to drought of rye

COMMENT: increasingly used in the UK, triticale replaces winter and spring feed barleys. It has a high level of disease resistance and a reduced demand for chemical fertilizer. The name is made up from the Latin words for wheat (Triticum) and rye (Secale)

Triticum ['trɪtɪkəm] *Latin for* wheat

trocar ['trəʊkɑː] *noun* pointed rod which slides inside a cannula to draw off liquid or to puncture an animal's stomach to let gas escape

Trondheim ['trɒndhaɪm] *see* BLACKSIDED TRONDHEIM

tropics ['trɒpɪks] *noun* tropical countries, hot areas of the world between the Tropic of Cancer or Tropic of Capricorn and the equator; *he lives in the tropics; disease which is endemic in the tropics*

◊ **tropical** ['trɒpɪkl] *adjective* referring to the tropics

trotter ['trɒtə] *noun* foot of a pig or sheep

trough [trɒf] *noun* (a) long narrow open wooden or metal container for holding water or feed for livestock (b) long narrow area of low pressure, leading away from the centre of a depression

truck farming ['trʌk 'fɑːmɪŋ] *noun US* term used to describe intensive vegetable cultivation at a considerable distance from the urban markets where the produce is sold

true [truː] *adjective* correct; **to breed true** = to reproduce all the characteristics of the species; *F_1 hybrids do not breed true*

trug [trʌg] *noun* low fruit or garden basket made of willow strips fastened to a strong framework of ash or chestnut

trunk [trʌŋk] *noun* main stem of a tree

truss [trʌs] *noun* (a) bundle of hay or straw (b) compact cluster or flowers or fruit (tomatoes, for example)

trypanosome [[trɪ'pænəsəʊm] *noun* parasite which causes sleeping sickness, transmitted by the tsetse fly

◊ **trypanosomiasis** [trɪpænəsəʊ'maɪəsɪs] *noun* sleeping sickness, a serious disease attacking both livestock and humans, spread by the tsetse fly

tsetse fly ['tsetsi 'flaɪ] *noun* African insect which passes trypanosomes into the bloodstream of humans and other animals, causing sleeping sickness

tsutsugamushi disease [tsuːtsəgə'muːʃi dɪ'ziːz] *noun* form of virus caused by Rickettsia bacteria, found in South East Asia

tuber ['tjuːbə] *noun* fat part of an underground stem or root, which holds nutrients, and which has buds from which shoots develop; *a potato is the tuber of a potato plant*

◊ **tuberous** ['tju:bərəs] *adjective* (i) like a tuber; (ii) (plant) which grows from a tuber

> COMMENT: potatoes, cassavas and sweet potatoes are all tubers from which new shoots develop

tubercle ['tju:bəkl] *noun* **(a)** small lump on the surface of a bone **(b)** small tuber

tuberculin [tju'bɜ:kjulɪn] *noun* substance which is derived from the culture of the tuberculosis bacillus and is used to test patients for the presence of tuberculosis; **tuberculin testing** = testing of cattle for the presence of bovine tuberculosis

tuberculosis [tjubɜ:kju'ləusɪs] *noun* infectious disease caused by the tuberculosis bacillus, where infected lumps form in tissue and which affects humans and other animals. Cattle and pigs are more commonly affected than other species

> COMMENT: tuberculosis in cattle has been eradicated in the UK, by a policy of tuberculin testing and the slaughter of animals which react to the test. Bovine tuberculosis is a notifiable disease and can affect certain wild animals, in particular badgers. Infected badgers are believed by some people to be able to transmit the disease to cattle though this has not been proved. In humans, tuberculosis can take many forms: the commonest form is infection of the lungs (pulmonary tuberculosis). Tuberculosis is caught by breathing in germs or by eating contaminated food, especially unpasteurized milk. It can be passed from one person to another, and the carrier usually shows no signs of having the disease

TUFF = TECHNICAL UPDATING FOR FARMERS

Tull [tʌl] Jethro Tull (1674–1740), an 18th century gentleman farmer, he invented the mechanical seed drill and the horse-drawn hoe

tundra ['tʌndrə] *noun* cold treeless arctic and alpine region which may be covered with low shrubs, grasses, mosses and lichens

tung tree ['tʌŋ 'tri:] *noun Aleurites montana,* the tree from which tung oil is obtained, used for paints and varnishes. The oil has quick drying properties

tunnel ['tʌnəl] *noun* long enclosure, covered with a semicircular roof; **lambing tunnel** = covered enclosure for ewes and lambs; **tunnel drying** = method of storage drying of hay, where the bales are stacked in the form of a tunnel over a central duct through which unheated air is blown; *see also* POLYTUNNEL

tup [tʌp] **1** *noun* uncastrated male sheep **2** *verb* to serve a ewe

turbary ['tɜ:bəri] *noun* place where turf or peat is dug for fuel

turf [tɜ:f] *noun* surface earth covered with grass, with its roots matted in the soil

turkey ['tɜ:ki] *noun Meleagris gallopavo,* a large poultry bird raised for meat (NOTE: the adult males are called **cocks** *or* **toms,** the adult females are **hens)**

turmeric ['tɜ:mərɪk] *noun* tropical plant *(Curcuma longa);* its tubers produce an oil which is used in making curry powder; mainly grown in India

turn out ['tɜ:n 'aut] *verb* to put animals out to pasture, after being kept indoors during the winter; *the ewes are turned out in March*

◊ **turn out time** ['tɜ:n aut 'taɪm] *noun* season, usually in the spring, when animals which have been kept indoors during the winter are let out to grass

turnip ['tɜ:nɪp] *noun* brassica plant which has a swollen root and is an important forage crop, and also used as a vegetable; **stubble turnips** = quick-growing types of turnip sown into cereal stubble and grown as catch crops

> COMMENT turnips can be harvested by machine and stored outdoors in clamps; in milder areas they can be left growing in the fields and used when needed. Turnips are often grazed off in the field

tussock grass ['tʌsək 'grɑ:s] *noun Deschampsia caespitosa,* a coarse grass growing in tufts

TVO = TRACTOR VAPORIZING OIL

twice-a-day ['twaɪsədeɪ] *noun* milking system where cows are milked two times a day

twig [twɪg] *noun* small woody growth from

the branch of a tree, bearing leaves, flowers or fruit

twinter ['twɪntə] *noun* two year old ewe

twin [twɪn] *noun* one of two young born at the same time to the same mother; **twin embryos** = embryos of twin young, used in ET; **twin lamb disease** = pregnancy toxaemia, a metabolic disorder affecting ewes during late pregnancy. Animals wobble and fall down, breathing is difficult, and death may follow. It is associated with lack of feed in late pregnancy

◊ **twinning** ['twɪnɪŋ] *noun* giving birth to twins; **induced twinning** = producing twin young after ET with twin embryos

twine [twaɪn] **1** *noun* strong string used for binding bales **2** *verb (of a tendril)* to coil round a support

twist [twɪst] *noun* disease of cereals and grasses which causes malformation of the leaves and stalks due to the growth of internal fungus. May prevent the ear

emerging from its sheath *(Dilophospora alopecuri)*

twitch [twɪtʃ] *see* COUCH GRASS; LOUPING-ILL

2,4-D ['tuːfɔː 'diː] selective translocated hormone herbicide especially effective against broad-leaved weeds growing in cereals

2,4,5-T ['tuːfɔːfaɪv 'tiː] trichloro-phenoxyacetic acid, a phenoxy hormone herbicide, effective against woody shrubs and similar vegetation

COMMENT: dioxin is formed as a by-product during the manufacturing process (this was the gas which escaped during the Seveso disaster in 1976)

two-sward system ['tuːswɔːd 'sɪstəm] *noun* grazing system where the area being grazed is kept separate from the area being conserved for cutting

two-tooth ['tuːtuːθ] *noun* sheep showing two permanent incisors (approximately 18 months old)

Uu

U Good (in the EUROP carcass classification system)

udder ['ʌdə] *noun* the mammary gland of an animal, which secretes milk; it takes the form of a bag under the body of the animal with teats from which the milk is sucked; **udder oedema** = livestock disease probably due to increased pressure in the milk vein around the time of calving. The udder becomes swollen. Sometimes caused by excess sodium in diet

> COMMENT: in the cow there are four glands, each with a teat. In the ewe there are two glands, each with a teat, and in the sow there are two glands per teat and from 12 to 20 teats

ugli ['ʌgli] *noun* citrus fruit similar to a grapefruit, but of uneven shape

UHT = ULTRA HEAT TREATED, ULTRA HIGH TEMPERATURE

UKASTA = UNITED KINGDOM AGRICULTURAL SUPPLY TRADE ASSOCIATION

UKDA = UNITED KINGDOM DAIRY ASSOCIATION

UKROFS = UNITED KINGDOM REGISTER ORGANIC FOOD STANDARD

UKEPRA = UNITED KINGDOM EGG PRODUCERS ASSOCIATION

ulcer ['ʌlsə] *noun* open sore in the skin *or* in mucous membrane, which is inflamed and difficult to heal

◊ **ulcerated** ['ʌlsəreɪtɪd] *adjective* covered with ulcers

◊ **ulceration** [ʌlsə'reɪʃn] *noun* (i) condition where ulcers develop; (ii) the development of an ulcer

ulluco [u'luːkəʊ] *noun Ullucus tuberosus,* tuber grown in South America and used as a vegetable

Ulmus ['ʌlməs] *Latin for* elm

Ulster White ['ʌlstə 'waɪt] *noun* breed of pig popular for bacon production in Northern Ireland; quite rare today, having been replaced by the Large White

ultra- ['ʌltrə] *prefix* meaning (i) further than; (ii) extremely

◊ **ultra heat treated milk (UHT milk)** ['ʌltrə 'hiːt triːtɪd 'mɪlk] *noun* milk which has been treated by sterilizing at temperatures above 135°C, and then put aseptically into containers (it has a much longer shelf-life than normal milk)

◊ **ultra high temperature (UHT)** ['ʌltrə 'haɪ 'temprətʃə] *adjective* process of sterilization at very high temperatures; **UHT sterilization** = sterilization of milk at very high temperatures; milk which has been treated in this way may be stored for periods of up to one year

◊ **ultramicroscopic** [ʌltrəmaɪkrə-'skɒpɪk] *adjective* so small that it cannot be seen using a normal microscope

◊ **ultrasonics** [ʌltrə'sɒnɪks] *noun* using high frequency sound waves to tell what is below the skin of a live animal: by using ultrasonics, it is possible to tell the amount of fat layers and the muscle area

umbel ['ʌmbəl] *noun* flower cluster in which the flowers all rise on stalks from the same point on the plant's stem (for example, the carrot or polyanthus)

◊ **umbellifer** [ʌm'belɪfə] *noun* plant belonging to the Umbelliferae

◊ **Umbelliferae** [ʌmbə'lɪfəriː] *noun* family of herbs and shrubs, including important food plants such as carrot, parsnip and celery

◊ **umbelliferous** [ʌmbə'lɪfərəs] *adjective* referring to a plant of the Umbelliferae

unavailable water [ʌnə'veɪləbl 'wɔːtə] *noun* water in the soil which is held in the smallest soil pores and so is not available for plants

uncastrated [ʌnkæ'streɪtɪd] *adjective* (male animal) which has not been castrated (NOTE: also called an **entire male**)

unchitted [ʌn'tʃɪtɪd] *adjective* (seed) which has not been chitted

uncultivated [ʌn'kʌltɪveɪtɪd] *adjective* (land) which is not cultivated

under- ['ʌndə] *prefix* meaning less than *or* not as strong as *or* below; **underfeeding** = giving an animal less feed than it needs; **underproduction** = producing less than normal

undercoat ['ʌndəkəʊt] *noun* coat of fine hair under the main coat of some animals

underdeveloped countries [ʌndədɪ'veləpd 'kʌntriz] *noun* countries which have not been industrialized

under-drain ['ʌndədreɪn] *noun* drain under the surface of the soil, such as a mole drain or pipe drain

undergrowth ['ʌndəgrəʊθ] *noun* shrubs and other plants growing under large trees

undersow ['ʌndəsəʊ] *verb* to sow a grass mixture after an arable crop has established, so that both develop at the same time; cereals are most often used as cover crops

understorey ['ʌndəstɔːri] *noun* lowest layer of trees and shrubs in a forest

undulant fever ['ʌndjʊlənt 'fiːvə] *noun* (*in humans*) brucellosis

COMMENT: most cases of brucellosis in humans occur through contact with infected cows or by drinking unpasteurized infected milk

unenclosed land [ʌnɪn'kləʊzd 'lænd] *noun* area of land without any walls or fences round it

Ungulata *or* **ungulates** [ʌŋɡju:'lɑːtə *or* 'ʌŋɡjuːleɪts] *noun* grazing animals which have hoofs, such as cattle, sheep, goats and horses

unheated [ʌn'hiːtɪd] *adjective* which is not heated; **unheated glasshouse** *or* **cold glasshouse** = glasshouse with no heating

unimproved [ʌnɪm'pruːvd] *adjective* (land) which has not been well looked after, and has not been improved by fertilizing and proper husbandry

union ['juːniən] *noun* point of contact between a scion and stock in a grafted fruit tree

unisexual [juːni'seksjuəl] *adjective* (plant) which has either male or female flowers, but not both

unit ['juːnɪt] *noun* (i) single part (as of a series of numbers); (ii) basic standard of measurement; **unit of account** = currency used for calculating the EU budget and farm prices; **unit cost** = the cost of one item (i.e. the total product cost divided by the number of units produced); **unit of fertilizer** = 1% of one hundredweight

COMMENT: the recommended rates of application of plant food to crops are shown in units such as kilograms per hectare (kg/ha); one unit per acre equals 1kg/ha

unpalatable [ʌn'pælətəbl] *adjective* not palatable, feedingstuff which has an unpleasant taste

unpasteurized [ʌn'pɑːstʃəraɪzd] *adjective* which has not been pasteurized; *unpasteurized milk can carry bacilli*

unploughed land [ʌn'plaʊd 'lænd] *noun* land which has not been ploughed

unsaturated fat [ʌn'sætʃəreɪtɪd 'fæt] *noun* fat which does not have a large amount of hydrogen, and so can be broken down more easily

unthrifty [ʌn'θrɪfti] *adjective* (*of animals*) not thriving *or* not growing well

untreated milk [ʌn'triːtɪd 'mɪlk] *noun* milk which has not been treated and which is sold direct from the farm to the public (in the UK this is called 'green top milk' and sales are now banned)

unweaned [ʌn'wiːnd] *adjective* (young animal) which has not yet been weaned

upgrade [ʌp'greɪd] *verb* to make improvements to a herd by repeated crossing with superior males

upland ['ʌplænd] *noun* high areas of land; *upland areas or uplands have different ecosystems from the lowlands* (NOTE: opposite is **lowland**)

uptake ['ʌpteɪk] *noun* taking in of trace elements or nutrients by an animal

QUOTE zinc, iron, lead or cadmium have depressed copper uptake
Farmers Weekly

urban fringe ['ɜːbən 'frɪnʒ] *noun* area of land use where the urban activities meet the rural (usually a source of conflict between townspeople and farmers)

urd [ɜːd] *noun* Indian name for black gram, a pulse plant yielding pods and seeds used as vegetables

urea [juː'rɪə] *noun* substance produced in the liver from excess amino acids, and excreted by the kidneys into the urine. It is widely used as a nitrogenous fertilizer, both in solid form and in aqueous solution, in which form it is easily leached. In the soil it is turned into ammonia and some is released as ammonia gas, and so is lost

uric acid ['juːrɪk 'æsɪd] *noun* chemical compound which is formed from nitrogen in waste products from an animal's body; uric acid is the principal nitrogenous excretory end-product in birds and land reptiles

urine ['jʊərɪn] *noun* liquids containing uric acid, secreted as waste from an animal's body

Urticaceae [ɜːtɪ'keɪsiɪ] *noun* family of shrubs and herbs with stinging leaves (such as the common stinging nettle *Urtica dioica*)

urticaria [ɜːtɪ'keərɪə] *noun* allergic reaction (to injections or certain foods) where the skin forms irritating reddish patches

uterus ['juːtərəs] *noun* hollow organ in a female animal linked to the ovaries; the lower end opens into the vagina. When an ovum is fertilized it becomes implanted in the wall of the uterus and develops into an embryo inside it

Vv

vaccinate ['væksɪneɪt] *verb* to use a vaccine to give an animal *or* person immunization against a specific disease (NOTE: you vaccinate **against** a disease)

◊ **vaccination** [væksɪ'neɪʃn] *noun* action of vaccinating; *mass vaccination of poultry has been carried out against Newcastle disease*

◊ **vaccine** ['væksiːn] *noun* substance which contains the germs of a disease, used to inoculate or vaccinate

◊ **vaccinia** [væk'sɪnɪə] *noun* cowpox, infectious disease of cattle, used to prepare vaccine to vaccinate against smallpox (NOTE: Originally the words **vaccine** and **vaccination** applied only to smallpox immunization, but they are now used for immunization against any disease)

> COMMENT: a vaccine contains antigens (the germs of the disease) sometimes alive and sometimes dead, and this is injected into the animal so that its body will develop immunity to the disease. The antigens provoke the body to produce antibodies, some of which remain in the bloodstream for a very long time and react against the same antigens if they enter the body naturally at a later date when the animal is exposed to the disease.

> QUOTE the Ministry of Agriculture and Water has announced the successful development of a new vaccine against a camel pox known as seed virus
> **Middle East Agribusiness**

vacuum silage ['vækjuːm 'saɪlɪdʒ] *noun* silage placed in large polythene bags, usually by a baler specially adapted for this purpose; air is excluded, so preventing the development of moulds and the green crop is conserved in succulent form

vagal indigestion ['veɪgl ɪndɪ'dʒestʃn] *noun* disease of livestock due to malfunction of the vagus nerve which controls the activity of the stomach and intestines

vanilla [və'nɪlə] *noun* tropical climbing plant *(Vanilla planiolia)* which produces long pods, used for flavouring in confectionery

VAPS = VARIABLE ANIMAL PREMIUM SCHEME

variant ['veərɪənt] *noun* specimen of a plant *or* animal which is different from the normal type

variegation [veərɪ'geɪʃn] *noun* phenomenon in some plants where two or more colours occur in patches on the leaves or flowers

◊ **variegated** ['veərɪgeɪtɪd] *adjective* (plant) with different coloured patches

variety [və'raɪəti] *noun* type of organism, especially a cultivated plant

◊ **varietal** [və'raɪətəl] *noun* variety of cultivated plant, such as a grapevine

> QUOTE as a result of varietal improvement trials conducted at the Indian Grassland and Fodder Research Institute, a new high-yielding variety of cowpea has been developed
> **Indian Farming**

vasectomy [və'sektəmi] *noun* operation to cut the duct which takes sperm from the testicles, so making the animal infertile

◊ **vasectomize** [və'sektəmaɪz] *verb* to perform a vasectomy on (an animal)

veal [viːl] *noun* meat of a young calf fed solely on a milk diet, slaughtered between three and fifteen weeks old; **veal crate system** = intensive method of veal production, where calves are kept in crates; abandoned in the UK, but still practised elsewhere

vector ['vektə] *noun* anything that carries a disease-causing organism from an infected organism to a healthy one, such as the mosquito which carries malaria

veer [vɪə] *verb (of wind)* to change

direction, moving in a clockwise direction
(NOTE: the opposite is **back**)

vegeculture ['vedʒɪkʌltʃə] *noun* type of
agriculture based on vegetatively
propagated crops

vegetable ['vedʒɪtəbl] *noun* plant or parts
of plants grown for food, not usually sweet
(including some fruits, such as tomatoes,
and some seeds, such as peas); **vegetable oils**
= oils obtained from plants and their seeds,
which are low in saturated fats; **vegetable
protein** = protein obtained from cereals,
oilseeds, pulses, green vegetables and roots,
which provides for the feeding
requirements of both humans and livestock

COMMENT: the main vegetable crops
being grown on a field scale are broad,
green and navy beans, vining and dried
peas, Brussels sprouts, cabbages, carrots,
cauliflowers, celery, onions, turnips and
swedes and potatoes. In 1993, 180,000
hectares of potatoes and 135,000 hectares
of other vegetables were cultivated for
human consumption in the UK. 126,000
hectares of vegetables were grown in the
open

vegetation [vedʒɪ'teɪʃn] *noun* (i) plants in
general; (ii) plants which are found in a
particular area

◊ **vegetative** ['vedʒətətɪv] *adjective*
referring to vegetation; **vegetative
propagation** *or* **vegetative reproduction** =
reproduction of plants by taking cuttings *or*
by grafting, not by seed

veldt [velt] *noun* extensive grasslands with
a scattering of trees, found in the east of the
interior of South Africa

venison ['venɪsən] *noun* meat from deer

ventilate ['ventɪleɪt] *verb* to provide fresh
air

◊ **ventilation** [ventɪ'leɪʃn] *noun* allowing
air to circulate; measures necessary to
remove polluted air from a building
without making too many draughts; good
ventilation is of prime importance in
housing livestock

◊ **ventilator** ['ventɪleɪtə] *noun* device
which circulates fresh air into a room *or*
building

verandah [və'rændə] *noun* type of
housing for poultry or pigs with a slatted or

wire floor, through which the droppings
fall

vermicide ['vɜːmɪsaɪd] *noun* substance
which kills worms

vermiculite [vɜː'mɪkjuːlaɪt] *noun*
substance which is used in soilless
gardening

COMMENT: vermiculite occurs naturally as
a form of silica and is capable of retaining
moisture. It is processed into small pieces
and used instead of soil in horticulture

vermifuge ['vɜːmɪfjuːdʒ] *noun* substance
used to get rid of parasitic worms in the
intestines of livestock

vermin ['vɜːmɪn] *noun* **(a)** organisms
which are looked upon as pests; *see also*
PEST **(b)** insects (such as lice) which live on
other animals as parasites

vernal ['vɜːnəl] *adjective* referring to the
late spring

◊ **vernalization** [vɜːnəlaɪ'zeɪʃn] *noun*
treatment of seeds to make them flower at a
set time; seed is made to germinate early by
refrigerating it for a time

vertebrate ['vɜːtɪbrət] *adjective & noun*
animal with a backbone (all animals raised
on farms are vertebrates, with the
exception of edible snails) (NOTE:
opposite is **invertebrate**)

verticillium wilt [vɜːtɪ'sɪliəm 'wɪlt] *noun*
plant disease caused by a fungus, which
makes leaves become yellow and wilt; a
notifiable disease of hops, and also a
serious disease in lucerne and clovers

vet [vet] *noun* = VETERINARY SURGEON

◊ **veterinarian** [vetərɪ'neəriən] *noun US* =
VETERINARY SURGEON

◊ **veterinary** ['vetrənəri] *adjective*
referring to the care of sick animals;
veterinary science = scientific study of
diseases of animals and their treatment;
veterinary service = branch of the
Agricultural Development and Advisory
Service. The service is responsible for the
control and eradication of notifiable
diseases, national disease control
programmes, investigation of new diseases,
meat hygiene, animal welfare, the import
and export of animals and animal
products, and the liaison with private

veterinary surgeons on livestock disease and production programmes; **veterinary surgeon** = person who is qualified in veterinary science

vetch [vetʃ] *noun* leguminous plant (*Vicia sativa*) also called 'tares'. Vetches can be sown with oats as an arable silage

V-graft ['viːgrɑːft] *noun* method of grafting, where the stem of the stock is trimmed to a point, and the stem of the cutting is split to allow it to be fitted over the point of the stock

vibrate [vaɪ'breɪt] *verb* to shake, making very small rapid movements

◊ **vibrojet** ['vaɪbrəʊdʒet] *noun* jet for irrigation, sending the water out in small sharp pulses (useful to control spray drift)

vice [vaɪs] *noun* bad habit in an animal, such as the habit of biting other animals' tails

Vicia ['vɪsɪə] *Latin name for* beans, such as broad beans

Victoria [vɪk'tɔːrɪə] *see* PLUM

VIDA = VETERINARY DIAGNOSTIC DATA

vigour ['vɪɡə] *noun* strength; **hybrid vigour** = increase in size *or* rate of growth *or* fertility *or* resistance to disease found in offspring of a cross between two species

◊ **vigorous** ['vɪɡərəs] *adjective* which grows strongly; *plants put out vigorous shoots in warm damp atmosphere*

VMD = VETERINARY MEDICINES DIRECTORATE

vine [vaɪn] *noun* **(a) grape vine** = plant of the genus *Vitis,* yielding grapes, either for eating fresh, for drying or for making into wine **(b) vine crops** = crops (*Cucurbitaceae*) such as cucumber, marrow, gourds and melons, which are annuals and produce long trailing shoots and heavy fleshy fruit

◊ **viner** ['vaɪnə] *noun* machine for harvesting vining peas

◊ **vineyard** ['vɪnjəd] *noun* plantation of grapevines

◊ **vining** ['vaɪnɪŋ] *noun* harvesting of peas for processing; **vining peas** = peas used for canning or freezing

violet root rot ['vaɪələt 'ruːt rɒt] *noun* common disease of sugar beet. A violet coloured fungus growth on surface of root. It lowers sugar content of the plant (*Helicobasadium purpureum*)

virgin ['vɜːdʒɪn] *adjective* untouched; **virgin land** = land which has never been cultivated; **virgin oil** = first pressing of olive oil, which produces oil of the best quality

virus ['vaɪrəs] *noun* tiny cell formed largely of genetic material and protein, which can only develop in other cells and often destroys them; all viruses are parasitic; **virus pneumonia** = disease affecting pigs and calves; the animals suffer from a hard cough and are unthrifty. In calves, shivering, loss of appetite and noisy breathing are symptoms; **virus yellows** = disease of sugar beet and mangolds; the leaves turn yellow and the sugar content is greatly reduced. Crops are most at risk when virus-carrying aphids infest the plants at the two-leaf stage

◊ **viral** ['vaɪrəl] *adjective* (disease) caused by a virus

COMMENT: viruses produce disease in man, animals and plants. Many common diseases such as measles or the common cold are caused by viruses; viral diseases cannot be treated with antibiotics (which only destroy bacteria). Viruses can be transmitted from one animal to another, reproducing the same disease. Insects, particularly aphids, transmit certain virus diseases in plants

viscera ['vɪsərə] *noun* internal organs (such as heart, lungs, stomach and intestines)

vitamin ['vɪtəmɪn] *noun* essential nutrient usually needed by animals in minute quantities for growth and health; **vitamin deficiency** = lack of necessary vitamins

COMMENT: vitamins are a group of chemical compounds found in a variety of foodstuffs, which are necessary for the healthy regulation of physical processes in an animal's body. Vitamins are classified by letters: *Vitamin A* is fat-soluble and is stored in the fat compounds in the body. It protects mucous membranes and is necessary for the growth and good health of the animal. It is formed in the animal's body from the yellow substance, carotene, which is available to animals in green foods, carrots, maize and milk. *Vitamin B* is water-soluble, and is a group of several different substances: riboflavin, thiamine, nicotinic acid, biotin, choline and pantothenic acid. It is necessary for the nervous system and for oxidation processes in the body. It is found in whole grains, grass, milk and yeast. Cattle supply their own needs for Vitamin B in the rumen. *Vitamin C* is ascorbic acid and is found in fruit and vegetables. *Vitamin D* is fat-soluble. It is necessary to young animals as it is important in bone formation. It is formed in the skin of an animal by the action of sunlight. *Vitamin E* is fat-soluble and is important for muscle development and normal reproduction. It is found in green food and cereals. *Vitamin K* is fat-soluble and prevents blood-clotting; it is available in green foodstuffs such as spinach. Vitamin deficiencies in animals can cause serious health problems and once identified, can be cured by the use of mineral supplements either as an individual simple substance or by complex mixtures added to normal rations and supplied according the animals' needs

viticulture ['vɪtɪkʌltʃə] *noun* cultivation of grapes

volatile oils ['vɒlətaɪl 'ɔɪlz] *noun* concentrated oils from a scented plant used in cosmetics or as antiseptics

voluntary restraints agreements (VRA) ['vɒlʌntəri rɪ'streɪnts ə'griːmənts] *noun* agreements by which farmers agree not to spray in windy conditions; such agreements are not legally binding

volunteer (plant) [vɒlən'tɪə] *noun* plant that has grown by natural propagation, as opposed to having been planted; *volunteer cereals are a problem in establishing oilseed rape*

VRA = VOLUNTARY RESTRAINTS AGREEMENTS

Vulgare *or* **Vulgaris** [vʌl'gɑːri *or* vʌl'gɑːrɪs] *Latin word meaning* 'common', often used in plant names

Ww

wages ['weɪdʒɪz] *noun* money paid to an employee for work done; wages for farm workers are set by the Agricultural Wages Board. In 1995 the minimum wage for farm workers was around £150 a week, although the average was £223

wall barley grass ['wɔːl 'bɑːli grɑːs] *noun* weed *(Hordeum murinum)* found in grassland

walnut ['wɔːlnʌt] *noun* hardwood tree of the genus *Juglans*, with edible nuts; the timber is used in furniture making

WAOS = WELSH AGRICULTURAL ORGANISATION SOCIETY

warble fly ['wɔːbl 'flaɪ] *noun* parasitic fly, whose larvae infest cattle

◊ **warbles** ['wɔːblz] *noun* swellings on the backs of cattle caused by the warble fly

> COMMENT: eggs of warble flies are laid on the legs or bellies of cattle. On hatching, the maggots burrow into the skin and cause swellings on the back of the animal. When adult, the maggots leave the body through the skin and fall to the ground to pupate. They cause severe irritation, loss of condition and in young animals may cause death

ware (potatoes) [weə] *noun* potatoes grown for human consumption, as opposed to those grown for seed; **ware growers** = farmers who grow potatoes for consumption, not for seed

> COMMENT: ware potatoes are used in many ways: for crisps, chips, canning and dehydration. Good quality ware potatoes should not be damaged or diseased, should not be green in colour, and should be between 40 and 80mm in size

> QUOTE a number of merchants have been selling small ware potatoes as seed. Although there is nothing to stop growers from selling small ware, it becomes illegal if the seller knows that it is going for planting as seed
> *Farmers Weekly*

warfarin ['wɔːfərɪn] *noun* substance used to poison rats; it acts as a blood anticoagulant

warping ['wɔːpɪŋ] *noun* farming practice which permits a river to flood low-lying land to cover it with silt in which crops will be grown

wart [wɔːt] *noun* common complaint in young housed cattle. Warts can occur in cattle up to two years old; they are found around the eyes and on the neck and body. Warts are infectious and disfigure the animal; **wart disease** = disease of potatoes. A notifiable disease, in which warts appear on the surface of the tubers, and develop into large eruptions which may become larger than the potatoes themselves; **wart-hog disease** = AFRICAN SWINE FEVER

waste [weɪst] **1** *adjective* useless *or* which has no use; *waste products are dumped in the sea; waste matter is excreted by the body in the faeces or urine;* **waste lime** = lime obtained from industrial concerns after it has been used as a purifying material; **waste product** = substance which is not needed in a process **2** *noun* unenclosed land used for pasture

◊ **wasteland** ['weɪstlænd] *noun* area of land which can no longer be used for agriculture, or for any other purpose; *overgrazing has produced wastelands in Central Africa*

water ['wɔːtə] **1** *noun* common liquid (H_2O) which forms rain, rivers, the sea, etc., and which makes up a large part of the bodies of organisms; **water balance** = (i) state where the water lost in an area by evaporation *or* by runoff is replaced by water received in the form of rain; (ii) state where the water lost by the body (in urine *or* perspiration, etc.) is balanced by water absorbed from food and drink; **water-salt balance** = state where the water in the soil balances the amount of salts in the soil; **waterbowl** = container for water in a stable or loose-box; **water buffalo** *see* BUFFALO **water meadow** = grassy field near a river, which is subject to flooding **2** *verb* to give water to (a plant); *plants may need watering two times a day in hot dry weather*

◊ **watercourse** ['wɔːtəkɔːs] *noun* stream or river

◊ **waterlogged** ['wɔːtəlɒgd] *adjective* (soil) in which the pore space is completely filled with water; waterlogged soil is not suitable for plant growth since the roots cannot obtain oxygen for respiration

◊ **watermill** ['wɔːtəmɪl] *noun* mill which is driven by the power of a stream of water which turns a large wheel

◊ **Water Pollution Act** act of 1989 which makes it an offence to cause a discharge of poisonous, noxious or polluting matter or solid matter to any controlled water under the responsibility of the National Rivers Authority. Controls are also in force to ensure that silage, slurry and fuel oil installations are of adequate standard

◊ **watershed** ['wɔːtəʃed] *noun* dividing line between two catchment areas

◊ **water-soluble** [wɔːtə'sɒljuːbl] *adjective* which can dissolve in water; *some vitamins are water-soluble*

◊ **water table** ['wɔːtə 'teɪbl] *noun* the upper surface of the ground water or the level below which the soil is permanently saturated

◊ **watery mouth** ['wɔːtəri 'maʊθ] *noun* disease affecting new-born lambs

COMMENT: water is essential to plant and animal life. Water pollution can take many forms: the most common are discharges from industrial processes, household sewage and the runoff of chemicals used in agriculture

waterwheel ['wɔːtəwiːl] *noun* wheel with containers which catch and hold water as the wheel turns

COMMENT: waterwheels have two purposes: placed over a moving stream the pressure of the moving current turns the wheel and so drives machinery (as in a watermill); in the second case, the wheel is also placed over a moving stream, but the water collects in the buckets on the wheel, which, as it turns, raises the full buckets and tips the water into an irrigation channel

wattle ['wɒtl] *noun* **(a)** species of Acacia tree, leguminous trees of tropical and subtropical origin. Many commercial products are derived from acacias, including dyes, perfumes and timber; **wattle bark** = bark from the wattle, an important source of tannin **(b)** rods and twigs woven together to make a type of fence **(c)** fleshy skin hanging down below the throat of some birds, such as the turkey

Watusi [wə'tuːsi] *noun* breed of cattle found in East Africa; the animals are coloured red, brown or black

wean [wiːn] *verb* to remove a young animal from the milk source of its mother; weaning is common at 5 weeks

◊ **weaner** ['wiːnə] *noun* young animal which has been weaned (especially used of young pigs)

weather ['weðə] **1** *noun* daily atmospheric conditions (e.g. sunshine, wind, precipitation) in a certain area, as opposed to climate which is the average weather for the area; **weather chart** *or* **weather map** = map which shows the weather in a certain area at a certain time; **weather forecast** = description of what the weather will be like for a period in the future, important for the farmer when carrying out certain jobs such as spraying; **weather forecasting** = scientific study of weather which allows forecasts of the weather to be made **2** *verb* to wear down (rock) under the influence of frost *or* wind *or* rain, etc.

◊ **weathering** ['weðərɪŋ] *noun* changing the state of soil *or* rock through the influence of the climate (rainfall, hot sun, frost, etc.) or by chemical pollutants present in the rain or the atmosphere

weatings *or* **wheatings** ['wiːtɪŋz] *noun* by-product of milling wheat; brans of various particle sizes and varying amounts of attached endosperm; used as a feedingstuff

wedge [wedʒ] *see* DORSET WEDGE

web conveyor ['web kən'veɪə] *noun* machine used to move material along a moving web; found on all types of harvesters and some processing machines

weed [wiːd] **1** *noun* wild plant which grows in cultivated land, and which the farmer *or* gardener tries to remove **2** *verb* to pull out weeds (usually by hand, or with a hoe)

◊ **weed beet** ['wiːd 'biːt] *noun* weed affecting sugar beet crops; the weed can harbour rhizomania. It is controlled by limiting bolters and so preventing cross-pollination. The most effective control is by hand-pulling bolters

COMMENT: weeds compete with crops for nutrients and water; the presence of weeds can lower the quality of a crop and often make it more difficult to harvest. Some weeds may taint milk when eaten by cows and some weeds are poisonous and can affect livestock. Weeds also harbour pests and diseases which can spread to crops. Chemical control of weeds is an additional cost but weeds can be controlled by good rotations and tillage treatment

QUOTE farmers need to check young trees for weed growth and plan strategies to prevent insect damage. Some follow-up weed control would be needed if weed control was inadequate before the trees were planted. Weeds will reduce tree growth rates
The West Australian

weedkiller ['wiːdkɪlə] *noun* substance used to kill weeds; **selective weedkiller** = substance which is meant to kill only certain types of weed

COMMENT: although weedkillers are widely used in farming and horticulture, they can have serious side-effects. They may blow away from the area where they are being sprayed; excess weedkiller enters the soil and is leached away by rain into watercourses, killing water plants

weevil ['wiːvɪl] *noun* kind of beetle which feeds on grain, nuts, fruit and leaves; the larvae of grain beetles feed on the stored grain where they also pupate; bean and pea weevils can be treated with BHC

Weil's disease ['veɪlz dɪ'ziːz] *noun* sometimes fatal disease of humans caused by Leptospira bacteria, caught from the urine of infected cattle or rats

well [wel] *noun* hole dug in the ground to the level of the water table, from which water can be extracted either by a pump or by a bucket; **artesian well** = well bored into a confined aquifer

Welsh [welʃ] *adjective* breed of pig, white in colour, with lop ears (one of the older breeds of British pig); **Welsh black** = hardy

dual-purpose breed of cattle formed when the northern Anglesey strain was bred with the Castlemartin strain. Welsh blacks produce a reasonable milk yield and very lean meat; **Welsh half bred** = cross between a border Leicester ram and a Welsh mountain ewe; **Welsh mountain** = hardy breed of sheep, well adapted to wet conditions; the animals are small with white faces and very fine fleece. Only the rams have horns; **Welsh mule** = cross between a Blue-faced Leicester and a ewe of one of the Welsh mountain breeds

Wensleydale ['wenslideɪl] *noun* **(a)** longwool breed of sheep; the animals are large and polled. The skin of the face, legs and ears is blue; now rare, but still found in Yorkshire **(b)** type of hard white cheese

West African Dwarf ['west 'æfrɪkən dwɔːf] *noun* breed of goat which is about half the height of other African breeds; it is mainly reared for meat production, and is extremely resistant to trypanosomiasis

Wessex Saddleback ['wesɪks 'sædlbæk] *noun* one of two saddleback breeds now joined with the Essex Saddleback to give the British Saddleback, a dual-purpose breed of pig, now rare

Western ['westən] *see* WILTSHIRE HORN

Westerwold ryegrass ['westəwəʊld 'raɪgrɑːs] *noun* annual type of ryegrass; a fast-growing summer crop

wether ['weðə] *noun* castrated male sheep

wetlands ['wetlændz] *noun* area of land which is often covered by water *or* which is very marshy; *there are two important wetland sites in the area*

◊ **wet rice** ['wet 'raɪs] *noun* cultivation of rice in water for most of the growing period; also called 'paddy rice'

WFU = WOMENS' FARMING UNION

wheat [wiːt] *noun* Triticum sp., the most important of the cereals grown in the temperate regions; **hard wheat** = wheat with a high protein content, used to make flour for bread; **soft wheat** = wheat with a lower protein content, used to make flour for cakes; **spring wheat** = varieties of wheat which are sown in spring and harvested towards the end of the summer; **winter wheat** = varieties of wheat sown in the autumn or early winter months and harvested the following summer (these are harder than the spring wheats); **wheat belt** = the wheat-growing regions of the North American prairies; **wheat blossom midge** = pest that affects wheat; **wheat bulb fly** = fly whose larvae feed on the roots of wheat; the central shoot turns yellow and dies; **wheat root rot** = common soil fungus *(Rhizoctonia solani)* which attacks the roots of seedlings and retards growth

◊ **wheatfeed** ['wiːtfiːd] *see* WHEAT OFFALS

◊ **wheatgerm** ['wiːtdʒɜːm] *noun* central part of the wheat seed, which contains valuable nutrients

◊ **wheatings** ['wiːtɪŋz] *see* WEATINGS

◊ **wheatmeal** ['wiːtmiːl] *noun* brown flour with a large amount of bran, but not as much as wholemeal

◊ **wheat offals** ['wiːt 'ɒfəlz] *noun* the embryo and seed coat of the wheat grain, used as animal feed

COMMENT: the two main species of wheat grown are *Triticum aestivum* which is grown for bread flour, and some varieties of which produce the most suitable flour for cakes and biscuits, and *Triticum durum* which is grown for pasta. Cereal drilling usually takes place in the UK between September and April. Winter wheat usually yields higher quantities than spring wheat and is harvested before it. Spring wheat varieties are grown in those areas with more extreme climates as they are the quick-maturing varieties. Spring wheat is grown in the prairie provinces of Canada, in the Dakotas and Montana in the USA, and in the more northerly parts of the steppe wheat belt in Russia. Winter varieties account for about three-quarters of the total output of wheat in the USA, and over 80% of exports. The state of Kansas grows 30% of the hard winter red winter wheat grown in the USA

QUOTE In the UK, the wheat acreage has risen above 2m hectares (5m acres) for the first time ever. Prompted by good autumn and winter drilling conditions and prices well above intervention levels, 20% more winter wheat went in. The overall wheat acreage is up 10% at 2.08m hectares (5.14m acres) in the year set-aside was introduced to cut the cereals acreage

Farmers Weekly

wheelwright ['wiːlraɪt] *noun* person who makes wooden wheels for farm carts

whey [weɪ] *noun* residue from milk after the casein and most of the fat have been removed; used as pig feed

whip [wɪp] *noun* short stick with a lash attached, used to control horses; also called a 'crop'; **whip and tongue cutting** = form of graft, where the stock and scion are cut diagonally to form large open surfaces with a small notch in each; the surfaces are bound together tightly with twine

◊ **whipworm** ['wɪpwɜːm] *noun* variety of worm affecting pigs, especially weaners

Whitbred shorthorn ['wɪtbred 'ʃɔːthɔːn] *noun* breed of white beef cattle

white clover ['waɪt 'kləʊvə] *noun* type of perennial clover *(Trifolium repens)*. There are several varieties including the large-leaved variety suitable for silage or hay and the small-leaved variety which is quick to establish and keeps out weeds and other grasses

◊ **White-faced Woodland** ['waɪtfeɪst 'wʊdlənd] *noun* large hill breed of sheep, with white face and legs and pinkish nostrils. The ram has heavy twisted horns. Found mainly in the South Pennines, it has been crossed with other hill breeds to give them its size and vigour. Also called the 'Penistone'

◊ **White Fulani** ['waɪt fʊ'lɑːni] *noun* breed of cattle from Northern Nigeria; the animals are usually white in colour, with distinctive long curved horns; used both as draught animals and for milk

◊ **whiteheads** ['waɪthedz] *see* TAKE-ALL

◊ **White Leghorn** ['waɪt 'leghɔːn] *noun* laying breed of poultry

◊ **white lupin** ['waɪt 'luːpɪn] *noun* new strain of lupin able to withstand cold. Seeds are 40% protein and at least 12% edible oil *(Lupinus albus)*

◊ **white mustard** ['waɪt 'mʌstəd] *noun* Sinapsis alba, crop grown to increase the organic content of the soil by using it as a green manure

◊ **White Park** ['waɪt 'pɑːk] *noun* rare breed of cattle; white in colour with either black or red muzzle, eyelids, ears and feet. One of the most ancient breeds of British cattle

◊ **White Plymouth Rock** ['waɪt plɪməθ 'rɒk] *noun* large heavy breed of table poultry

◊ **white rot** ['waɪt 'rɒt] *noun* fungal disease of onions and leeks; the leaves turn yellow and a white mass appears on the bulb

◊ **white scour** ['waɪt 'skaʊə] *noun* disease affecting young calves

◊ **White Wyandotte** ['waɪt 'waɪəndɒt] *noun* dual-purpose breed of poultry

whole crops ['həʊl 'krɒps] *noun* crops used for silage which do not need wilting

◊ **wholefood** ['həʊlfuːd] *noun* naturally *or* organically grown food, which has not been processed

◊ **wholegrain** ['həʊlgreɪn] *noun* food (such as rice) of which the whole of the seed is eaten

◊ **wholemeal** ['həʊlmiːl] *noun* flour which has had nothing removed or added to it and contains a large proportion of the original wheat seed, including the bran

◊ **wholesale** ['həʊlseɪl] *adjective & adverb* selling in bulk to shops, who then sell in smaller quantities to individual buyers; **wholesale seed merchant** = merchant who sells seed in bulk

wild [waɪld] *adjective* not domesticated; **wild boar** = *Sus scrota*, species of feral pig, common in parts of Europe, but extinct in the UK. Wild boars are preserved for hunting, but are now bred on farms; their meat is dark, with very little fat, and is of high value; **wild crop** = crop which is harvested by man, but not cultivated (such as wild berries, herbs, etc.); **wild oats** = several species of annual weeds *(Avena fatua)* found among cereal crops, and now largely controlled by selective herbicides, although manual weeding or roguing is used; **wild onion** = *Allium vineale,* perennial weed affecting cereal crops, beans and rape; also called 'crow garlic'; **wild rice** = species of grass which is found naturally in North America and which is similar to rice; **wild white clover** = *Trifolium repens,* variety of small-leaved white clover; slow to get established, but an essential part of a long ley. It is drought resistant and very productive

◊ **wildlife** ['waɪldlaɪf] *noun* any wild animals and birds which have not been domesticated; *plantations of conifers are poorer for wildlife than mixed or deciduous woodlands; the effects of the open-cast mining scheme would be disastrous on wildlife, particularly on moorland birds*

William ['wɪljəm] *see* PEAR

willow ['wɪləʊ] *noun* strong temperate hardwood (genus *Salix*), often used for coppicing or pollarding

wilt [wɪlt] **1** *noun* **(a)** drooping of leaves and stems, which become limp through lack of moisture **(b)** plant disease characterized by wilting **2** *verb* **(a)** *(of plant leaves, etc.)* to hang limp **(b)** to allow silage plants to become limp before harvesting; *the cut grass is left for a few days to wilt*

◊ **wilting** ['wɪltɪŋ] *noun* limpness found when plant tissues do not contain enough water; **(permanent) wilting point** = percentage of water remaining in the soil after a test plant has wilted under controlled conditions, so that it will not recover unless it receives water

Wiltshire cure ['wɪltʃə 'kjʊə] *noun* special method of mild curing and smoking sides of bacon over wood fires

◊ **Wiltshire horn** ['wɪltʃə 'hɔːn] *noun* distinctive white-faced breed of sheep, with curled horns; it grows a coat of thick matted hair. Found in the Midlands and Anglesey. It is a hardy breed, producing rapid-growing lambs. Also called 'Western'

wind [wɪnd] *noun* air which moves in the lower atmosphere; *the weather station has instruments to measure the speed of the wind;* **wind dispersal** = scattering of spores, pollen and seeds by the wind; **wind-driven** *or* **wind-powered** = (machine) that is powered by the wind; **wind erosion** = erosion of soil *or* rock by wind; **wind farm** *or* **wind park** = group of large windmills, built to harness the wind to produce electricity; **wind pollination** = pollination of flowers when the pollen is blown by the wind; **wind power** = power generated by using wind to drive machine (as in a windmill) *or* to drive a turbine which creates electricity; **wind pump** = small pump driven by the wind, used to pump water from wells

◊ **windbreak** ['wɪndbreɪk] *noun* hedge or line of trees, planted to give protection from the wind to land with growing crops

◊ **wind chill factor** ['wɪnd 'tʃɪl 'fæktə] *noun* way of calculating the risk of exposure in cold weather by adding the speed of the wind to the number of degrees of temperature below zero

windmill ['wɪndmɪl] *noun* building with sails which are turned by the wind, so driving a machine; used originally to mill grain and pump water

> COMMENT: windmills were originally built to grind corn or to pump water from marshes. Large modern windmills are used to harness the wind to produce electricity

windrow ['wɪndrəʊ] *noun* single row of hay or other crop formed from two more cut swaths; row of a root crop such as potatoes, which are lifted and left lying on the soil surface; **windrow pickup** = pickup mechanism which lifts a crop into a harvester

◊ **windrowed** ['wɪndrəʊd] *adjective* (crop) which has been lifted and left in a swath

◊ **windrower** ['wɪndrəʊə] *noun* machine which lifts a crop (such as potatoes) and leaves it in a swath on the surface of the soil

wing [wɪŋ] *noun* lower part of the ploughshare behind the point

winnow ['wɪnəʊ] *verb* to separate grain from chaff; originally this was done by throwing the grain and chaff up into the air, the lighter chaff being blown away by the wind

winter ['wɪntə] **1** *noun* the cold season of the year, when the days are short, when plants do not flower *or* produce new shoots, and when some animals hibernate; **winter burn** = leaf burn in winter; **winter feeding** = system of feeding livestock during the winter months, giving them feeds of hay, silage and concentrates; **winter greens** = hardy Brassicas which are grown for use during the winter; **winter kill** = death of plants in winter; **winter solstice** = 21st December, the shortest day in the Northern Hemisphere, when the sun is furthest north of the equator; **winter wash** = egg-killing spray applied to fruit trees in the dormant winter period; tar oil is the commonest winter spray; **winter wheat** = varieties of wheat that are sown in the autumn or early winter and harvested the following summer **2** *verb* to spend the winter in a place

wireweed ['waɪəwiːd] *noun* common name for 'knotgrass' (*Agriotes lineatus*)

wireworm ['waɪəwɜːm] *noun* shiny thin hard-bodied larvae of the click beetle, which feeds on the roots of cereals and other plants; they can be treated with BHC

wither ['wɪðə] *verb* (*of plants, leaves, flowers*) to shrivel and die

withers ['wɪðəz] *noun* ridge between the shoulder blades of an animal

withstand [wɪθ'stænd] *verb* not to be affected by; *some plants can withstand very low temperatures*

WLE = WELSH LAMB ENTERPRISE

WOAD = WELSH OFFICE AGRICULTURE DEPARTMENT

wolds [wəʊldz] *noun* open rolling countryside found in chalk areas (such as Yorkshire and Lincolnshire); wolds are characterized by having few hedges and no surface water

wood [wʊd] *noun* **(a)** hard tissue which forms the main body of a shrub or tree; **new wood** = growth made during the current year; **old wood** = growths made during previous years **(b)** building material that comes from trees; **wood alcohol** = methanol, alcohol manufactured from waste wood, which can be used as fuel; **wood ash** = ash from burnt wood, a source of potash; *see also* HARDWOOD, SOFTWOOD

◊ **woodlands** *or* **woods** ['wʊdləndz *or* wʊdz] *noun* land covered with trees with clear spaces between them (i.e. not as dense as a forest); **woodland management** = controlling an area of woodland so that it is productive (by regular felling, coppicing, planting, etc.)

◊ **woodlot** ['wʊdlɒt] *noun* small area of land planted with trees

◊ **woody** ['wʊdi] *adjective* plant tissue which is like wood *or* which is becoming wood

> QUOTE three-quarters of our breeding land birds, half of our butterflies and one sixth of our flowers need woodland habitats
> *Sunday Times*

wool [wʊl] *noun* soft curly hair, the coat of the domesticated sheep; also produced by goats and rabbits; **wool ball** = mass of wool found in the first or fourth stomachs of lambs; small amounts of wool swallowed by the lamb collect to form a ball which can increase in size until it blocks the stomach and causes death; **wool fat** = lanolin, fat which covers the fibres of sheep's wool

◊ **British Wool Marketing**

Board **(BWMB)** government body responsible for the buying, grading and marketing of home-produced wool

> COMMENT: the British Wool Marketing Board is responsible for the specifications by which wool is graded. These are based on the length of staple, lustre, softness, springiness, strength, colour and fineness. The diameter of the wool fibre (measured in microns) is used to measure fineness, and in Britain ranges from 28 (the thickest) to 58 (the thinnest).In 1994 income from wool accounted for only some 10% of gross return per ewe on lowland farms, but up to 20% in many hill situations

◊ **woolsorter's disease** ['wulsɔːtəz dɪ'ziːz] see ANTHRAX

work [wɜːk] verb to cultivate land; **work days** = the number of days when land can be worked with acceptable risk of damage to soil structure during the main activities of tillage, drilling and harvesting. Heavy soils have fewer work days than light soils; in general, good ground conditions exist when the soil is below field capacity

◊ **workability** [wɜːkə'bɪlɪti] noun the ability of soil to be cultivated; it is an interaction between climatic conditions and the physical condition of the soil

World Food Programme part of the Food and Agriculture Organization of the United Nations; the programme is intended to give international aid in the form of food from countries with food surpluses

World Trade Organization (WTO) international organization set up with the aim of reducing restrictions in trade between countries (replacing GATT)

World Wide Fund for Nature (WWF) international organization, set up in 1961, to protect endangered species of wildlife and their habitats, and now also

involved with projects to control pollution and promote policies of sustainable development

worm [wɜːm] **1** noun **(a)** earthworm, any of several species of invertebrate animals with a long thin shape **(b) worms** = parasitic animals which can cause disease in animals; see also ROUNDWORM, TAPEWORM, WHIPWORM **2** verb to remove worms from an animal

◊ **wormer** ['wɜːmə] noun substance used to worm cattle

> COMMENT: 1. earthworms provide a useful service in aerating the soil. It is also believed that they encourage root development in plants by a substance which they secrete. 2 parasitic worms infest most animals, but especially cattle and sheep, and can be removed with anthelmintics. Wormed cattle may give higher yields of milk than untreated animals, but tests are not conclusive. Various substances are used in worming, such as thiabendazole or fenbendazole

> QUOTE worming is essential. Worm on housing for the winter. Worm three weeks after turn out in the spring and at three week intervals after that. If the infestation is low, then worming could probably stop after three or four doses
> *Smallholder*

worrying ['wʌrɪɪŋ] noun chasing of sheep and other livestock, by dogs which are not controlled by their owners

WTO = WORLD TRADE ORGANIZATION

WWF = WORLD WIDE FUND FOR NATURE

Wyandotte ['waɪəndɒt] see WHITE WYANDOTTE

wych elm see ELM

Xx Yy Zz

X chromosome ['ekskrəʊməsəʊm] *noun* sex chromosome

> COMMENT: most animals have a series of pairs of chromosomes, one of which is always an X chromosome; a normal female has one pair of XX chromosomes, while a male has one XY pair; these are the sex chromosomes

xeno- ['zenəʊ] *prefix* meaning different

◊ **xenobiotics** [zenəʊbaɪ'ɒtɪks] *noun* chemical compounds that are foreign to an organism (such as some food additives)

xero- ['zɪərəʊ] *prefix* meaning dry

◊ **xeromorphic** [zɪərəʊ'mɔːfɪk] *adjective* (plant) which can prevent water loss from its stems during hot weather

◊ **xerophilous** [zɪ'rɒfɪləs] *adjective* (plant) which lives in very dry conditions

◊ **xerophyte** ['zɪərəfaɪt] *noun* plant which is adapted to living in very dry conditions

◊ **xerosere** ['zɪərəʊsiːə] *noun* succession of communities growing in dry conditions

xylem ['zaɪləm] *noun* the solid tissue in plants which takes water from the roots to the rest of the plant; the xylem is formed of tubes made from dead cells (as opposed to the phloem, which is formed of living cells), thickened with lignum; they occur both in stems and roots

yak [jæk] *noun* domesticated animal found in Tibet; a draught animal which also yields milk. The animals are black or brown in colour, with long thick hair and large black curved horns; they can be crossbred with domestic cattle

yam [jæm] *noun* (i) term applied to several tropical root crops; and in the southern USA also applied to sweet potatoes; (ii) in particular, the thick tubers of a tropical plant *(Dioscorea* sp) which is mainly grown as a subsistence crop

> COMMENT: yams are mainly grown in Nigeria, Cote d'Ivoire and Ghana

yard [jɑːd] *noun* open space in a farm, surrounded on three sides by barns, stables and farm buildings; **yard and parlour** = system of housing dairy cattle in yards and bringing them through a parlour for milking

yarr [jɑː] *noun* another name for 'corn spurrey'

yarrow ['jærəʊ] *noun* common weed *(Achillea millifolium)* which can cause taints in milk; also called 'milfoil'

Y chromosome ['waɪkrəʊməsəʊm] *noun* male chromosome

> COMMENT: the Y chromosome has male characteristics and does not form part of the female genetic structure. A normal male has an XY pair of chromosomes. See also the note at X CHROMOSOME

yearling ['jɜːlɪŋ] *noun* animal aged between one and two years

yeast [jiːst] *noun* fungus which is used in the fermentation of alcohol and in making bread and is a valuable source of vitamin B

yellow ['jeləʊ] *adjective* **yellow dwarf virus** = fungal disease affecting barley, and also wheat and grass; the leaves turn red and yellow and yields are reduced; the disease is carried by wingless aphids; the common name for it is 'BYDV'; **yellow rattle** = annual weed *(Rhinanthus minor)* found in grasslands; **yellow rust** = *Puccinia glumarum,* a fungal disease of cereals, mainly affecting wheat and barley; yellow pustules form on leaves, stems and ears; also called 'stripe rust'

◊ **yellowing** ['jeləʊɪŋ] *noun* chlorosis, a condition where the leaves of plants turn yellow, caused by lack of light; also a sign of disease or of nutrient deficiency; **interveinal yellowing** = condition of plants caused by magnesium deficiency, where the surface of the leaves turns yellow and the veins stay green; **lethal yellowing (LY)** = disease which attacks and kills coconut palms; **speckled yellowing** = disease affecting the leaves of sugar beet, cause by manganese deficiency

◊ **yellows** ['jeləuz] *noun* **(a)** general term for any plant disease in which the leaves become yellow **(b)** jaundice in animals

> COMMENT: yellow diseases are often caused by viruses, but may also be caused by bacteria and fungi. Yellowing is a common symptom when there is a deficiency of elements which are important to chlorophyll production (such as iron and magnesium)

yelt [jelt] *noun* young female pig (NOTE: commonly called a **gilt**)

yeoman farming ['jəumən 'fɑ:mɪŋ] *noun* system of farming by a class of small freeholders which resulted from the enclosure movement in the Middle Ages

Yerba de Maté ['jɜrbæ deɪ 'mæteɪ] *noun* tree found mainly in Brazil; its leaves are used for making a beverage like tea (maté); the drink is popular throughout South America

yew [ju:] *noun* coniferous tree or large shrub *(Taxus baccata)* **yew poisoning** = poisoning through eating yew berries or leaves

> COMMENT: all varieties of the British yew tree are poisonous and stock which eat the leaves or berries suffer vomiting and in severe cases die

YFC = YOUNG FARMERS' CLUB

yield [ji:ld] **1** *noun* quantity of a product produced from a plant or animal *or* from an area of land: *the normal yield is 2 tonnes per hectare; the green revolution increased rice yields in parts of Asia* **2** *verb* to produce a crop *or* a product: *the rice can yield up to 2 tonnes per hectare*

> QUOTE oilseed rape yielded an estimated 3.25t/ha
> *Farmers Weekly*

> QUOTE research has shown what must have been known to stockmen for years, that kindness to cows increases their milk yield, yet the pharmaceutical industry is determined to increase milk production by injecting cows with the hormone bovine somatotropin (BST)
> *Ecologist*

> QUOTE the new cotton variety 'F414' is a high-yielding potential variety with an average yield of 18.5 quintals per hectare with a maximum yield potential of 30 quintals per hectare under good management conditions
> *Indian Farming*

yoghurt *or* **yogurt** *or* **yoghourt** ['jɒgət] *noun* soured milk in which fermentation is accelerated by the introduction of specific bacterial microorganisms

yoke [jəuk] *noun* **(a)** wooden crosspiece fastened over the necks of two oxen **(b)** a pair of oxen

yolk [jəuk] *noun* **(a)** yellow central part of an egg; **yolk sac** = membrane which encloses the yolk in embryo fish, reptiles and birds **(b)** greasy material present in sheep wool

Yorkshire ['jɔ:kʃə] *noun* breed of large white pig, similar to the Large White (the name is not much used in the UK)

◊ **Yorkshire fog** ['jɔ:kʃə 'fɒg] *noun* weed grass *(Holcus lanatus)* able to grow under poor conditions; it is unpalatable and of little value

Young Farmers' Club (YFC) social organization for young farmers

yuca ['ju:kə] *noun* Spanish name for tapioca, the flour from the cassava plant

Zadoks scale ['zeɪdɒks 'skeɪl] *noun* scale of 10 used to show the growth stages of a plant from germination to ripening

Zea [zɪə] *Latin name for* maize or corn

zebu ['zi:bu:] *noun* humped cattle of the tropics; a domesticated Asiatic cattle breed with a pronounced shoulder hump and prominent dewlap; in the USA, called a 'Brahman'

zero grazing ['zɪərəu 'greɪzɪŋ] *noun* practice of harvesting forage crops and taking the green material to feed housed livestock

◊ **zero tillage** ['zɪərəu 'tɪlɪdʒ] *noun* technique using herbicides instead of tilling the soil before sowing an arable crop by direct drilling

zigzag harrow ['zɪgzæg 'hærəʊ] *noun* light harrow used for final seedbed work, and also for covering sown seeds; the frames are zigzag in shape, with short tines bolted to them

zinc [zɪŋk] *noun* white metallic trace element, essential to biological life (NOTE: chemical symbol is **Zn;** atomic number is 30)

COMMENT: zinc deficiency in plants prevents the expansion of leaves and internodes; in animals zinc forms part of certain enzyme systems and is present in crystallized insulin

zoo- [zəʊ] *prefix* meaning animals

◊ **zoonoses** [zəʊ'nəʊsiːz] *plural noun* diseases transmitted between farm animals and humans; for example, tuberculosis and disorders caused by Salmonella bacteria

QUOTE zoonoses are diseases which are communicable from animals to humans and include conditions like ringworm, leptospirosis and certain types of food poisoning

Smallholder

zoophyte ['zəʊəfaɪt] *noun* animal which looks like a plant

zucchini [zuˈkiːni] *noun* Italian name for courgettes, the fruit of the marrow at a very immature stage in its development, cut when between 10 and 20cm long; it may be green or yellow in colour

zygote ['zaɪgəʊt] *noun* fertilized ovum; the first stage of development of an embryo, it is the product of the union of two gametes

SUPPLEMENT

Weights and Measures

Metric Measures

Length

1 millimetre (mm)			= 0.0394 in
1 centimetre (cm)	=	10 mm	= 0.3937 in
1 metre (m)	=	100 cm	= 1.0936 yds
1 kilometre (km)	=	1000 m	= 0.6214 mile

Weight

1 milligramme (mg)			= 0.0154 grain
1 gramme (g)	=	1000 mg	= 0.0353 oz
1 kilogramme (kg)	=	1000 g	= 2.2046 lb
1 tonne (t)	=	1000 kg	= 0.9842 ton

Area

1 cm²	=	100 mm²	= 0.1550 sq. in
1 m²	=	10 000 cm²	= 1.1960 sq. yds
1 are (a)	=	100 m²	= 119.60 sq. yds
1 hectare (ha)	=	100 ares	= 2.4711 acres
1 km²	=	100 hectares	= 0.3861 sq. mile

Capacity

1 cm³			= 0.0610 cu. in
1 dm³	=	1000 cm³	= 0.0351 cu. ft
1 m³	=	1000 dm³	= 1.3080 cu. yds
1 litre	=	1 dm³	= 0.2200 gallon
1 hectolitre	=	100 litres	= 2.7497 bushels

Imperial Measures

Length

1 inch			= 2.54 cm
1 foot	=	12 inches	= 0.3048 m
1 yard	=	3 feet	= 0.9144 m
1 rod	=	5.5 yards	= 4.0292 m
1 chain	=	22 yards	= 20.117 m
1 furlong	=	220 yards	= 201.17 m
1 mile	=	1760 yards	= 1.6093 km

Weights and Measures (cont'd)

Weight

1 ounce	=	437.6 grains	=	28.350 g
1 pound	=	16 ounces	=	0.4536 kg
1 stone	=	14 pounds	=	6.3503 kg
1 hundredweight	=	112 pounds	=	50.802 kg
1 ton	=	20 cwt	=	1.0161 tonnes

Area

1 sq. inch			=	6.4516 cm^2
1 sq. foot	=	144 sq. ins	=	0.0929 m^2
1 sq. yard	=	9 sq. ft	=	0.8361 m^2
1 acre	=	4840 sq. yds	=	4046.9 m^2
1 sq. mile	=	640 acres	=	259.0 hectares

Capacity

1 cu. inch			=	16.387 cm^3
1 cu. foot	=	1728 cu. ins	=	0.0283 m^3
1 cu. yard	=	27 cu. ft	=	0.7646 m^3
1 pint	=	4 gills	=	0.5683 litre
1 quart	=	2 pints	=	1.1365 litres
1 gallon	=	8 pints	=	4.5461 litres
1 bushel	=	8 gallons	=	36.369 litres
1 fluid ounce	=	8 fl. drachms	=	28.413 cm^3
1 pint	=	20 fl. oz	=	568.26 cm^3

US Measures

Dry Measures

1 pint	=	0.9689 UK pt	=	0.5506 litre
1 bushel	=	0.9689 UK bu	=	35.238 litres

Liquid Measures

1 fluid ounce	=	1.0408 UK fl. oz	=	0.0296 litre
1 pint (16 oz)	=	0.8327 UK pt	=	0.4732 litre
1 gallon	=	0.8327 UK gal	=	3.7853 litres

Conversion Tables

AREA

hectares	hectares or acres	acres
0.41	1	2.47
0.81	2	4.94
1.21	3	7.41
1.62	4	9.88
2.02	5	12.36
2.43	6	14.83
2.83	7	17.30
3.24	8	19.77
3.64	9	22.24
4.05	10	24.71
8.09	20	49.42
12.14	30	74.13
16.19	40	98.84
20.23	50	123.56
24.28	60	148.27
28.33	70	172.98
32.38	80	197.69
36.42	90	222.40
40.47	100	247.11

CAPACITY

litres	litres or gallons	gallons
4.55	1	0.22
9.09	2	0.44
13.64	3	0.66
18.18	4	0.88
22.73	5	1.10
27.28	6	1.32
31.82	7	1.54
36.37	8	1.76
40.91	9	1.98
45.46	10	2.20
90.92	20	4.40
136.38	30	6.60
181.84	40	8.80
227.31	50	11.00
272.77	60	13.20
318.23	70	15.40
363.69	80	17.60
409.15	90	19.80
454.61	100	22.00

LENGTH

centimetres	cm or inches	inches
2.54	1	0.39
5.08	2	0.79
7.62	3	1.18
10.16	4	1/58
12.70	5	1.97
15.24	6	2.36
17.78	7	2.76
20.32	8	3.15
22.86	9	3.54
25.40	10	3.94
50.80	20	7.87
76.20	30	11.81
101.60	40	15.75
127.00	50	19.69
162.40	60	23.62
177.80	70	27.56
203.20	80	31.50
228.60	90	35.43
254.00	100	39.37

Conversion Tables

WEIGHT

kilogrammes	kg or pounds	pounds	tonnes	tonnes or tons	tons
0.45	1	2.20	1.02	1	0.98
0.91	2	4.41	2.03	2	1.97
1.36	3	6.61	3.05	3	2.95
1.81	4	8.82	4.06	4	3.94
2.27	5	11.02	5.08	5	4.92
2.72	6	13.23	6.10	6	5.91
3.18	7	15.43	7.11	7	6.89
3.63	8	17.64	8.13	8	7.87
4.08	9	19.84	9.14	9	8.86
4.54	10	22.05	10.16	10	9.84
9.07	20	44.09	20.32	20	19.68
13.61	30	66.14	30.48	30	29.53
18.14	40	88.19	40.64	40	39.37
22.68	50	110.23	50.80	50	49.21
27.22	60	132.28	60.96	60	59.05
31.75	70	154.32	71.12	70	68.89
36.29	80	176.37	81.28	80	78.74
40.82	90	198.41	91.44	90	88.58
45.36	100	200.46	101.60	100	98.42

Periodic Table of the Elements

IA	IIA	IIIB	IVB	VB	VIB	VIIB	VIII			IB	IIB	IIIA	IVA	VA	VIA	VIIA	O
1 H																	2 He
3 Li	4 Be											5 B	6 C	7 N	8 O	9 F	10 Ne
11 Na	12 Mg											13 Al	14 Si	15 P	16 S	17 Cl	18 Ar
19 K	20 Ca	21 Sc	22 Ti	23 V	24 Cr	25 Mn	26 Fe	27 Co	28 Ni	29 Cu	30 Zn	31 Ga	32 Ge	33 As	34 Se	35 Br	36 Kr
37 Rb	38 Sr	39 Y	40 Zr	41 Nb	42 Mo	43 Tc	44 Ru	45 Rh	46 Pd	47 Ag	48 Cd	49 In	50 Sn	51 Sb	52 Te	53 I	54 Xe
55 Cs	56 Ba	57* La	72 Hf	73 Ta	74 W	75 Re	76 Os	77 Ir	78 Pt	79 Au	80 Hg	81 Tl	82 Pb	83 Bi	84 Po	85 At	86 Rn
87 Fr	88 Ra	89† Ac															

← Transition metals →

*Lanthanides	57 La	58 Ce	59 Pr	60 Nd	61 Pm	62 Sm	63 Eu	64 Gd	65 Tb	66 Dy	67 Ho	68 Er	69 Tm	70 Yb	71 Lu
†Actinides	89 Ac	90 Th	91 Pa	92 U	93 Np	94 Pu	95 Am	96 Cm	97 Bk	98 Cf	99 Es	100 Fm	101 Md	102 No	103 Lr

Chromosome numbers in some common species of plants and animals

Plants

Apple 34, 51	Oats *(Avena sativa)* 42	
Asparagus 20	Onion 16, 32	
Banana 22, 33, 44	Orange 18	
Barley 14	Pea 14	
Bean (broad) 12	Pear 34, 51	
Bean (French) 22	Potato 48	
Beet 18	Potato (Sweet) 90	
Carrot 18	Radish 18	
Coffee 22, 44	Rice 24	
Cucumber 14	Rye 14	
Garlic 16	Soybean 20	
Grape vine 38	Spinach 12	
Groundnut 40	Strawberry 56	
Hemp 20	Sunflower 34	
Lemon 18, 36	Tangerine 18	
Lettuce 18	Tobacco 48	
Loganberry 42	Tomato 24	
Maize 20	Watermelon 22	
Maple (sugar) 26	Wheat *(Triticum aestivum)* 42	
Millet 14		

Animals

Cattle *(Bos taurus)* 60
Chicken 78
Dog 78
Duck 78-80 (approx.)
Goat 60
Goose 80
Horse 64
Rabbit 44
Sheep 54
Swine 38
Turkey 80 (approx.)

MAFF Land Use Classification

Grade 1:   land completely suitable for all types of agriculture, including horticulture

Grade 2:   land with minor limitations as regards use, but suitable for most crops

Grade 3:   land with limitations as regards use, due to the soil type, relief, climate, etc. Used generally for grass or cereals

Grade 4:   land with major limitations, due to soil, relief or climate. Suitable only for very limited crops

Grade 5:   land used for rough grazing, or not used for vegetation

Land Capability Classification

	Limitations	*Suitable Crops*
Class I:	none or very minor	All crops including horticultural crops
Class II:	minor	All crops, though horticultural crops may fail
Class III:	moderate	Cereals, forage crops, potatoes
Class IV:	moderate-severe	Grass, and some cereals, such as oats
Class V:	severe	Grass; also forestry
Class VI:	very severe	No arable crops; grazing for sheep and cattle; also forestry
Class VII:	extremely severe	Grazing for sheep and deer; forestry not possible

Nitrogen Fertilizers

Ammonium Nitrate: 33.5-34.5% N The commonest nitrogen fertilizer.

Ammonium Nitrate lime: 21-26% N.

Urea: 45% N The most concentrated solid Nitrogen fertilizer.

Sulphate of ammonia: seldom used.

Nitrate of soda: 16% N

Calcium nitrate: 15.5% N

Anhydrous ammonia: 82% N. Ammonia gas liquefied under high pressure.

Aqueous ammonia: 28% N

Organic fertilizers:

Hoof and Horn: 13% N

Shoddy (wool waste): Up to 15% N

Dried Blood: 10-13% N

HORSE

nose

nostril

cheek

neck

shoulder

chest

knee

hoof

mane

back

flank

croup

pastern

tail

fetlock

COW

muzzle

neck

brisket

hoof

back

udder

teat

rump

tail

PLANT

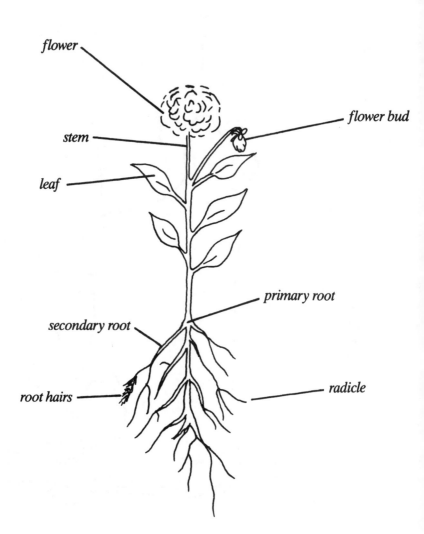

flower

flower bud

stem

leaf

primary root

secondary root

radicle

root hairs

EARS OF CEREALS

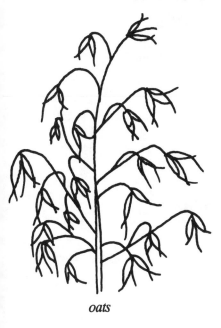

oats

barley

wheat

rye

Gestation Periods

	average period	
	months	*days*
cow	9	283-284
ewe and goat	4-5	144-150
sow	3	114
mare	11	340

Oestrus cycles

	interval between oestrus	length of oestrus
cow	21 days	18 hours
ewe	17 days	36 hours
goat	20 days	40 hours
sow	21 days	45 hours
mare	21 days	120 hours

Zadoks Scale

0	*Germination*		5	*Ear emergence*
00	dry seed		52	¼ of ear visible
05	root emerges		54	½ of ear visible
07	shoot emerges		56	¾ of ear visible
			58	ear complete
1	*Seedling*			
11	first leaf unfolds		6	*Flowers*
12	second leaf unfolds		69	flowering complete
13	third leaf unfolds			
			7	*Milk development*
2	*Tillering*			
21	main shoot + one side shoot		8	*Dough development*
22	main shoot + two side shoots			
			9	*Seed ripening*
3	*Stem*		91	grain hard
31	one node		93	grain loose
32	two nodes		95	dormant seed
			96	viable seed
4	*Boots*			
43	boots visible			
49	awns visible			

World Commodity Markets

Argentina	Bolsa de Cereales, Buenos Aires	grains
Australia	Sydney Futures Exchange	wool, live cattle, gold
Austria	Wiener Börsekammer	timber, textiles, leather
Brazil	Bolsa de Mercadorias de São Paulo	cotton, coffee, soya beans
Canada	Winnipeg Commodity Exchange	grain, gold
France	International Market of Robusta Coffee	coffee
	Cocoa Terminal Market (part of Paris Commodity Exchange)	cocoa beans
	International Market for White Sugar	sugar
Germany	Berliner Produktenbörse	fodder, flour, grain
	Frankfurter Getreide und Produktenbörse	fodder, eggs, grain, flour, potatoes, fuel oil
	Niedersächsische Getreide und Produktenbörse	grain, fodder, flour, fertilizer, potatoes
	Getreide und Warenbörse Rhein-Ruhr	grain, fodder, flour
	Wormser Getreide und Produktenbörse	grain, fodder, seeds, fertilizer
Hong Kong	Hong Kong Commodity Exchange	cotton, sugar
India	Tobacco Board, Andhra Pradesh	tobacco
	Coffee Board, Bangalore	coffee
	Central Silk Board, Bombay	silk
	Tea Board of India, Calcutta	tea
	Cardamom Board, Cochin	cardamoms
	Coir Board, Cochin	coir
	Rubber Board, Kerala	rubber
Italy	Borsa Merci di Bologna	agricultural produce, flour, cereals
	Borsa Merci di Firenze	cereals, fertilizers, flour, vegetable oil, wheat, wine
	Borsa Merci di Padova	alcohol, live cattle, chickens, eggs, fertilizers, fruit, grain, olive oil, timber
	Borsa Merci di Parma	cereals, ham, eggs, flour, pork, potatoes, chickens, tomato concentrate

Japan	Hokkaido Grain Exchange	starch, beans
	Kobe Grain Exchange	grain
	Kobe Raw Silk Exchange	raw silk
	Kobe Rubber Exchange	rubber
	Nagoya Grain Exchange	starch, beans
	Nagoya Textile Exchange	cotton, rayon, wool yarn
	Osaka Grain Exchange	starch, beans
	Osaka Sampin Exchange	cotton yarn
	Osaka Sugar Exchange	raw sugar, white sugar
	Kanmon Commodity Exchange	starch, beans, sugar
	Tokyo Grain Exchange	starch, beans
	Tokyo Rubber Exchange	rubber
	Tokyo Sugar Exchange	raw sugar, white sugar
	Tokyo Textile Commodities Exchange	cotton yarn, wool yarn
	Tokohashi Dried Cocoon Exchange	silk cocoons
Kenya	Coffee Board of Kenya	coffee
	East African Tea Trade Association	tea
	Kenya Tea Development Authority	tea
Malaysia	Malaysian Rubber Exchange	rubber
	Straits Tin Market, Penang	tin
Netherlands	Egg Terminal Market, Amsterdam	eggs
	Pork Terminal Market	pork
	Potato Terminal Market	potatoes
Norway	Oslo Fur Auctions	fur skins
Singapore	Rubber Association of Singapore	rubber
	Gold Exchange of Singapore	gold
United Kingdom	Liverpool Cotton Association	raw cotton
	London Cocoa Terminal Market	cocoa
	London Grain Futures Market	EC barley, EC wheat
	London Metal Exchange	aluminium, copper, lead, silver, nickel, tin, zinc
	London Rubber Terminal Market	rubber
	London Soya Bean Meal Futures Market	soya beans
	London Vegetable Oil Terminal Market	palm oil, soya bean oil
	Coffee Terminal Market	coffee
	Federation of Oils, Seeds and Fats Association	vegetable oil, animal fat, oil seed
	Tea Brokers' Association	tea

	United Terminal Sugar Market	raw sugar, white sugar
	London Jute Association	jute
	London Wool Terminal Market	Australian & New Zealand wool
	Rubber Trade Association	rubber
United States	Mid-American Commodity Exchange	corn, oats, soya beans, wheat, cattle, hogs, gold, silver
	Board of Trade of Kansas City	wheat, grains
	Commodity Exchange, New York	copper, gold, silver, zinc
	New York Cocoa Exchange	cocoa
	New York Mercantile Exchange	butter, beef, potatoes, gold, platinum, palladium, Swiss francs, Deutschmarks, Canadian dollars, pounds, yen
	Chicago Board of Trade	corn, oats, soya bean oil, frozen chickens, wheat, plywood, soya beans, gold, silver, Treasury bonds, Government National Mortgage Association certificates
	Chicago Mercantile Exchange (main market)	cattle, hogs, pork bellies, beef, hams
	(International Monetary Market)	pounds, yen, Canadian dollars, Deutschmarks, French francs, guilders, Mexican pesos, Swiss francs, gold, copper, US Treasury bills
	(Associate Mercantile Division)	eggs, timber, frozen turkeys, sorghum, butter, potatoes
	Minneapolis Grain Exchange	wheat, barley, oats, corn, flax seed, rye, soya beans
	Amex Commodities Exchange	Financial futures, GNMA futures
	Citrus Associates (NY Exchange)	frozen orange juice
	NY Cocoa Exchange (Rubber)	natural rubber
	NY Coffee and Sugar Exchange	coffee & sugar futures
	NY Cotton Exchange	cotton